# Molecular Pharmacology and Interventions in Cardiovascular Disease

# Molecular Pharmacology and Interventions in Cardiovascular Disease

Guest Editor

**Magdalena Zabielska-Kaczorowska**

MDPI

Basel • Beijing • Wuhan • Barcelona • Belgrade • Novi Sad • Cluj • Manchester

*Guest Editor*
Magdalena Zabielska-Kaczorowska
Department of Physiology
Medical University of Gdansk
Gdansk
Poland

*Editorial Office*
MDPI AG
Grosspeteranlage 5
4052 Basel, Switzerland

This is a reprint of the Special Issue, published open access by the journal *International Journal of Molecular Sciences* (ISSN 1422-0067), freely accessible at: https://www.mdpi.com/journal/ijms/special_issues/LHWKNUUCV5.

For citation purposes, cite each article independently as indicated on the article page online and as indicated below:

| |
|---|
| Lastname, A.A.; Lastname, B.B. Article Title. *Journal Name* **Year**, *Volume Number*, Page Range. |

**ISBN 978-3-7258-5295-6 (Hbk)**
**ISBN 978-3-7258-5296-3 (PDF)**
**https://doi.org/10.3390/books978-3-7258-5296-3**

# Contents

# About the Editor

**Magdalena Zabielska-Kaczorowska**

Magdalena Zabielska-Kaczorowska is a notable researcher in the field of experimental cardiology. She earned her MSc in molecular biology from Warmia and Mazury University in 2009. In 2019, she graduated from the Medical University of Gdansk with a PhD in medical sciences. Her research with the Physiology Department team focuses on understanding the mechanisms that underlie cardiovascular diseases by using innovative experimental approaches. She has contributed to advancing knowledge in areas such as myocardial function, heart energy metabolism, vascular biology, and the molecular basis of cardiovascular diseases, such as ischemic heart disease or atherosclerosis. She uses multi-omic approaches, such as metabolomic and proteomic analyses, to discover biomarkers of disease progression. Magdalena Zabielska-Kaczorowska's research efforts aim to develop new therapeutic strategies and improve outcomes for patients with cardiovascular conditions. Her collaboration with interdisciplinary teams has further enhanced the potential impact of her findings, fostering a holistic approach to treatment.

# Preface

The incidence of cardiovascular disorders (CVDs), which are intricate and multifaceted, escalates in the aging population. The severity of the CVD is a vital factor when deciding the treatment, as it is associated with heart failure and mortality. Efforts are focused on developing novel pharmaceuticals through the physicochemical modification of drug molecules and the design and synthesis of new compounds, emphasizing drug delivery, controlled release, and targeted administration to enhance therapeutic efficacy. Researchers are investigating novel methodologies, encompassing contemporary pharmacology and gene therapy, to enhance these delivery systems. Although several pre-clinical investigations have demonstrated that specific pharmacologic drugs and therapeutic interventions dramatically diminish CVDs' severity beyond the effects of reperfusion alone, only a limited number of these therapies have successfully transitioned into clinical trials or normal clinical practice. The optimization of pharmacological substances is essential for the further development of diagnostic, therapeutic, and preventive techniques in cardiology. Promoting the endogenous regeneration of the heart is crucial for enhancing the prognosis of patients with cardiac damage and developing appropriate therapeutic options. The Special Issue titled "Molecular Pharmacology and Interventions in Cardiovascular Disease" is addressed to researchers, clinicians, and professionals involved in the fields of cardiovascular medicine, pharmacology, and biomedical research. Its goal is to compile and share the most recent research results, creative methods, and treatment approaches related to molecular mechanisms and interventions in cardiovascular diseases.

**Magdalena Zabielska-Kaczorowska**
*Guest Editor*

International Journal of
*Molecular Sciences*

MDPI

*Editorial*

# Editorial for the Special Issue Titled "Molecular Pharmacology and Interventions in Cardiovascular Disease"

**Magdalena Zabielska-Kaczorowska**

Department of Physiology, Medical University of Gdansk, 80-211 Gdansk, Poland;
magdalena.zabielska@gumed.edu.pl; Tel.: +48-58-349-15-20

## 1. Introduction

Cardiovascular diseases (CVDs) include coronary artery disease, hypertension, and various cardiac disorders, among other conditions that impact the heart and vascular system. These diseases are commonly linked to a variety of risk factors, including diabetes, high blood pressure, and high cholesterol [1]. As the population's mean age rises, so does the prevalence of cardiovascular disorders, which are intricate and multifaceted [2]. With a focus on drug delivery, controlled drug release, and targeting to the site of action to maximise the therapeutic effect, efforts are being made to develop novel pharmaceutical and non-pharmaceutical modulators in CVDs [3]. These advancements aim not only to enhance the efficacy of treatments but also to minimise side effects, ultimately improving patient outcomes. Researchers are exploring innovative approaches, including modern pharmacology and gene therapy, to further refine these delivery systems [4].

This special issue of the International Journal of Molecular Sciences (IJMS), entitled "Molecular Pharmacology and Interventions in Cardiovascular Disease", contains 4 reviews and 6 original research papers written by a panel of experts who highlighted recent advances, discussed potential drug targets, and aimed to elucidate pharmacologic strategies that may be employed to promote cardiac repair after injury. These contributions provide valuable insights into the intricate mechanisms underlying cardiovascular pathology and offer a glimpse into future therapeutic avenues. The collective findings underscore the importance of continued research in this critical area of medical science to achieve therapeutic gain and enhance the efficacy of protocols for cardiac therapy.

## 2. Research

The original articles cover a range of topics from the use of novel inhibitors to analyses of vascular permeability mediators in cardiovascular disorders. Two publications are focused on the treatment of an aneurysm. Increased levels of matrix metalloproteinases (MMPs) and destabilisation of the vessel wall due to the breakdown of the extracellular matrix's (ECM) structural elements, primarily collagen and elastin, are caused by accelerated inflammatory processes during aneurysm development. ECM proteolysis is inhibited by tissue inhibitors of metalloproteinases (TIMPs), which directly control MMP activity. Here, Golombek et al. suppressed MMP-9 by synthetic TIMP-1 encoding mRNA exogenously delivered into the aorta, which triggered the expression of TIMP-1 protein. These findings suggested that TIMP-1 mRNA administration is promising as an aneurysm therapy method [5]. In order to prevent aneurysm formation in Marfan syndrome (MFS), strategies to block downstream effectors of angiotensin II type 1 receptor (AT1R)-mediated signalling were investigated in the work of Callejon et al. It was found that inhibiting the progression of aneurysms in MFS may require total blocking of AT1R function, which can be accomplished by administering a high dosage of losartan [6]. The next set of two investigations is related to the inflammation strongly involved in cardiovascular disease. Inhibition of the interleukin-6 receptor (IL-6R) has demonstrated efficacy in alleviating myocardial damage and decreasing the levels of the prothrombotic and inflammatory mediator, neutrophil extracellular traps (NETs). A major contributor to the development of NET consists of the enzyme peptidylarginine deiminase 4 (PAD4). In STEMI patients, a correlation between PAD4, IL-6R, and troponin release was shown by Kindberg et al. [7]. The pathophysiology of atherosclerosis and the development of in-stent restenosis (ISR) are significantly influenced by inflammation. It was found by Hytönen et al. that a novel synthetic flavonoid, 3,7-dihydroxy-isoflav-3-ene (DHIF) showed promise in limiting ISR through antioxidant action. DHIF treatment did not postpone endothelial healing following stenting injury and markedly decreased inflammation and proliferation in the restenotic lesions [8]. Investigation of Luecht et al. was conducted into the inflammatory response in macrophages triggered by perioperative serum samples taken from congenital heart defect (CHD) patients undergoing cardiopulmonary bypass (CPB) heart surgery. By activating STAT3 through IL-6 and IL-8, CPB triggered an increase in the production of cytokines and mediators of vascular permeability. Treating those patients with Stattic increased TNF$\alpha$ expression while attenuating all mediators examined [9]. In the work of Brosinsky et al. using the inducible cardiomyocyte-specific monoamine oxidase B (cmMAO-B) knockout mouse model, the effects of ROS generation mediated by MAO-B on RV dilatation and function in response to pulmonary artery banding were examined [10]. During pulmonary hypertension (PH), the elevated production of reactive oxygen species (ROS) in the mitochondria is crucial for the development of right ventricular hypertrophy (RVH) and failure (RVF). Compared to their littermates, cmMAO-B KO mice were shielded against RV dilatation, hypertrophy, and dysfunction after RV pressure overload. The hypothesis that cmMAO-B plays a significant role in RV hypertrophy and failure during PH was validated by these findings.

## 3. Review

Review articles present several up-to-date aspects of the pharmacological interventions in CVDs. Kulpa et al., reviewed data showing that research on catestatin (CTS) is encouraging and suggests that it may be used to treat a variety of CVDs. It was found that CTS stimulates endothelial cells and cardiomyocytes to produce more NO. Additionally, for patients with mildly decreased or retained HF, CTS may serve as a biochemical marker. However, there are some conditions, including atherosclerosis, where studies regarding the impact of CTS on vessel wall remodelling were conflicting, and its significance in these conditions is still unclear [11]. Liu et al. made an insightful review about the therapeutic strategies of cirrhotic cardiomyopathy (CCM). Patients with CCM may benefit from treatments including antioxidants, anti-inflammatory, and anti-apoptotic drugs. Such strategies are now primarily restricted to animal studies, hence carefully planned clinical trials are required to validate these compounds. Another possible therapeutic agent are non-selective beta-blockers (NSBBs), which should, in theory, have therapeutic effects on CCM. Nevertheless, there is inconsistency among the outcomes from various studies [12]. Menezes et al. summarised the current state of treatment for hypertrophic cardiomyopathy (HCM), a complex heart disease characterised by alterations in a number of biological processes, such as ROS control, ionic homeostasis, structural remodelling, and metabolic balance. It was proposed that mitochondrial redox signalling and genetic mutations are the key regulating variables in HCM [13]. Szczepanska-Sadowska et al. explored the interactions between the hormones of the hypothalamo–pituitary–adrenal axis and vasopressin (AVP). This review summarises recent knowledge about steroid hormones and AVP that are often released together and act closely together to control behaviour, metabolism, blood pressure, and water-electrolyte balance. Significant changes occur in the interactions between AVP and HPA during inflammation, neurogenic stress, and metabolic, respiratory, and cardiovascular disorders. In metabolic diseases, inappropriate interactions between AVP and steroids may cause or accelerate cardiovascular diseases [14].

This Special Issue underlines the richness but also the complexity of approaches in CVDs treatment. The present challenge is to accelerate the exploration of unique pharmacological targets for clinical applications. As the Guest Editor, I hope that the findings included in this Special Issue will inspire further investigations in this challenging field.

**Acknowledgments:** The guest editor would like to express gratitude to the reviewers for their insightful criticism and to all of the authors for their outstanding contributions.

**Conflicts of Interest:** The author declares no conflicts of interest.

**List of Contributions:**

1. Kindberg, K.M.; Nordeng, J.; Langseth, M.S.; Schandiz, H.; Roald, B.; Solheim, S.; Seljeflot, I.; Stokke, M.K.; Helseth, R. IL-6R Signaling Is Associated with PAD4 and Neutrophil Extracellular Trap Formation in Patients with STEMI. *Int. J. Mol. Sci.* **2025**, *26*, 5348. https://doi.org/10.3390/IJMS26115348.
2. Luecht, J.; Pauli, C.; Seiler, R.; Herre, A.L.; Brankova, L.; Berger, F.; Schmitt, K.R.L.; Tong, G. Prolonged Extracor-poreal Circulation Leads to Inflammation and Higher Expression of Mediators of Vascular Permeability Through Activation of STAT3 Signaling Pathway in Macrophages. *Int. J. Mol. Sci.* **2024**, *25*, 12398. https://doi.org/10.3390/IJMS252212398.
3. Hytönen, J.P.; Leppänen, O.; Taavitsainen, J.; Ylä-Herttuala, S. Synthetic Flavonoid 3,7-Dihydroxy-Isoflav-3-Ene (DHIF) Reduces In-Stent Restenosis in an Atherosclerotic Watanabe Heritable Hyperlipidemic Rabbit Stent Model. *Int. J. Mol. Sci.* **2024**, *25*, 11530. https://doi.org/10.3390/IJMS252111530.
4. Golombek, S.; Doll, I.; Kaufmann, L.; Lescan, M.; Schlensak, C.; Avci-Adali, M. A Novel Strategy for the Treatment of Aneurysms: Inhibition of MMP-9 Activity through the Delivery of TIMP-1 Encoding Synthetic MRNA into Arteries. *Int. J. Mol. Sci.* **2024**, *25*, 6599. https://doi.org/10.3390/IJMS25126599.
5. Brosinsky, P.; Heger, J.; Sydykov, A.; Weiss, A.; Klatt, S.; Czech, L.; Kraut, S.; Schermuly, R.T.; Schlüter, K.D.; Schulz, R. Does Cell-Type-Specific Silencing of Monoamine Oxidase B Interfere with the Development of Right Ventricle (RV) Hypertrophy or Right Ventricle Failure in Pulmonary Hypertension? *Int. J. Mol. Sci.* **2024**, *25*, 6212. https://doi.org/10.3390/IJMS25116212.
6. Valdivia Callejon, I.; Buccioli, L.; Bastianen, J.; Schippers, J.; Verstraeten, A.; Luyckx, I.; Peeters, S.; Danser, A.H.J.; Van Kimmenade, R.R.J.; Meester, J.; et al. Investigation of Strategies to Block Downstream Effectors of AT1R-Mediated Signalling to Prevent Aneurysm Formation in Marfan Syndrome. *Int. J. Mol. Sci.* **2024**, *25*, 5025. https://doi.org/10.3390/IJMS25095025.
7. Kulpa, J.; Paduch, J.; Szczepanik, M.; Gorący-Rosik, A.; Rosik, J.; Tchórz, M.; Pawlik, A.; Gorący, J. Catestatin in Cardiovascular Diseases. *Int. J. Mol. Sci.* **2025**, *26*, 2417. https://doi.org/10.3390/IJMS26062417.
8. Szczepanska-Sadowska, E.; Czarzasta, K.; Bogacki-Rychlik, W.; Kowara, M. The Interaction of Vasopressin with Hormones of the Hypothalamo–Pituitary–Adrenal Axis: The Significance for Therapeutic Strategies in Cardio-vascular and Metabolic Diseases. *Int. J. Mol. Sci.* **2024**, *25*, 7394. https://doi.org/10.3390/IJMS25137394.
9. Menezes Junior, A. da S.; França-e-Silva, A.L.G. de; Oliveira, H.L. de; Lima, K.B.A. de; Porto, I. de O.P.; Pedroso, T.M.A.; Silva, D. de M. e.; Freitas, A.F. Genetic Mutations and Mitochondrial

Redox Signaling as Modulating Factors in Hypertrophic Cardiomyopathy: A Scoping Review. *Int. J. Mol. Sci.* **2024**, *25*, 5855. https://doi.org/10.3390/IJMS25115855.
10. Liu, H.; Ryu, D.; Hwang, S.; Lee, S.S. Therapies for Cirrhotic Cardiomyopathy: Current Perspectives and Future Possibilities. *Int. J. Mol. Sci.* **2024**, *25*, 5849. https://doi.org/10.3390/IJMS25115849.

## References

1. Prousi, G.S.; Joshi, A.M.; Atti, V.; Addison, D.; Brown, S.A.; Guha, A.; Patel, B. Vascular Inflammation, Cancer, and Cardiovascular Diseases. *Curr. Oncol. Rep.* **2023**, *25*, 955–963. [CrossRef]
2. Gaidai, O.; Cao, Y.; Loginov, S. Global Cardiovascular Diseases Death Rate Prediction. *Curr. Probl. Cardiol.* **2023**, *48*, 101622. [CrossRef] [PubMed]
3. Stern, C.S.; Lebowitz, J. Latest Drug Developments in the Field of Cardiovascular Disease. *Int. J. Angiol.* **2010**, *19*, e100–e105. [CrossRef] [PubMed]
4. Plowright, A.T.; Engkvist, O.; Gill, A.; Knerr, L.; Wang, Q.D. Heart Regeneration: Opportunities and Challenges for Drug Discovery with Novel Chemical and Therapeutic Methods or Agents. *Angew. Chem.—Int. Ed.* **2014**, *53*, 4056–4075. [CrossRef] [PubMed]
5. Golombek, S.; Doll, I.; Kaufmann, L.; Lescan, M.; Schlensak, C.; Avci-Adali, M. A Novel Strategy for the Treatment of Aneurysms: Inhibition of MMP-9 Activity through the Delivery of TIMP-1 Encoding Synthetic MRNA into Arteries. *Int. J. Mol. Sci.* **2024**, *25*, 6599. [CrossRef] [PubMed]
6. Valdivia Callejon, I.; Buccioli, L.; Bastianen, J.; Schippers, J.; Verstraeten, A.; Luyckx, I.; Peeters, S.; Danser, A.H.J.; Van Kimmenade, R.R.J.; Meester, J.; et al. Investigation of Strategies to Block Downstream Effectors of AT1R-Mediated Signalling to Prevent Aneurysm Formation in Marfan Syndrome. *Int. J. Mol. Sci.* **2024**, *25*, 5025. [CrossRef] [PubMed]
7. Kindberg, K.M.; Nordeng, J.; Langseth, M.S.; Schandiz, H.; Roald, B.; Solheim, S.; Seljeflot, I.; Stokke, M.K.; Helseth, R. IL-6R Signaling Is Associated with PAD4 and Neutrophil Extracellular Trap Formation in Patients with STEMI. *Int. J. Mol. Sci.* **2025**, *26*, 5348. [CrossRef] [PubMed]
8. Hytönen, J.P.; Leppänen, O.; Taavitsainen, J.; Ylä-Herttuala, S. Synthetic Flavonoid 3,7-Dihydroxy-Isoflav-3-Ene (DHIF) Reduces In-Stent Restenosis in an Atherosclerotic Watanabe Heritable Hyperlipidemic Rabbit Stent Model. *Int. J. Mol. Sci.* **2024**, *25*, 11530. [CrossRef] [PubMed]
9. Luecht, J.; Pauli, C.; Seiler, R.; Herre, A.L.; Brankova, L.; Berger, F.; Schmitt, K.R.L.; Tong, G. Prolonged Extracorporeal Circulation Leads to Inflammation and Higher Expression of Mediators of Vascular Permeability Through Activation of STAT3 Signaling Pathway in Macrophages. *Int. J. Mol. Sci.* **2024**, *25*, 12398. [CrossRef] [PubMed]
10. Brosinsky, P.; Heger, J.; Sydykov, A.; Weiss, A.; Klatt, S.; Czech, L.; Kraut, S.; Schermuly, R.T.; Schlüter, K.D.; Schulz, R. Does Cell-Type-Specific Silencing of Monoamine Oxidase B Interfere with the Development of Right Ventricle (RV) Hypertrophy or Right Ventricle Failure in Pulmonary Hypertension? *Int. J. Mol. Sci.* **2024**, *25*, 6212. [CrossRef] [PubMed]
11. Kulpa, J.; Paduch, J.; Szczepanik, M.; Gorący-Rosik, A.; Rosik, J.; Tchórz, M.; Pawlik, A.; Gorący, J. Catestatin in Cardiovascular Diseases. *Int. J. Mol. Sci.* **2025**, *26*, 2417. [CrossRef] [PubMed]
12. Liu, H.; Ryu, D.; Hwang, S.; Lee, S.S. Therapies for Cirrhotic Cardiomyopathy: Current Perspectives and Future Possibilities. *Int. J. Mol. Sci.* **2024**, *25*, 5849. [CrossRef] [PubMed]
13. da Menezes Junior, A.S.; de França-e-Silva, A.L.G.; de Oliveira, H.L.; de Lima, K.B.A.; de Porto, I.O.P.; Pedroso, T.M.A.; de Silva, D.M.E.; Freitas, A.F. Genetic Mutations and Mitochondrial Redox Signaling as Modulating Factors in Hypertrophic Cardiomyopathy: A Scoping Review. *Int. J. Mol. Sci.* **2024**, *25*, 5855. [CrossRef] [PubMed]
14. Szczepanska-Sadowska, E.; Czarzasta, K.; Bogacki-Rychlik, W.; Kowara, M. The Interaction of Vasopressin with Hormones of the Hypothalamo–Pituitary–Adrenal Axis: The Significance for Therapeutic Strategies in Cardiovascular and Metabolic Diseases. *Int. J. Mol. Sci.* **2024**, *25*, 7394. [CrossRef] [PubMed]

International Journal of
*Molecular Sciences*

MDPI

*Article*

# IL-6R Signaling Is Associated with PAD4 and Neutrophil Extracellular Trap Formation in Patients with STEMI

**Kristine Mørk Kindberg [1,2,*], Jostein Nordeng [1], Miriam Sjåstad Langseth [3], Hossein Schandiz [2,4], Borghild Roald [2,5], Svein Solheim [1], Ingebjørg Seljeflot [1], Mathis Korseberg Stokke [6] and Ragnhild Helseth [1]**

[1] Oslo Center for Clinical Heart Research, Department of Cardiology Ullevaal, Oslo University Hospital, 0450 Oslo, Norway
[2] Faculty of Medicine, Institute of Clinical Medicine, University of Oslo, 0450 Oslo, Norway
[3] Department of Cardiology Rikshospitalet, Oslo University Hospital, 0450 Oslo, Norway
[4] Department of Oncology, Akershus University Hospital, 1478 Lørenskog, Norway
[5] Department of Pathology, Oslo University Hospital Ullevaal, 0450 Oslo, Norway
[6] Institute for Experimental Medical Research, Oslo University Hospital and University of Oslo, 0450 Oslo, Norway
* Correspondence: kromoe@ous-hf.no

**Abstract:** Inflammation contributes to myocardial injury in ST-elevation myocardial infarction (STEMI). Interleukin-6 receptor (IL-6R) inhibition has been shown to mitigate myocardial injury and reduce levels of the prothrombotic and inflammatory mediator, neutrophil extracellular traps (NETs). The enzyme peptidylarginine deiminase 4 (PAD4) is central in NET formation. We hypothesized that PAD4 links IL-6R activation and NET formation. Methods: We conducted thrombus aspiration and peripheral blood sampling in 33 STEMI patients. In thrombi and leukocytes, we quantified the mRNA of IL-6, IL-6R, and PAD4. In peripheral blood, the protein levels of IL-6, IL-6R, PAD4, dsDNA, H3Cit, MPO-DNA, and troponin T were quantified. Results: In thrombi and circulating leukocytes, PAD4 mRNA was associated with IL-6R mRNA (thrombi: $\beta = 0.34$, 95% CI [0.16–0.53], $p = 0.001$, circulating leukocytes: $\beta = 0.92$, 95% CI [0.07–1.77], $p = 0.036$). There were no correlations between PAD4 and IL-6 in thrombi and leukocytes. The protein levels of IL-6R were associated with the NET marker H3Cit ($r_s = 0.40$, $p = 0.02$). In thrombi, PAD4 mRNA was associated with high levels of troponin T ($\beta = 1.15$ 95% CI [0.27–2.04], $p = 0.013$). Conclusion: We demonstrate an association between PAD4, IL-6R, and troponin release in STEMI patients. Our findings indicate a PAD4-mediated connection between IL-6R and NET formation and highlight PAD4 as a potential treatment target for mitigating inflammation and myocardial injury in STEMI.

**Keywords:** coronary thrombus; NETs; STEMI; inflammation; IL-6R; PAD4

## 1. Introduction

Circulating levels of interleukin (IL)-6 increase during the acute phase of ST-elevation myocardial infarction (STEMI) and are associated with infarct size and adverse clinical outcomes [1,2]. A central source of IL-6, which binds to the IL-6 receptor (IL-6R), is the NOD-, LRR-, and pyrin domain-containing protein 3 (NLRP3) inflammasome axis, a key mediator of inflammation [3]. A randomized, placebo-controlled trial, ASSessing the effect of Anti-IL-6 treatment in Myocardial Infarction (ASSAIL-MI), demonstrated that IL-6R inhibition with tocilizumab administered during PCI in patients with STEMI reduced the inflammatory response and myocardial damage [4]. Activation of IL-6R can mediate a vast array of both pro- and anti-inflammatory effects in cardiovascular diseases [5]. We recently reported that tocilizumab attenuates the formation of neutrophil extracellular traps (NETs) in the ASSAIL-MI trial and suggested that this could explain part of the IL-6R-mediated cardioprotective effect of tocilizumab [6].

NETs are extracellular web-like structures composed of double-stranded DNA (dsDNA), citrullinated histones, and various neutrophil proteins. They are released from activated neutrophils during STEMI in a process called NETosis [7,8]. NETs contribute to atherothrombosis through their clot-stabilizing, cytotoxic, and proinflammatory properties [9–12]. Circulating NET markers have repeatedly been linked to myocardial damage and clinical outcomes in STEMI [10,13], and NETs have been observed in coronary thrombi [14–18]. During NETosis, chromatin decondensation is initiated by the intracellular protein peptidylarginine deiminase 4 (PAD4) [19]. Inhibition of PAD4 has shown cardioprotective effects in preclinical models of myocardial infarction, making it a promising target for further studies [20]. We have previously identified an association between increased IL-6R gene expression and markers of myocardial damage in aspirated coronary thrombi from patients with STEMI [21]. In the present study, we hypothesized that IL-6R and NETosis are linked by PAD4. To test this, we analyzed aspirated coronary thrombi and circulating blood from patients with STEMI and assessed the associations between IL-6R, IL-6, PAD4, and NETs.

## 2. Results

### 2.1. Baseline Characteristics

The baseline characteristics of the study cohort are outlined in Table 1. The mean age was 60 years, and the female inclusion rate was 9%. The majority (79%) presented with total coronary occlusion prior to PCI, defined as TIMI flow 0. Twelve patients (36%) had retrograde flow to the culprit coronary artery, suggesting longstanding atherosclerotic disease. The available markers with their respective locations are listed in Table 2.

**Table 1.** Baseline characteristics.

| | **Total $n$ = 33** |
|---|---|
| Age, years | 60 (±11) |
| Females | 3 (9%) |
| Current or previous smoker | 27 (84%) |
| BMI, kg/m$^2$ | 27.7 (23.4, 28.6) |
| sBP, mmHg | 127 (±31) |
| dBP, mmHg | 82 (±19) |
| Heart rate | 70 (65, 90) |
| Prior myocardial infarction | 1 (3%) |
| Prior hypertension | 11 (33%) |
| Prior diabetes mellitus type 2 | 4 (12%) |
| Prior medication: | |
| Acetylsalicylic acid | 6 (18%) |
| P2Y12-inhibitor | 2 (6%) |
| Anticoagulation | 3 (9%) |
| RAAS inhibitors | 5 (15%) |
| Betablocker | 4 (12%) |
| Aldosterone antagonist | 0 (0%) |

**Table 1.** *Cont.*

| | Total $n$ = 33 |
|---|---|
| Statins | 6 (18%) |
| Symptom start to PCI, min | 152 (122, 343) |
| Culprit artery: | |
| LAD | 16 (48%) |
| Cx | 6 (18%) |
| RCA | 11 (33%) |
| Troponin T peak, μg/L | 3434 (1250, 6967) |
| CRP, mg/L | 2.71 (0.96, 6.03) |

Values are given as mean (±SD), median (IQR 25, 75), or proportions (%). BMI = body mass index, sBP = systolic blood pressure, dPB = diastolic blood pressure, RAAS = renin–angiotensin–aldosterone system, PCI = percutaneous coronary intervention, LAD = left anterior descending artery, Cx = circumflex artery, RCA = right coronary artery, CRP = C-reactive protein.

**Table 2.** Available markers and their locations.

| | Gene Expression in Thrombi | Gene Expression in Circulating Leukocytes | Circulating Levels |
|---|---|---|---|
| IL-6 | √ | √ | √ |
| IL-6R | √ | √ | √ |
| PAD4 | √ | √ | √ |
| dsDNA | x | x | √ |
| MPO-DNA | x | x | √ |
| H3Cit | x | x | √ |

All circulating proteins were quantified by ELISA in serum, except MPO-DNA, which was quantified in EDTA plasma. dsDNA was stained with PicoGreen in EDTA plasma and quantified with fluorometry. IL-6 = intereleukin-6, IL-6R = interleukin-6 receptor, PAD4 = peptidylarginine deiminase 4, dsDNA = double-stranded DNA, MPO-DNA = myeloperioxidase–DNA, H3Cit = citrullinated histone H3.

## *2.2. Investigations in Coronary Thrombi*

### 2.2.1. Associations Between PAD4, IL-6R, and IL-6 mRNA Levels

PAD4 mRNA was detected in 27 of the 33 coronary thrombi. A correlation between mRNA for PAD4 and IL-6R in coronary thrombi was strongly suggested, although not statistically significant (Figure 1a). This association persisted, now statistically significant, in a multivariate, linear regression model adjusted for age, sex, and symptom duration (Table 3). In contrast, no association was found between PAD4 and IL-6 mRNA.

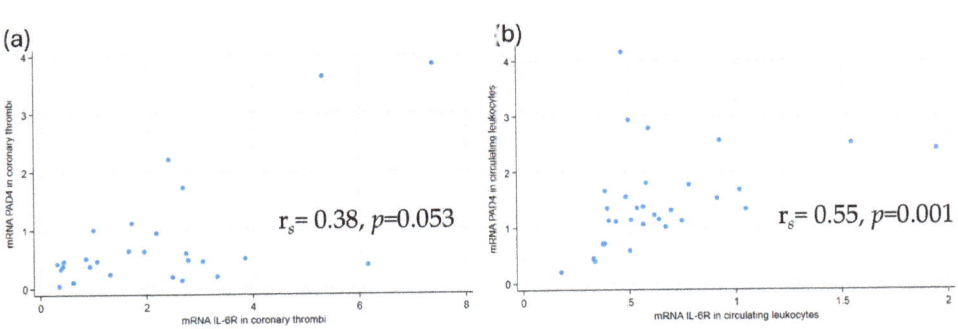


**Figure 1.** Scatterplots with Spearman's Rho of PAD4 mRNA and IL-6R mRNA levels in (**a**) coronary thrombi and (**b**) circulating leukocytes. mRNA levels were measured by qPCR. PAD4 = peptidylarginine deiminase 4, IL-6R = interleukin-6 receptor.

**Table 3.** Associations between PAD4 mRNA, IL-6R mRNA, and troponin T in thrombi and PAD4 mRNA and IL-6R mRNA in leukocytes.

| | Univariate | | | Multivariate | | |
|---|---|---|---|---|---|---|
| | β | CI | *p*-value | β | CI | *p*-value |
| Association with PAD4 mRNA in coronary thrombi | | | | | | |
| IL-6R mRNA | 0.34 | 0.16–0.51 | 0.001 | 0.34 | 0.16–0.53 | 0.001 * |
| Circulating peak troponin T, quartile q4 vs. 1–3 | 1.0 | 0.20–1.80 | 0.016 | 1.15 | 0.27–2.04 | 0.013 * |
| Association with PAD4 mRNA in circulating leukocytes | | | | | | |
| IL-6R mRNA | 0.98 | 0.18–1.77 | 0.017 | 0.92 | 0.07–1.77 | 0.036 * |

Univariate and multivariate linear regression models of associations between mRNA in coronary thrombi (orange) and circulating leukocytes (green). mRNA levels were measured by qPCR. PAD4 = peptidylarginine deiminase 4, IL-6R = interleukin-6 receptor. * Adjusted for age, sex, and symptom duration.

#### 2.2.2. Associations Between PAD4 mRNA Levels and Troponin T

The level of PAD4 mRNA in coronary thrombi was associated with high peak troponin T in a multivariate, linear regression model adjusted for age, sex, and symptom duration (Table 3). In contrast, PAD4 mRNA levels in circulating leukocytes, PAD4 protein levels, and the other circulating NET markers (dsDNA, MPO-DNA, and H3Cit) were not associated with troponin T.

#### 2.2.3. Immunofluorescence Investigations of NETosis

The NET markers, histone H3 and neutrophil elastase, were detected in 27 and 29 of 33 thrombi, respectively. The presence of NETs in coronary thrombi is illustrated in Figure 2 with co-localization of two NET markers, histone H3 and neutrophil elastase, alongside extracellular DNA.

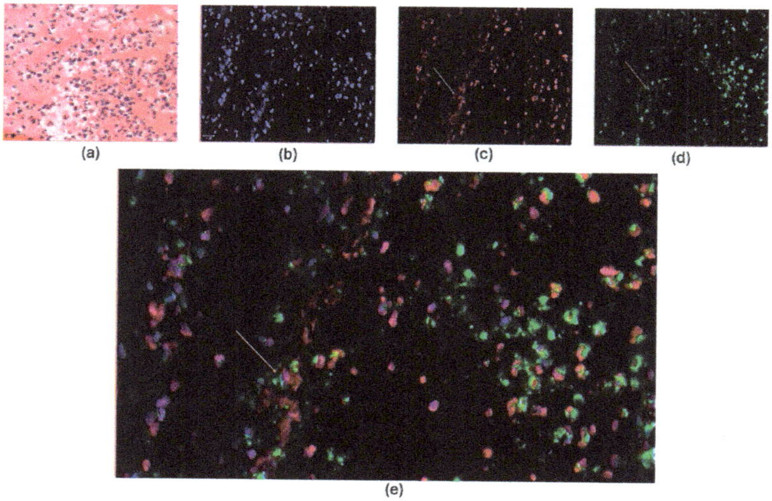


**Figure 2.** Co-localization of NET markers in coronary thrombi. (**a**): The HE-stained image displayed a cluster of abundant inflammatory cells, predominantly neutrophil granulocytes, with fibrin appearing as eosinophilic and amorphous deposit filaments. b-e: Immunofluorescence-stained thrombus, as in (**a**). (**b**): The nuclei and extracellular DNA were stained with DAPI (blue). (**c**): The nuclei were stained with histone H3 (red). On the left, web-like structures show extracellular deposits of histone H3. (**d**): Green-stained neutrophil elastase (NE) was abundant in the cytoplasm, cell membrane, and extracellular granular debris. (**e**): Co-staining showed the co-localization of DAPI (blue), NE (green), and histone H3 (red). Magnification was 100×, and the scale bar length was 20 μm. Arrows indicate extracellular staining of the chosen NET markers.

### 2.3. Investigations in Peripheral Blood

#### 2.3.1. Associations Between PAD4 mRNA, IL-6, IL-6R, and NETs

As for the coronary thrombi, the mRNA of PAD4 and IL-6R mRNA were correlated in circulating leukocytes (Figure 1b). This association remained statistically significant in a multivariate, linear regression model adjusted for age, sex, and symptom duration (Table 3). The serum protein levels of the NET marker H3Cit were correlated with IL-6R protein levels ($r_s = 0.40$, $p = 0.02$). No other correlations were observed between NET markers and IL-6R protein levels, or between NET markers and IL-6 protein levels.

#### 2.3.2. Comparison of Coronary PAD4 mRNA and Circulating Markers

PAD4 mRNA levels in coronary thrombi correlated with protein levels of PAD4 in peripheral blood measured in serum ($r_s = 0.39$, $p = 0.048$). PAD4 mRNA levels in thrombi also correlated with dsDNA ($r_s = 0.47$, $p = 0.014$) measured in plasma, but not with the other NET markers in peripheral blood.

## 3. Discussion

In coronary thrombi from patients with STEMI, we found that PAD4 mRNA levels correlated with IL-6R mRNA expression, as well as myocardial injury measured as troponin T. We also observed correlations between PAD4 mRNA and IL-6R mRNA in circulating leukocytes, and between protein levels of IL-6R and H3Cit in peripheral blood. Together, these findings might suggest that PAD4 links IL-6R signaling and NETosis in STEMI.

Individually, both NETs and IL-6 signaling have been associated with the formation and stabilization of thrombi in myocardial infarction [3,22]. The web-like structure of NETs coated with neutrophil proteins (Figure 2) has prothrombotic and procoagulant effects and hampers fibrinolysis [10,11,23–25]. The release of NETs, called NETosis, occurs from minutes to hours after the neutrophil cell is activated, and local NETosis within the coronary thrombi during acute myocardial infarction participates in the prevention of spontaneous reperfusion of the coronary artery [8,26–28]. In previous studies, the levels of NETs in coronary thrombi have been correlated with infarct size and clinical outcomes, and our finding that PAD4 mRNA is associated with high levels of troponin T corroborates the importance of NETosis in coronary thrombi for myocardial injury in STEMI [8,14]. Importantly, this is further supported by the correlation between PAD4 mRNA and circulating dsDNA, as both NETosis and myocardial cell damage are potential sources of circulating dsDNA after STEMI. As part of IL-6 signaling, our group has previously shown that the NLRP3-activated pathway, with IL-1β and IL-6 as downstream proteins, is also highly regulated in coronary thrombi, and that local IL-6R mRNA in thrombi is linked to myocardial damage [21]. It has recently been shown that neutrophil cells themselves also release NLRP3 in the early inflammatory phase of myocardial ischemia, and that NLRP3 is not only a precursor but also a stimulus for NETosis [29,30]. Although the IL-6-PAD4 axis has been described in other acute and chronic diseases [31,32], this is, to the best of our knowledge, the first time the interplay between NETosis and IL-6R signaling in coronary thrombi has been investigated. Our demonstration of an association between PAD4 mRNA and IL-6R mRNA aligns with our recent finding that IL-6R inhibition reduces the neutrophil cell's ability to undergo NETosis and marks PAD4 as an emerging target to mitigate local inflammation [6]. The concept of PAD4 inhibition is supported by experimental models that have shown reductions in NET-induced damage in myocardial infarction by PAD4 inhibition [20,33]. Considering the findings from this study, exploring the effects of PAD4 or IL-6R inhibition on coronary thrombus composition is warranted. Such interventions may not only promote thrombus destabilization but also mitigate ischemia–reperfusion injury in patients with STEMI. As ischemia–reperfusion injury is a significant contributing factor to myocardial damage without effective treatment, it represents an ongoing challenge in modern cardiology [34–36].

We observed, like the findings in thrombi, an association between PAD mRNA and IL-6R mRNA in circulating leukocytes, where neutrophils represent the predominant subtype. This observation might suggest that the findings in coronary thrombi were derived from neutrophils, although admixture from other inflammatory cells like macrophages cannot be ruled out. Activated circulating neutrophil cells also play an active role during ischemia–reperfusion injury [36]. NETosis might thus play a role in this detrimental process. In this study, peripheral blood samples were collected simultaneously with thrombus aspiration, i.e., as reperfusion started and before the full effect was presumably reached. We have previously published that coronary thrombi and circulating leukocyte mRNA levels of IL-6 and IL-6R did not correlate with the protein levels of IL-6 and IL-6R in peripheral blood [21], suggesting that peak concentrations of mRNA and protein may occur at distinct time points. Nevertheless, we did find a correlation between the protein levels of the NET marker H3Cit and IL-6R in the circulation. H3Cit is the direct product of PAD4 activity, which was upregulated both in coronary thrombi and circulating leukocytes, and is considered as a specific NET marker [37,38].

*Limitations*

The sample size in our study is suitable for hypothesis-generating studies but limits subgroup analysis and increases the risk of type II statistical errors. We only included patients that underwent thrombus aspiration at the discretion of the PCI-operator. These patients might represent a subgroup with higher propensity for thrombus formation compared to the general STEMI population, and their thrombi composition might differ from that of most other patients with STEMI. In addition, the low female study participation rate of 9% limits the generalizability of our results. Moreover, the chosen marker of NETs can be discussed, as NET composition might vary depending on the initial stimuli [39,40], and more ideal markers for STEMI-related research cannot be ruled out. Also, IL-6R exists in two forms, as membrane-bound (mIL-6R) and soluble (sIL-6R). In this study, only sIL-6R was measured in the circulation. In addition, quantification of the immunofluorescence-stained factors in coronary thrombi would have added valuable insight; however, we lacked the necessary expertise and facilities to perform it at the time of the study. Furthermore, the timing of blood sampling may be relevant to our findings. All samples were obtained at the catheterization lab shortly after hospital admission, but the ischemic period ranged from 60 to 1440 min. The processes of translation from mRNA to protein, as well as the elimination of proteins, could be in different stages. Lastly, this is a cross-sectional study, and causality cannot be determined. Future studies are needed to establish the value of these markers in the context of myocardial infarction.

## 4. Materials and Methods

### 4.1. Study Design

The material analyzed in this study was collected for the Thrombus Aspiration in ST-elevation myocardial Infarction (TASTI) trial (registered at clinicaltrials.gov, NCT02746822), as previously described [21]. Briefly, 33 patients with STEMI treated with percutaneous coronary intervention (PCI) and thrombus aspiration were consecutively included between August 2015 and January 2019. Thrombi and peripheral blood samples were obtained simultaneously in the catheterization lab, and an additional peripheral blood sample was drawn the following morning.

*4.2. Thrombi Preparation*

Intracoronary thrombi were obtained using a standard aspiration catheter, rinsed with saline, and divided into two equal portions. One portion was preserved in 10% buffered formalin, processed chemically, and embedded in paraffin for subsequent histological and immunofluorescence analyses. The second portion was snap-frozen in RNA-later solution (Qiagen, Hilden, Germany) and stored at −80 °C to facilitate later RNA extraction and gene expression analysis.

The formalin-fixed, paraffin-embedded (FFPE) thrombi were serially sectioned at 3.5 μm. Sections 1 and 7 were stained with hematoxylin and eosin (HE). For immunofluorescence analysis, the FFPE thrombi were immunofluorescence-stained for histone H3 and neutrophil elastase (Supplementary Tables S1 and S2). Double immunofluorescence was performed sequentially using a Ventana Discovery Ultra automated slide stainer (Ventana Medical System, 750–601, Roche, Basel, Switzerland). Monoclonal mouse anti-human neutrophil elastase (M752, Dako, Glostrup, Denmark) diluted 1:50 in Discovery Antibody Diluent (05266319001, Roche Diagnostics, Basel, Switzerland) was incubated for 32 min, followed by incubation with OmniMap anti-mouse HRP (5269652001, Roche) for 12 min and Discovery FITC (07259212001, Roche Diagnostics, Basel, Switzerland) for 12 min.

Monoclonal rabbit anti-histone H3 (4499S, Cell Signaling Technology, Inc., Danvers, MA, USA) diluted 1:400 in Discovery Antibody Diluent (05266319001, Roche Diagnostics, Basel, Switzerland) was incubated for 32 min, followed by OmniMap anti-Rabbit HRP (05269679001, Roche Diagnostics, Basel, Switzerland) for 12 min and Discovery Rhodamine (07259883001, Roche Diagnostics, Basel, Switzerland) for 12 min. Heat-induced epitope retrieval (HIER) was performed before the second cycle using Discovery CC1 (06414575001, Roche Diagnostics, Basel, Switzerland) for 48 min at 95 °C, and peroxidase was inactivated using a Discovery inhibitor (7017944001, Roche Diagnostics, Basel, Switzerland). Cell nuclei were stained with Discovery QD 4′,6-Diamidino-2-Phenylindole (DAPI) (5268826001, Roche Diagnostics, Basel, Switzerland) for 8 min, and the sections were mounted with ProLong Glass Antifade Mountant (Molecular Probes, Thermo Fisher Scientific, Waltham, MA, USA).

*4.3. Blood Sampling Protocol*

Peripheral venous blood was drawn into tubes without additives for serum analysis and tubes containing EDTA for plasma analysis. Tubes without additives were kept at room temperature for 30–60 min to ensure complete coagulation, followed by centrifugation at 2100× *g* for 10 min at room temperature. EDTA tubes were immediately placed on ice and centrifuged within 30 min at 2500 g for 20 min at 4 °C to procure platelet-poor plasma. Following centrifugation, serum and plasma samples were divided into aliquots and stored at −80 °C for subsequent analyses. BD PAXgene™ Blood RNA tubes were maintained at room temperature for 2–72 h to stabilize leukocyte RNA prior to storage at −80 °C.

*4.4. Laboratory Analysis*

For mRNA quantification, we extracted RNA from the snap-frozen thrombi samples utilizing the High Pure RNA Tissue Kit (Roche Diagnostics GmbH, Mannheim, Germany), supplemented with Proteinase K Solution and stabilized with lysing buffer. Homogenization was achieved using a Thermomixer (Eppendorf AG, Hamburg, Germany) and stainless steel grinding balls (Qiagen GmbH, Hilden, Germany). We isolated RNA from PAXgene tubes with the PAXgene® Blood RNA Kit (PreAnalytix, Qiagen GmbH) and further performed purification using RNeasy® MinElute® Cleanup Kit (Qiagen). RNA quality and concentration (ng/μL) were assessed with the NanoDrop™ 1000 Spectrophotometer (Thermo

Scientific, Wilmington, DE, USA). Complementary DNA was synthesized from equal amounts of RNA using the qScript™ cDNA SuperMix (Quanta Biosciences, Gaithersburg, MD, USA). Real-time PCR was conducted on a ViiA™ 7 system (Applied Biosystems, Foster City, CA, USA) with TaqMan® Universal PCR Master Mix (P/N 4324018) and TaqMan® assays for IL-6 (Hs00174131_m1), IL-6R (Hs01075664_m1), and PAD 4 (Hs01057483_m1).

In peripheral blood drawn at admission, dsDNA was quantified in EDTA plasma using the fluorescent nucleic acid stain Quant-iT PicoGreen (Invitrogen, Paisley, UK) and measured by fluorometry (Fluoroskan Ascent, Thermo Fisher Scientific Oy, Vantaa, Finland). Myeloperoxidase–DNA (MPO-DNA) complexes were assessed in undiluted EDTA plasma using an enzyme-linked immunosorbent assay (ELISA) technique previously described by Kessenbrock et al. [41]. Briefly, microplates were coated with the capture antibody anti-MPO (Bio-Rad, Hercules, CA, USA) and incubated overnight at 4 °C. After blocking with bovine serum albumin, patient samples and a peroxidase-labeled anti-DNA antibody (Cell Death Detection kit, Roche Diagnostics GmbH, Mannheim, Germany) were added and incubated for two hours. Following incubation, a peroxidase substrate was introduced, and absorbance was measured and expressed as optical density (OD) units. H3Cit levels were analyzed in serum in a 1:2 dilution with ELISA buffer using a commercial sandwich ELISA kit (Cayman Chemical, Ann Arbor, MI, USA). PAD4, IL-6, and IL-6R were quantified in serum using commercially available ELISA kits (PAD4 (human), Cayman Chemical, Ann arbor, MI, USA and Quantikine® HS ELISA, R&D Systems®, Minneapolis, MN, USA). All samples were analyzed on one plate of the specific assays. The intra-assay coefficients of variability were 5.8% (dsDNA), 7.9% (MPO-DNA), 13.2% (H3Cit), 1.4% (PAD4), 10.6% (IL-6), and 3.6% (IL-6R).

Cardiac troponin T was measured in serum collected at admission and on the following day with commercial electrochemiluminescence immunoassay (third-generation cTnT, Elecys 2010, Roche, Mannheim, Germany). The inter-assay coefficient of variability was 7%.

### *4.5. Statistical Methods*

The demographic data were given as median (25% and 75% percentiles), mean (SD), or numbers (%) as appropriate. All variables were evaluated for normality by histograms and qq-plots. Correlation analyses were performed by Spearman's Rho. Linear associations were analyzed with linear regression. Multivariable models were adjusted for the covariates: age, sex, and symptom duration. We chose not to adjust for inflammatory-linked variables like CRP and troponin T. The level of statistical significance was set to two-sided $p \leq 0.05$. All statistical analyses were performed on STATA v.18 SE (StataCorp LLC, College Station, TX, USA).

## 5. Conclusions

We demonstrate an association between PAD4, IL-6R, and troponin release in STEMI patients. This indicates a PAD4-mediated link between IL-6R and NET formation and highlights PAD4 as a potential treatment target for mitigating inflammation and myocardial injury in STEMI. Further mechanistic and clinical studies are needed to confirm a causal relationship and clarify the therapeutic potential of PAD4 inhibition in STEMI.

**Supplementary Materials:** The following supporting information can be downloaded at https://www.mdpi.com/article/10.3390/ijms26115348/s1.

**Author Contributions:** Conceptualization, S.S., R.H. and I.S.; methodology; S.S., R.H. and I.S.; formal analysis, K.M.K.; investigation, J.N.; resources, J.N. and M.S.L.; writing—original draft preparation,

K.M.K.; writing—review and editing, R.H., S.S., M.S.L., I.S., J.N., H.S. and M.K.S.; visualization, K.M.K., H.S., B.R. and R.H.; supervision, R.H., I.S. and M.K.S.; project administration, S.S. and R.H.; funding acquisition, S.S., R.H., M.S.L., I.S. and B.R. All authors have read and agreed to the published version of the manuscript.

**Funding:** This research was supported by grants from Stein Erik Hagens Foundation for Clinical Heart Research; the Blix Foundation for the Promotion of Medical Research; Ada og Hagbart Waages Humanitære og Veldedige Stiftelse; the South-Eastern Norway Regional Health Authority (PhD grant K.M.K., grant number 2022012); Marie Stenbergs Legat; and the National Health Association (Postdoctoral Fellowship grant R.H., grant number 43600). All grants were unrestricted and did not impact the study results.

**Institutional Review Board Statement:** The study was approved by the Regional Committee of Medical Research Ethics in South-Eastern Norway (2015/169) and conducted according to the Declaration of Helsinki.

**Informed Consent Statement:** Written informed consent was collected from all patients.

**Data Availability Statement:** Data will be available upon reasonable request.

**Acknowledgments:** The authors would like to thank Sissel Åkra, Sheryl Palermo, and Hogne Røed Nilsen for excellent laboratory work and technical assistance.

**Conflicts of Interest:** The authors declare no conflicts of interest.

## Abbreviations

The following abbreviations are used in this manuscript:

| | |
|---|---|
| STEMI | ST-elevation myocardial infarction |
| Il-6 | Interleukin-6 |
| IL-6R | Interleukin-6 receptor |
| NETs | Neutrophil extracellular traps |
| PAD4 | Peptidylarginine deiminase 4 |
| dsDNA | Double-stranded deoxyribonucleic acid |
| H3Cit | Citrullinated histone H3 |
| MPO-DNA | Myeloperioxidase bound to deoxyribonucleic acid |
| NLRP3 | NOD-, LRR-, and pyrin domain-containing protein 3 |
| ASSAIL-MI | Assessing the effect of anti-IL-6 treatment in myocardial infarction |
| PCI | Percutaneous coronary intervention |
| FFPE | Formalin-fixed, paraffin-embedded |
| ELISA | Enzyme-linked immunosorbent assay |

## References

1. Tollefsen, I.M.; Shetelig, C.; Seljeflot, I.; Eritsland, J.; Hoffmann, P.; Andersen, G.O. High levels of interleukin-6 are associated with final infarct size and adverse clinical events in patients with STEMI. *Open Heart* **2021**, *8*, e001869. [CrossRef]
2. Groot, H.E.; Al Ali, L.; van der Horst, I.C.C.; Schurer, R.A.J.; van der Werf, H.W.; Lipsic, E.; van Veldhuisen, D.J.; Karper, J.C.; van der Harst, P. Plasma interleukin 6 levels are associated with cardiac function after ST-elevation myocardial infarction. *Clin. Res. Cardiol.* **2019**, *108*, 612–621. [CrossRef]
3. Libby, P. Targeting Inflammatory Pathways in Cardiovascular Disease: The Inflammasome, Interleukin-1, Interleukin-6 and Beyond. *Cells* **2021**, *10*, 951. [CrossRef]
4. Broch, K.; Anstensrud, A.K.; Woxholt, S.; Sharma, K.; Tollefsen, I.M.; Bendz, B.; Aakhus, S.; Ueland, T.; Amundsen, B.H.; Damas, J.K.; et al. Randomized Trial of Interleukin-6 Receptor Inhibition in Patients With Acute ST-Segment Elevation Myocardial Infarction. *J. Am. Coll. Cardiol.* **2021**, *77*, 1845–1855. [CrossRef]
5. Feng, Y.; Ye, D.; Wang, Z.; Pan, H.; Lu, X.; Wang, M.; Xu, Y.; Yu, J.; Zhang, J.; Zhao, M.; et al. The Role of Interleukin-6 Family Members in Cardiovascular Diseases. *Front. Cardiovasc. Med.* **2022**, *9*, 818890. [CrossRef]

6. Kindberg, K.M.; Broch, K.; Andersen, G.O.; Anstensrud, A.K.; Akra, S.; Woxholt, S.; Tollefsen, I.M.; Ueland, T.; Amundsen, B.H.; Klow, N.E.; et al. Neutrophil Extracellular Traps in ST-Segment Elevation Myocardial Infarction: Reduced by Tocilizumab and Associated With Infarct Size. *JACC Adv.* **2024**, *3*, 101193. [CrossRef]
7. Brinkmann, V.; Reichard, U.; Goosmann, C.; Fauler, B.; Uhlemann, Y.; Weiss, D.S.; Weinrauch, Y.; Zychlinsky, A. Neutrophil extracellular traps kill bacteria. *Science* **2004**, *303*, 1532–1535. [CrossRef]
8. Mangold, A.; Alias, S.; Scherz, T.; Hofbauer, M.; Jakowitsch, J.; Panzenbock, A.; Simon, D.; Laimer, D.; Bangert, C.; Kammerlander, A.; et al. Coronary neutrophil extracellular trap burden and deoxyribonuclease activity in ST-elevation acute coronary syndrome are predictors of ST-segment resolution and infarct size. *Circ. Res.* **2015**, *116*, 1182–1192. [CrossRef]
9. Doring, Y.; Soehnlein, O.; Weber, C. Neutrophil Extracellular Traps in Atherosclerosis and Atherothrombosis. *Circ. Res.* **2017**, *120*, 736–743. [CrossRef]
10. Nappi, F.; Bellomo, F.; Avtaar Singh, S.S. Worsening Thrombotic Complication of Atherosclerotic Plaques Due to Neutrophils Extracellular Traps: A Systematic Review. *Biomedicines* **2023**, *11*, 113. [CrossRef]
11. Varju, I.; Longstaff, C.; Szabo, L.; Farkas, A.Z.; Varga-Szabo, V.J.; Tanka-Salamon, A.; Machovich, R.; Kolev, K. DNA, histones and neutrophil extracellular traps exert anti-fibrinolytic effects in a plasma environment. *Thromb. Haemost.* **2015**, *113*, 1289–1298. [CrossRef]
12. Folco, E.J.; Mawson, T.L.; Vromman, A.; Bernardes-Souza, B.; Franck, G.; Persson, O.; Nakamura, M.; Newton, G.; Luscinskas, F.W.; Libby, P. Neutrophil Extracellular Traps Induce Endothelial Cell Activation and Tissue Factor Production Through Interleukin-1alpha and *Cathepsin G*. *Arterioscler. Thromb. Vasc. Biol.* **2018**, *38*, 1901–1912. [CrossRef]
13. Helseth, R.; Shetelig, C.; Andersen, G.O.; Langseth, M.S.; Limalanathan, S.; Opstad, T.B.; Arnesen, H.; Hoffmann, P.; Eritsland, J.; Seljeflot, I. Neutrophil Extracellular Trap Components Associate with Infarct Size, Ventricular Function, and Clinical Outcome in STEMI. *Mediat. Inflamm.* **2019**, *2019*, 7816491. [CrossRef]
14. Blasco, A.; Coronado, M.J.; Vela, P.; Martin, P.; Solano, J.; Ramil, E.; Mesquida, A.; Santos, A.; Cozar, B.; Royuela, A.; et al. Prognostic Implications of Neutrophil Extracellular Traps in Coronary Thrombi of Patients with ST-Elevation Myocardial Infarction. *Thromb. Haemost.* **2022**, *122*, 1415–1428. [CrossRef]
15. Novotny, J.; Chandraratne, S.; Weinberger, T.; Philippi, V.; Stark, K.; Ehrlich, A.; Pircher, J.; Konrad, I.; Oberdieck, P.; Titova, A.; et al. Histological comparison of arterial thrombi in mice and men and the influence of Cl-amidine on thrombus formation. *PLoS ONE* **2018**, *13*, e0190728. [CrossRef]
16. Riegger, J.; Byrne, R.A.; Joner, M.; Chandraratne, S.; Gershlick, A.H.; Ten Berg, J.M.; Adriaenssens, T.; Guagliumi, G.; Godschalk, T.C.; Neumann, F.J.; et al. Histopathological evaluation of thrombus in patients presenting with stent thrombosis. A multicenter European study: A report of the prevention of late stent thrombosis by an interdisciplinary global European effort consortium. *Eur. Heart J.* **2016**, *37*, 1538–1549. [CrossRef]
17. Fuchs, T.A.; Brill, A.; Duerschmied, D.; Schatzberg, D.; Monestier, M.; Myers, D.D., Jr.; Wrobleski, S.K.; Wakefield, T.W.; Hartwig, J.H.; Wagner, D.D. Extracellular DNA traps promote thrombosis. *Proc. Natl. Acad. Sci. USA* **2010**, *107*, 15880–15885. [CrossRef]
18. Gould, T.J.; Vu, T.T.; Swystun, L.L.; Dwivedi, D.J.; Mai, S.H.; Weitz, J.I.; Liaw, P.C. Neutrophil extracellular traps promote thrombin generation through platelet-dependent and platelet-independent mechanisms. *Arter. Thromb. Vasc. Biol.* **2014**, *34*, 1977–1984. [CrossRef]
19. Thiam, H.R.; Wong, S.L.; Qiu, R.; Kittisopikul, M.; Vahabikashi, A.; Goldman, A.E.; Goldman, R.D.; Wagner, D.D.; Waterman, C.M. NETosis proceeds by cytoskeleton and endomembrane disassembly and PAD4-mediated chromatin decondensation and nuclear envelope rupture. *Proc. Natl. Acad. Sci. USA* **2020**, *117*, 7326–7337. [CrossRef]
20. Du, M.; Yang, W.; Schmull, S.; Gu, J.; Xue, S. Inhibition of peptidyl arginine deiminase-4 protects against myocardial infarction induced cardiac dysfunction. *Int. Immunopharmacol.* **2020**, *78*, 106055. [CrossRef]
21. Nordeng, J.; Schandiz, H.; Solheim, S.; Akra, S.; Hoffman, P.; Roald, B.; Bendz, B.; Arnesen, H.; Helseth, R.; Seljeflot, I. The Inflammasome Signaling Pathway Is Actively Regulated and Related to Myocardial Damage in Coronary Thrombi from Patients with STEMI. *Mediat. Inflamm.* **2021**, *2021*, 5525917. [CrossRef] [PubMed]
22. Thalin, C.; Hisada, Y.; Lundstrom, S.; Mackman, N.; Wallen, H. Neutrophil Extracellular Traps: Villains and Targets in Arterial, Venous, and Cancer-Associated Thrombosis. *Arter. Thromb. Vasc. Biol.* **2019**, *39*, 1724–1738. [CrossRef]
23. Bonaventura, A.; Vecchie, A.; Abbate, A.; Montecucco, F. Neutrophil Extracellular Traps and Cardiovascular Diseases: An Update. *Cells* **2020**, *9*, 231. [CrossRef] [PubMed]
24. Farkas, A.Z.; Farkas, V.J.; Gubucz, I.; Szabo, L.; Balint, K.; Tenekedjiev, K.; Nagy, A.I.; Sotonyi, P.; Hidi, L.; Nagy, Z.; et al. Neutrophil extracellular traps in thrombi retrieved during interventional treatment of ischemic arterial diseases. *Thromb. Res.* **2019**, *175*, 46–52. [CrossRef] [PubMed]
25. Longstaff, C.; Varju, I.; Sotonyi, P.; Szabo, L.; Krumrey, M.; Hoell, A.; Bota, A.; Varga, Z.; Komorowicz, E.; Kolev, K. Mechanical stability and fibrinolytic resistance of clots containing fibrin, DNA, and histones. *J. Biol. Chem.* **2013**, *288*, 6946–6956. [CrossRef]

26. Yipp, B.G.; Kubes, P. NETosis: How vital is it? *Blood* **2013**, *122*, 2784–2794. [CrossRef]
27. McGill, C.J.; Lu, R.J.; Benayoun, B.A. Protocol for analysis of mouse neutrophil NETosis by flow cytometry. *STAR Protoc.* **2021**, *2*, 100948. [CrossRef]
28. Stakos, D.A.; Kambas, K.; Konstantinidis, T.; Mitroulis, I.; Apostolidou, E.; Arelaki, S.; Tsironidou, V.; Giatromanolaki, A.; Skendros, P.; Konstantinides, S.; et al. Expression of functional tissue factor by neutrophil extracellular traps in culprit artery of acute myocardial infarction. *Eur. Heart J.* **2015**, *36*, 1405–1414. [CrossRef]
29. Heger, L.A.; Schommer, N.; Van Bruggen, S.; Sheehy, C.E.; Chan, W.; Wagner, D.D. Neutrophil NLRP3 promotes cardiac injury following acute myocardial infarction through IL-1beta production, VWF release and NET deposition in the myocardium. *Sci. Rep.* **2024**, *14*, 14524. [CrossRef]
30. Munzer, P.; Negro, R.; Fukui, S.; di Meglio, L.; Aymonnier, K.; Chu, L.; Cherpokova, D.; Gutch, S.; Sorvillo, N.; Shi, L.; et al. NLRP3 Inflammasome Assembly in Neutrophils Is Supported by PAD4 and Promotes NETosis Under Sterile Conditions. *Front. Immunol.* **2021**, *12*, 683803. [CrossRef]
31. Yahagi, A.; Saika, T.; Hirano, H.; Takai-Imamura, M.; Tsuji, F.; Aono, H.; Iseki, M.; Morita, Y.; Igarashi, H.; Saeki, Y.; et al. IL-6-PAD4 axis in the earliest phase of arthritis in knock-in gp130F759 mice, a model for rheumatoid arthritis. *RMD Open* **2019**, *5*, e000853. [CrossRef] [PubMed]
32. Kang, L.; Yu, H.; Yang, X.; Zhu, Y.; Bai, X.; Wang, R.; Cao, Y.; Xu, H.; Luo, H.; Lu, L.; et al. Neutrophil extracellular traps released by neutrophils impair revascularization and vascular remodeling after stroke. *Nat. Commun.* **2020**, *11*, 2488. [CrossRef]
33. Yang, K.; Gao, R.; Chen, H.; Hu, J.; Zhang, P.; Wei, X.; Shi, J.; Chen, Y.; Zhang, L.; Chen, J.; et al. Myocardial reperfusion injury exacerbation due to ALDH2 deficiency is mediated by neutrophil extracellular traps and prevented by leukotriene C4 inhibition. *Eur. Heart J.* **2024**, *45*, 1662–1680. [CrossRef]
34. Hausenloy, D.J.; Yellon, D.M. Myocardial ischemia-reperfusion injury: A neglected therapeutic target. *J. Clin. Investig.* **2013**, *123*, 92–100. [CrossRef]
35. Yellon, D.M.; Hausenloy, D.J. Myocardial reperfusion injury. *N. Engl. J. Med.* **2007**, *357*, 1121–1135. [CrossRef]
36. Shah, M.; Yellon, D.M.; Davidson, S.M. The Role of Extracellular DNA and Histones in Ischaemia-Reperfusion Injury of the Myocardium. *Cardiovasc. Drugs Ther.* **2020**, *34*, 123–131. [CrossRef] [PubMed]
37. Saisorn, W.; Santiworakul, C.; Phuengmaung, P.; Siripen, N.; Rianthavorn, P.; Leelahavanichkul, A. Extracellular traps in peripheral blood mononuclear cell fraction in childhood-onset systemic lupus erythematosus. *Sci. Rep.* **2024**, *14*, 23177. [CrossRef]
38. Paues Goranson, S.; Thalin, C.; Lundstrom, A.; Hallstrom, L.; Lasselin, J.; Wallen, H.; Soop, A.; Mobarrez, F. Circulating H3Cit is elevated in a human model of endotoxemia and can be detected bound to microvesicles. *Sci. Rep.* **2018**, *8*, 12641. [CrossRef] [PubMed]
39. Chen, T.; Li, Y.; Sun, R.; Hu, H.; Liu, Y.; Herrmann, M.; Zhao, Y.; Munoz, L.E. Receptor-Mediated NETosis on Neutrophils. *Front. Immunol.* **2021**, *12*, 775267. [CrossRef]
40. Petretto, A.; Bruschi, M.; Pratesi, F.; Croia, C.; Candiano, G.; Ghiggeri, G.; Migliorini, P. Neutrophil extracellular traps (NET) induced by different stimuli: A comparative proteomic analysis. *PLoS ONE* **2019**, *14*, e0218946. [CrossRef]
41. Kessenbrock, K.; Krumbholz, M.; Schonermarck, U.; Back, W.; Gross, W.L.; Werb, Z.; Grone, H.J.; Brinkmann, V.; Jenne, D.E. Netting neutrophils in autoimmune small-vessel vasculitis. *Nat. Med.* **2009**, *15*, 623–625. [CrossRef] [PubMed]

International Journal of
*Molecular Sciences*

MDPI

*Article*

# Prolonged Extracorporeal Circulation Leads to Inflammation and Higher Expression of Mediators of Vascular Permeability Through Activation of STAT3 Signaling Pathway in Macrophages

**Jana Luecht [1,†], Camila Pauli [2,†], Raphael Seiler [1], Alexa-Leona Herre [2], Liliya Brankova [2], Felix Berger [1], Katharina R. L. Schmitt [3,‡] and Giang Tong [3,*,‡]**

1 Department of Congenital Heart Disease/Pediatric Cardiology, Deutsches Herzzentrum der Charité, 13353 Berlin, Germany; jana.luecht@dhzc-charite.de (J.L.); rseiler@dhzc-charite.de (R.S.); felix.berger@dhzc-charite.de (F.B.)

2 Department of Developmental Pediatric Cardiology, Charité—Universitätsmedizin Berlin, 10117 Berlin, Germany; camila.pauli@charite.de (C.P.); alexa-leona.herre@charite.de (A.-L.H.); lyliya.brankova@charite.de (L.B.)

3 Department of Developmental Pediatric Cardiology, Deutsches Herzzentrum der Charité, 13353 Berlin, Germany; katharina.schmitt@dhzc-charite.de

* Correspondence: giang.tong@dhzc-charite.de; Tel.: +49-(0)30-4593-2806

† These authors contributed equally to the manuscript.

‡ These authors contributed equally to the manuscript.

**Abstract:** Congenital heart defects (CHDs) are one of the most common congenital malformations and often require heart surgery with cardiopulmonary bypass (CPB). Children undergoing cardiac surgery with CPB are especially at greater risk of post-operative complications due to a systemic inflammatory response caused by innate inflammatory mediators. However, the pathophysiological response is not fully understood and warrants further investigation. Therefore, we investigated the inflammatory response in macrophages initiated by peri-operative serum samples obtained from patients with CHD undergoing CPB cardiac surgery. Human differentiated THP-1 macrophages were pretreated with Stattic, a STAT3 (Tyr705) inhibitor, before stimulation with serum samples. STAT3 and NF-κB activation were investigated via a Western blot, IL-1β, TNFα, IL-10, mediators for vascular permeability (VEGF-A, ICAM), and SOCS3 gene expressions via RT-qPCR. CPB induced an inflammatory response in macrophages via the activation of the STAT3 but not NF-κB signaling pathway. Longer duration on the CPB correlated with increased cytokine, VEGF, and ICAM expressions, relative to individual pre-operation levels. Patients that did not require CPB showed no significant immune response. Pretreatment with Stattic significantly attenuated all inflammatory mediators investigated except for TNFα in the macrophages. CPB induces an increased expression of cytokines and mediators of vascular permeability via the activation of STAT3 by IL-6 and IL-8 in the serum samples. Stattic attenuates all mediators investigated but promotes TNFα expression.

**Keywords:** congenital heart defects; cardiopulmonary bypass; macrophages; sterile inflammation; vascular permeability; STAT3; NF-κB p65; Stattic; DAMPs; cytokines

## 1. Introduction

Congenital heart defects (CHDs) are one of the most common congenital malformations and often require heart surgery with cardiopulmonary bypass (CPB) [1]. Among others, the operation itself and CPB-induced ischemia/reperfusion injury lead to the release of inflammatory mediators and damage-associated patterns (DAMPs) into the bloodstream, activating various cascades (coagulation cascade, complement system, immune system) and inducing an acute inflammatory response, which may lead to post-operative complications such as systemic inflammatory response syndrome (SIRS) [2,3] and capillary leak syndrome (CLS) [4,5]. Despite technological progress, the harmful influences of CPB in neonates and children are often pronounced due to their immature tissue, organ function, and immune system. Moreover, the disparity in size between the CPB circuit and patients' cardiovascular system surface increases the risk for complications [6–8], resulting in higher morbidity with prolonged stays in the intensive care unit (ICU), prolonged ventilation times and sedation, increased drainage losses, and an increased need for diuretics and catecholamines [9,10]. Furthermore, the group of surviving adults with congenital heart defects (ACHDs) is growing thanks to medical advances and treatments. ACHDs represent an unique patient group with distinct physiological characteristics that may impact both their surgical outcomes and inflammatory responses to CPB [11,12].

To minimize complications in heart surgery, a multimodal approach is necessary due to the complex pathogenesis [13]. There have been further developments in various areas, including neuroprotective strategies, improved kidney and blood coagulation management, and measures in infection prevention specifically aimed at reducing the inflammatory response [14]. CPB systems optimized for children with shorter tubes of a smaller diameter, coating of the tubes with biocompatible substances such as heparin, and further development of surgical techniques that enable shorter operating times have led to a significant improvement in the surgical treatment of congenital heart defects and safer pediatric cardiac surgery with lower complication rates [13,15–17].

However, measures that intervene directly with the immune response such as the pre-operative administration of glucocorticoids, which is established in many centers, are viewed critically. There are currently no standardized guidelines. The optimal dosage, time of administration, and specific patient groups that may profit from glucocorticoids and many other factors remain unclear [18–22]. Moreover, there is a general lack of data from both pediatric and ACHD patients since previous studies have mainly focused on adults with non-congenital heart defects.

Regarding cellular signaling pathways, the emphasis of many former studies has been on the transcription factor NF-κB [3,23,24]. Furthermore, the pathophysiology of the triggered inflammatory reaction and the associated complications is still not fully understood. In our study, we focus on the evolutionarily conserved JAK-STAT3 pathway, which can be activated by inflammatory mediators such as IL-6 [25]. The JAK-STAT3 pathway plays a major role in regulating the immune response by modulating the expression of various cytokines and mediators for vascular permeability [26,27]. Moreover, multiple crosstalk mechanisms are being described between the STAT3- and NFκB-signaling pathway. Lee et al. demonstrate a constitutive NFκB activity maintained by STAT3 in cancer cells via complex formation with NFκB subunits and the direct modulation of the gene expression of components important for NFκB signaling [28]. IL-6 is released by different cell types including macrophages, T cells, and endothelial cells, and is an important mediator in acute inflammatory reactions by modulating the balance between pro- and anti-inflammatory mediators; by interfering in the cell growth, survival, and differentiation; or by promoting the production of acute-phase proteins [29,30]. Particularly in pediatric patients, IL-6 is already an established biomarker for Sepsis [31,32].

The goal of our exploratory study is to better understand the heterogenous patient population who underwent corrective or palliative heart surgery for CHD at our hospital, as well as to shed light on the complex pathophysiology of CPB-induced inflammation in this cohort. Therefore, we optimized THP-1 human macrophages differentiated from acute monocytic leukemia cells by the PMA model to investigate the in vitro inflammatory response induced by the secreted mediators in peri-operatively collected serum samples from our patient cohort with CHD who underwent cardiac surgery with and without CPB. Moreover, we aim to investigate if there is a correlation between CPB duration and inflammatory response with a focus on the activation of the JAK-STAT3 pathway.

## 2. Results

In the current study, we enrolled 56 consecutive patients whose demographic and clinical data are summarized in Table 1. The study cohort consisted of 31 males of a median age of 2.3 years (range: 0 to 59 years) and 25 females of a median age of 3.6 years (range: 0 to 65 years). Our heterogeneous cohort included patients with a broad range of ages (47 patients between 0 and 18 years old and 9 patients between 18 and 65 years old), congenital heart diseases, and a complexity of anomalies treated in our hospital. In order to collect a representative cohort for patients with CHD, we included patients receiving their first and corrective operation such as closure of simple atrial or ventricular septal defects, as well as patients undergoing palliative surgery such as Norwood I, pulmonal atrial banding, or Fontan procedures. No neonatal patients were included in the study and all patients survived until hospital discharge.

**Table 1.** Demographic and clinical data.

| | Sex | |
|---|---|---|
| | **Male** | **Female** |
| Number of patients: 56 | 31 (55%) | 25 (45%) |
| Demographic data (median; range) | | |
| Age (years) | 2.3 (0–59) | 3.6 (0–65) |
| Weight (kg) | 11.6 (2.8–101) | 15.2 (3.2–104) |
| Characteristics of operation and CBP (median; range) | | |
| RACHS-1 | 3 (1–4) | 2 (1–6) |
| Warnes | 2 (1–3) | 2 (1–3) |
| Operation time (min) | 361 (93–900) | 313 (153–804) |
| CBP time (min) | 180 (0–540) | 120 (60–540) |
| Aortic cross-clamp time (min) | 79 (0–518) | 56 (0–266) |

RACHS-1 = Risk Adjustment for Congenital Heart Surgery 1; CPB = cardiopulmonary bypass.

To quantify the complexity of the surgical procedure based on risk adjustment concerning in-hospital mortality, we used Risk Adjustment for Congenital Heart Surgery 1 (RACHS-1) [33]. RACHS-1 could not be assessed for one 2-year-old patient as the cardiac surgery was performed to implant a two-chamber pacemaker in the patient with a congenitally corrected transposition of the great arteries and third-degree congenital atrioventricular block, and this kind of operation is not included in this risk score.

Severity of CHD was classified according to the three categories suggested by Warnes et al. [12]: complex (e.g., the transposition of the great arteries, pulmonary atresia, tricuspid atresia, hypoplastic left heart syndrome), moderate (e.g., atrioventricular canal defects, tetralogy of Fallot, the coarctation of the aorta, ventricular septal defects with associated lesions), and mild (e.g., isolated small atrial septal defects or isolated small ventricular septal defect without associated lesions, isolated congenital aortic valve disease).

### 2.1. *Schedule for Peri-Operative Serum Sample Collection*

Serum samples from our patient cohort were collected as previously described [34]. Briefly, blood samples were obtained via the central venous line pre-operatively after the induction of anesthesia (T0), post-operatively upon arrival in the pediatric intensive care unit (PICU) (T1), 6 h after the operation (T2), and 24 h after the operation (T3), as illustrated in Figure 1.

### 2.2. *CPB Induces Activation of STAT3 but Not NF-κB p65 Signaling Pathway in THP-1 Macrophages In Vitro*

As previously reported, we detected significantly elevated levels of IL-6, IL-8, and IL-10 concentrations in the serum samples from patients with CHD undergoing cardiac surgery with CPB at all post-operative timepoints investigated (T1–T3), relative to individual baseline values (T0) [34]. Since both IL-6 and IL-8 are known to phosphorylate the STAT3 and IL-10 is known to inhibit the NF-κB p65 signaling pathways, we investigated the inflammatory signaling pathways induced in the THP-1 macrophages by these inflammatory mediators in our CHD cohort. We observed a significant increase in STAT3 phosphorylation at the Tyr705 site post-operatively, relative to individual baseline values, as shown in Figure 2a. Post-operative serum-mediated STAT3 phosphorylation was significantly increased immediately after surgery upon arrival in the pediatric intensive care unit (PICU) (T1, $p = 0.0363$), then gradually returned to individual baseline values over the 24 h observation period after the operation. No significant increase in STAT3 phosphorylation at the Ser727 site was observed. Moreover, pretreatment with 5 µM Stattic significantly attenuated STAT3 activation at T1 ($p = 0.0248$) in the macrophages. Interestingly, we did not observe any significant increase in NF-κB p65 phosphorylation induced by post-operative CPB serum samples, as shown in Figure 2b. In fact, post-operative values were, in general, lower than individual baseline values, and pretreatment with Stattic significantly reduced NF-κB p65 phosphorylation 6 h after the operation (T2, $p = 0.0162$). Moreover, we also observed a significant attenuation of NF-κB p65 phosphorylation by Stattic in our LPS-stimulated experimental positive control ($p = 0.0038$).

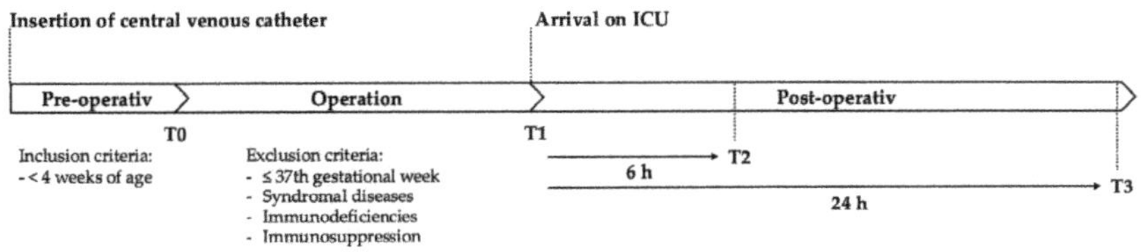


**Figure 1.** Patients' blood samples were obtained via the central venous line pre-operatively after the induction of anesthesia (T0), post-operatively upon arrival in the pediatric intensive care unit (PICU) (T1), and both 6 h (T2) and 24 h (T3) after the operation.

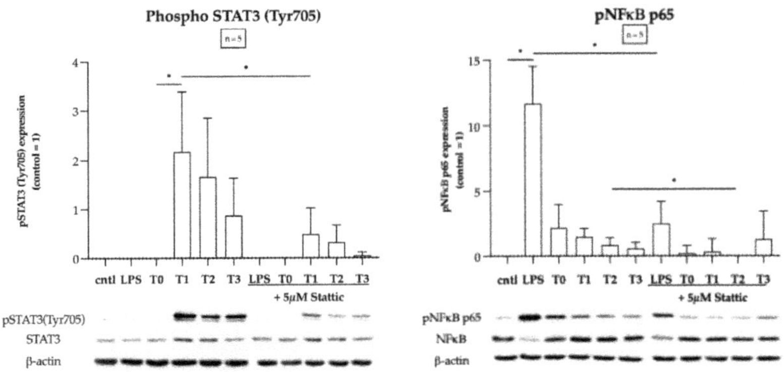


**Figure 2.** (**a**) CPB induces the activation of STAT3 at phosphorylation site Tyr705 that was attenuated by pretreatment with 5 µM Stattic in THP-1 macrophages in vitro. LPS had no effect on STAT3 phosphorylation. (**b**) CPB had no effect on the NF-κB p65 signaling pathway in THP-1 macrophages in vitro. Pretreatment with Stattic attenuated LPS-induced NF-κB p65 phosphorylation. Data from five separate experiments ($n = 5$) are represented as box plots. Statistical test analysis: RM one-way ANOVA with Dunnett's multiple comparisons post-test relative to individual pre-operative control (T0), a paired t-test relative to an untreated sample of the same time point, or an unpaired $t$-test relative to control; * $p \leq 0.05$ was considered significant.

### *2.3. Prolonged Duration of CPB Correlated with Induced Inflammatory Response and Increased Expression of Mediators of Vascular Permeability in THP-1 Macrophages In Vitro*

We observed that the peri-operative kinetics of the expression of cytokines and mediators of vascular permeability induced by patients' serum in THP-1 macrophages in vitro can be correlated with the duration of CPB, as shown in Figure 3. The patient cohort was classified into four distinct groups based on the duration of CPB as follows: 0 h ($n = 5$), 1–3 h ($n = 32$), 4–6 h ($n = 13$), and 7–9 h ($n = 6$). Patients undergoing cardiac surgery without CPB (0 h) did not elicit a significant immune response post-operatively in the macrophages for all targets investigated. Increased IL-1β expression in the macrophages correlated with an increased duration of CPB immediately after operation upon arrival in the pediatric intensive care unit (PICU) that was significantly higher for the 7–9 h CPB group (T1, $p = 0.0283$) relative to patients operated on without CPB (0 h), as shown in Figure 3a. A moderate duration of CPB, 1–3 h and 4–6 h, elicited a continuous increase in IL-1β expression throughout the post-operative phase that ultimately peaked at 24 h in both groups but did not reach significancy. Similarly, we observed that induced TNFα expression in the macrophages also continuously increased throughout the post-operative phase and peaked at 24 h in both 1–3 h and 4–6 h CPB groups but did not reach significancy relative to patients without CPB, as shown in Figure 3b. Interestingly, TNFα expression in the 7–9 h CPB group did not increase and was similar to the 0 h CPB group. Induced IL-10 expression for the 7–9 h CPB group was significantly higher than patients operated on without CPB immediately after the operation at T1 ($p = 0.0291$) and remained significantly higher 6 h after the operation (T2, $p = 0.0284$) before returning to baseline, as shown in Figure 3c. IL-10 expression for the 4–6 h CPB group continuously increased throughout the post-operative phase and reached significancy at both 6 h (T2, $p = 0.0243$) and 24 h (T3, $p = 0.0254$).

The elicit expression of the mediators of vascular permeability, including intracellular adhesion molecule-1 (ICAM) and vascular endothelial growth factor (VEGF), as well as the suppressor of cytokine signaling 3 (SOCS3), in the macrophages also showed similar correlation with the duration of CPB. We observed an immediate significant spike in SOCS3 expression relative to patients without CPB at T1 ($p = 0.0015$) that decreased towards baseline but remained significant 6 h after the operation (T2, $p = 0.0375$), as shown in Figure 3d. Interestingly, we did not observe any induced ICAM expression in the macrophages in our entire cohort, as shown in Figure 3e. ICAM expression in the 4–6 h CPB group continuously increased throughout the post-operative phase, which peaked at 24 h but did not reach significancy. Induced VEGF expression kinetics in the macrophages were similar to that of SOCS3, with a significant spike in the 7–9 h CPB group immediately after the operation at T1 ($p = 0.0005$) that decreased back to baseline but remained significant 6 h after the operation (T2, $p = 0.0009$), as shown in Figure 3f.

In summary, increasing the duration of CPB from 1 to 6 h elicited an immune response that gradually increased post-operatively upon reaching significancy relative to individual baseline values 24 h after the operation. In general, the 4–6 h CPB group showed a higher immune response than the shorter 1–3 h group. On the other hand, an extreme duration of CPB lasting from 7 to 9 h resulted in an immediate induced immune response in the macrophages upon arrival at the ICU (T1) that gradually decreased towards baseline.

### *2.4. Stattic Attenuates CPB-Induced Cytokines and Mediators of Vascular Permiability but Promotes TNFα Expression in THP-1 Macrophages In Vitro*

Cardiac surgery requiring the assistance of CPB to correct congenital heart defects is known to induce a systemic inflammatory response in patients. Based on our experimental observations and previous reports of the role of STAT3 in post-operative systemic inflammation [35], we investigated the role of CPB in the induction of the expression of cytokines and mediators of vascular permeability, as well as the effect of pretreatment with 5 μM Stattic on this inflammatory response, in an in vitro model of human THP-1 macrophages. Serum samples collected from patients were categorized according to the duration of CPB: no CPB (0 h) and 1–3 h, 4–6 h, and 7–9 h CPB. In general, we observed that serum samples from patients undergoing heart operation with the assistance of CPB elicited higher cytokine expressions than patients that did not require CPB, as shown in Figure 4. Additionally, the data are also shown in sub-groups with respect to patients' age, children (0–18 yrs old) and adults (18–65 yrs old), in Figure S1 of the supplementary file. Interestingly, serum-induced IL-1β and IL-10 expressions were significantly lower upon arrival at the ICU after the operation (T1) then individual baseline values before the operation (T0) in patients not requiring CPB, as shown in Figures 4a and 4c, respectively. We observed a significant decrease in IL-1β expression in patients not requiring CPB (0 h) at T1 ($p = 0.0260$), as shown in Figure 4a. However, for patients undergoing CPB, we observed a significant increase in IL-1β expression in the 1–3 h CPB group at T3 ($p = 0.0268$) and in the 4–6 h group at T3 ($p = 0.0216$). Stattic significantly attenuated IL-1β expression in the 4–6 h CPB group at T0 ($p = 0.0231$), T1 ($p = 0.0197$), and T3 ($p = 0.0158$).

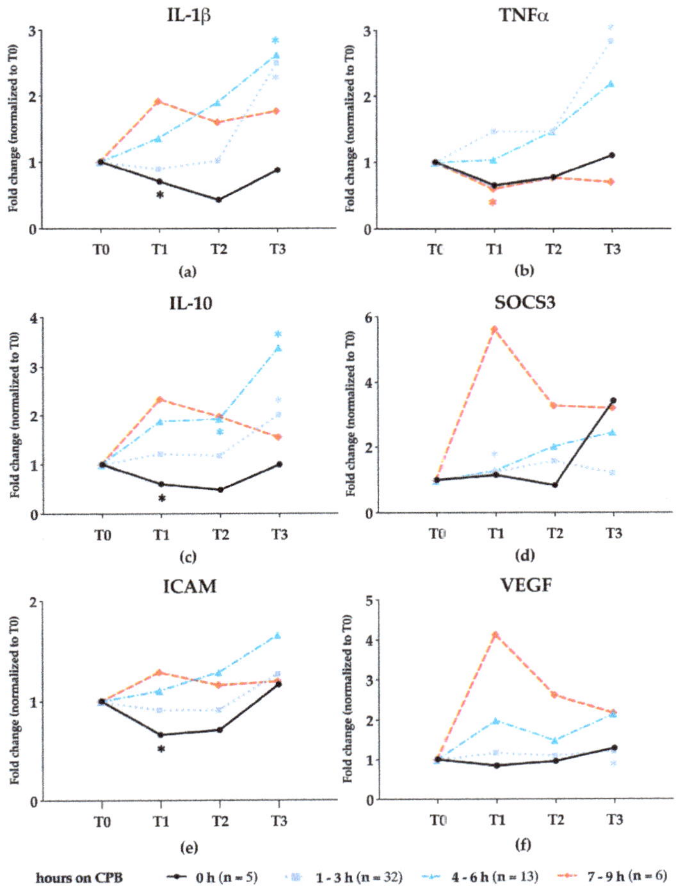


**Figure 3.** A prolonged duration of CPB correlates with expressions of serum-induced cytokines and mediators of vascular permeability in THP-1 macrophages in vitro, including (**a**) IL-1β, (**b**) TNFα, (**c**) IL-10, (**d**) SOCS3, (**e**) ICAM, and (**f**) VEGF. Data from 56 patients are represented as line graphs. Statistical test analysis: RM one-way ANOVA with Dunnett's multiple comparisons post-test relative to patients not undergoing CPB (0 h); * $p \leq 0.05$ was considered significant.

We also observed a significant increase in TNFα expression in the 1–3 h CPB group at T3 ($p = 0.0097$) but a significant decrease in the 1–9 h CPB group at T1 ($p = 0.0273$), as shown in Figure 4b. Interestingly, Stattic significantly augmented TNFα expression in the 1–3 h CPB group at all time points investigated, T0 ($p < 0.0001$), T1 ($p = 0.0029$), T2 ($p < 0.0001$), and T3 ($p = 0.0002$); in the 4–6 h CPB group at T0 ($p = 0.0015$), T1 ($p = 0.0332$), and T3 ($p = 0.0220$); and in the 7–9 h CPB group only at T1 ($p = 0.0237$).

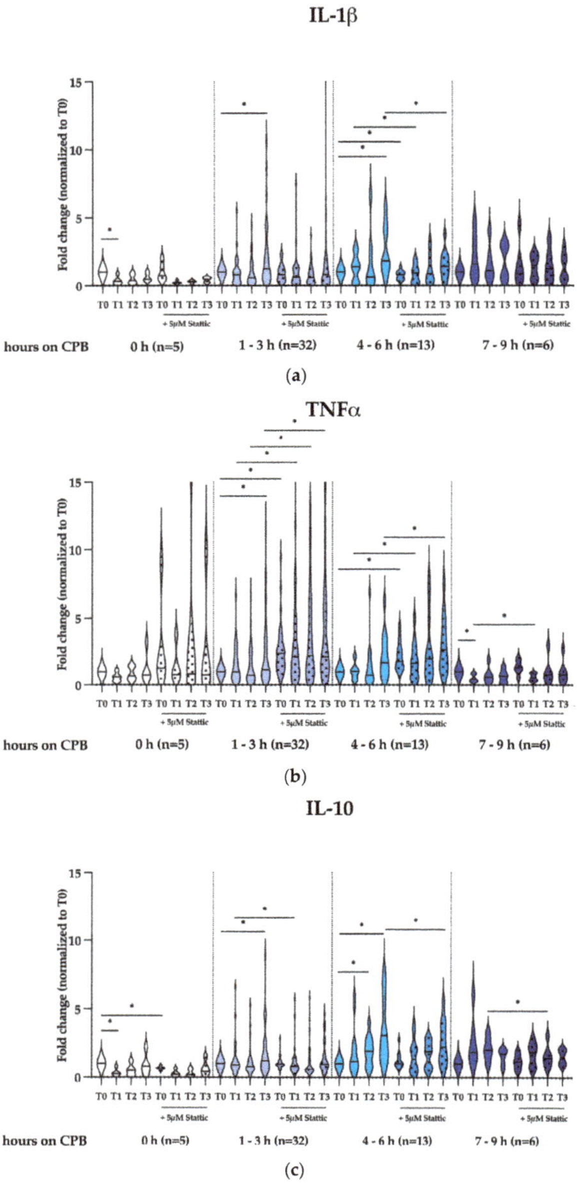


**Figure 4.** Stattic attenuates CPB-induced IL-1β and IL-10 expressions but augments TNFα expression in THP-1 macrophages in vitro. Serum samples from patients undergoing cardiac surgery without and with CPB for 1–3 h ($n$ = 32) and 4–6 h ($n$ = 13) induced significantly higher (**a**) IL-1β and (**c**) IL-10 expressions that could be attenuated by pretreatment with Stattic. No significant increases were observed in patients who did not require CPB (0 h, $n$ = 5) nor in patients with an extremely long duration of CPB (7–9 h, $n$ = 6). CPB for 1–3 h also elicited a significant increase in (**b**) TNFα expression that was augmented by pretreatment with Stattic. Data from 56 patients are represented as violin plots. Statistical test analysis: RM one-way ANOVA with Dunnett's multiple comparisons post-test relative to individual pre-operative control (T0) or a paired $t$-test relative to an untreated sample of the same time point; * $p \leq 0.05$ was considered significant.

IL-10 expression showed similar expression kinetics to IL-1β expression. We observed a significant decrease in IL-10 expression in patients not requiring CPB (0 h) at T1 ($p$ = 0.0128), as shown in Figure 4c. Moreover, for patients undergoing CPB, we observed a significant increase in IL-10 expression in the 1–3 h CPB group at T3 ($p$ = 0.0299) and in the 4–6 h group at T2 ($p$ = 0.0193) and T3 ($p$ = 0.0086). Stattic significantly attenuated IL-10 expression in the 0 h CPB group at T0 ($p$ = 0.0049), in the 1–3 h CPB group at T1 ($p$ = 0.0058), in the 4–6 h CPB group at T3 ($p$ = 0.0173), and in the 7–9 h CPB group at T2 ($p$ = 0.0030).

We further investigate the expression of the mediators of vascular permeability, ICAM and VEGF, as well as the inhibitor of the Stat3 activation molecule, SOCS3. We observed a significant decrease in ICAM expression in patients not requiring CPB (0 h) at T1 ($p$ = 0.0051), but no significant changes in patients undergoing CPB, as shown in Figure 5a. Interestingly, Stattic significantly increased ICAM expression in the 1–3 h CPB group at T0 ($p$ = 0.0430) and T2 ($p$ = 0.0207) but decreased ICAM expression in the 4–6 h CPB group at T1 ($p$ = 0.0368). For VEGF expression, we observed a significant increase only in the 1–3 h CPB group at T3 ($p$ = 0.0252), as shown in Figure 5b. Stattic significantly attenuated VEGF expression in the 1–3 h CPB group at T1 ($p$ = 0.0004) and in the 4–6 h CPB group at T1 ($p$ = 0.0057) and at T3 ($p$ = 0.0091). SOCS3 expression was significantly upregulated in the 1–3 h CPB group at T1 ($p$ = 0.0465), as shown in Figure 5c. Stattic increased SOCS3 in the 1–3 h CPB group at T1 ($p$ = 0.0252). Additionally, the data are also shown in sub-groups with respect to patients' age, children (0–18 yrs old) and adults (18–65 yrs old), in Figure S2 of the supplementary file.

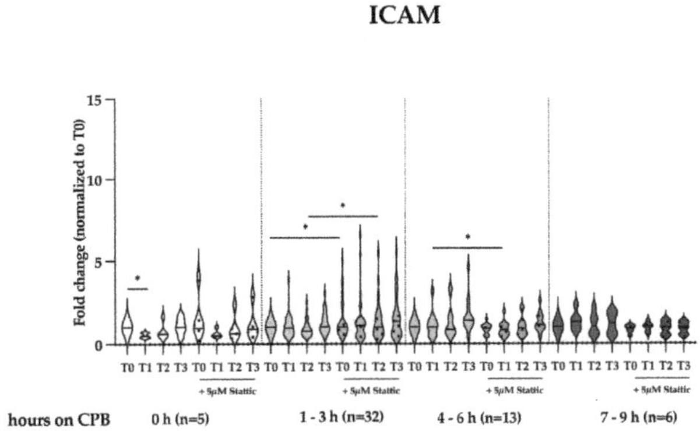


(a)

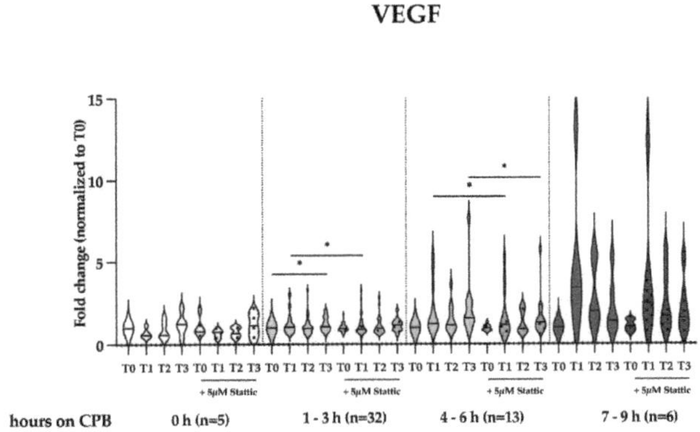


(b)

**Figure 5.** *Cont.*

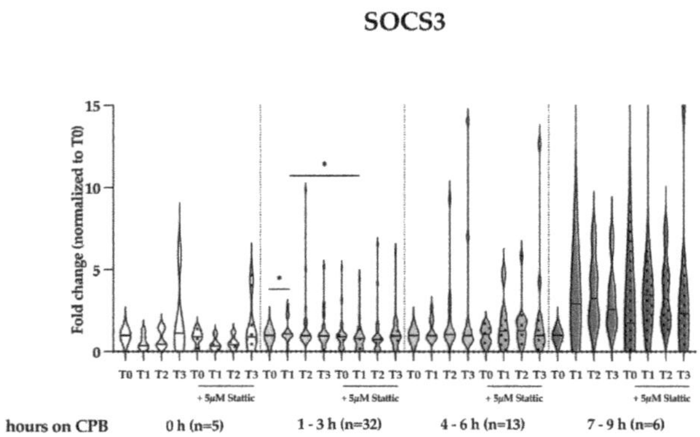


(c)

**Figure 5.** Effects of Stattic on expressions of mediators of vascular permeability, (**a**) ICAM, (**b**) VEGF, and (**c**) the inhibitor of STAT3 activation (SOCS3) in THP-1 macrophages in vitro, stimulated with serum samples from patients undergoing cardiac surgery without and with CPB. The patient cohort was classified into four distinct groups based on the duration of CPB as follows: 0 h ($n$ = 5), 1–3 h ($n$ = 32), 4–6 h ($n$ = 13), and 7–9 h ($n$ = 6). Statistical test analysis: RM one-way ANOVA with Dunnett's multiple comparisons post-test relative to individual pre-operative control (T0) or a paired $t$-test relative to an untreated sample of the same time point; * $p \leq 0.05$ was considered significant.

## 3. Discussion

As previously reported, we detected significantly elevated levels of serum IL-6, IL-8, and IL-10 at all post-operative timepoints investigated (T1–T3) in patients with congenital heart defects undergoing cardiac surgery involving CPB [34]. Other clinical studies have also reported similar significant elevations in plasma cytokine levels in patients undergoing CPB [36,37]. Moreover, CPB-induced release of DAMPs such as mitochondrial DNA (mtDNA), which can also elicit a systemic immune response via toll-like receptor-9 signaling, has also been reported [38]. In our effort to investigate the underlying molecular mechanisms of inflammation in our in vitro model of CPB-induced macrophage activation, the significant increase in Stat3 and lack of NF-κB p65 activation indicate cytokines and not DAMPs as the dominant mediators of inflammation after the operation in our cohort. In fact, we observed that NF-κB p65 phosphorylation was lower immediately after the operation (T1) and continued to decrease relative to individual baseline levels before CPB (T0) (Figure 2a and Figure 2b, respectively).

The combined effect on the signaling pathways resulted in a significant increase in IL-1β, TNFα, and IL-10 expressions 24 h post-operatively (T3) for patients who underwent moderate durations of CPB. Blocking Stat3 activation by Stattic partially attenuated IL-1β and IL-10 expressions at some observed time points but significantly augmented TNFα expression at basically all investigated time points after the operation after stimulation with serum samples from the same patients (Figure 3a, Figure 3b, and Figure 3c, respectively). Our findings correlate with those of de Jong et al., who reported that the abrogation of the IL-10/STAT3 pathway restored in vitro in LPS-induced TNF-α production in human PBMC isolated from a pediatric cohort undergoing CPB-assisted surgery to correct simple congenital heart defects was independent of p38 MAPK attenuation or IκB-α degradation [35].

During the acute phase reaction following CPB, surgical trauma, ischemia–reperfusion injury, contact with the surface of the external circulatory system, and other triggers can induce the secretion of various pro- and anti-inflammatory cytokines and adhesion molecules, which is mediated by NF-kB signaling. Justifiably so, various therapeutic strategies emphasize the attenuation of NF-kB activation to minimize the systemic inflammatory response. However, Stat3 has also been shown to play a crucial role in the anti-inflammatory pathway [39,40]. Specifically, the anti-inflammatory properties of IL-10 and its effect on macrophage deactivation have been reported [38]. However, the direct effect of IL-10 on NF-κB activity is controversial. Murray proposed that IL-10 can indirectly inhibit NF-κB-driven gene transcription via the activation of Stat3 to reduce the overall transcriptional rate of specific genes [41]. Moreover, Stat3-deficient macrophages have been reported to have high sensitivity to a lipopolysaccharide, a potent mediator of NF-κB-driven inflammation. The authors also further concluded that the IL-10 activation of Stat3 is critically involved in the deactivation of macrophages and neutrophils [39].

Both VEGF and ICAM-1 not only play an important role in promoting tumor angiogenesis but they are also potent inducers of vascular permeability, which play a major role in the induction of endothelial dysfunction and the associated capillary leak syndrome. It is well known that VEGF activates the STAT3 signaling pathway, inducing a positive feedback loop [42–44]. Furthermore, it has been shown that IL-6 promotes the expression of VEGF, and thereby the inhibition of the IL6R-STAT3 signaling pathway leads to a reduced production of VEGF [45,46]. Regarding ICAM, STAT3 is known to promote its expression by binding to the ICAM promoter after being activated by inflammatory mediators such as IL-6 or TNFα [42,47].

Our findings suggest that STAT3 signaling plays a vital role in the systemic inflammatory response commonly observed after cardiac surgery requiring CPB. Moreover, the STAT3 modulation of ICAM-1 and VEGF is crucial in determining not only the inflammatory response, but also vascular health and tissue repair mechanisms during and after CPB. Similarly to our previous findings, we continue to observe large inter-individual differences in the inflammatory response in our patients' cohort for all targets investigated [48]. Furthermore, we analyzed the influence of age in our previously published feasibility study including 19 patients ranging from 0 to 18 years of age [48], as well as in our follow-up study with 105 patients ranging from 0 to 65 years of age (data currently under preparation for publication). In both studies, we observed that patients' age and sex had no significant influence on the investigated biomarkers for post-operative inflammation. Minimizing the post-operative sterile inflammatory response induced by cardiopulmonary bypass may significantly reduce the patient's risk and improve the outcome. A multimodal approach is most likely needed for any therapeutic strategy and a focus on post-operative cytokine-induced STAT3 signaling should be included.

## 4. Materials and Methods

### 4.1. Protocol for Serum Sample Collection

The study was approved by the Ethics Committee of Charité—Universitätsmedizin Berlin, Germany (decision EA2/180/19). Serum samples from our patient cohort were collected as previously described [34]. Briefly, blood samples were obtained via the central venous line pre-operatively after the induction of anesthesia (T0), post-operatively upon arrival in the pediatric intensive care unit (PICU) (T1), and both 6 h (T2) and 24 h (T3) after the operation (Figure 1). We collected 1 mL blood from patients ≤15 kg and 2 mL of blood from patients >15 kg at each time point in Serum-Gel Microvette® 500 (20.1344 Sarstedt, Nümbrecht, Germany). Samples were centrifuged at 10,000× *g* for 10 min and temporarily stored at −8 °C until transfer to final storage at −80 °C.

### 4.2. Cell Culture

THP-1 cells represent a human monocytic cell line that can be transformed to macrophages by stimulating with PMA, as described further below. By optimizing a standard transformation and activation protocol in our laboratory, the use of THP-1 cells allows us to obtain reproducible results that mimic human inflammatory responses under controlled conditions [49,50]. Immortalized human THP-1 cells were purchased from ATCC (American Type Culture Collection: TIB-202) and grown in suspension in an RPMI 1640 (Gibco, Thermo Fisher, Karlsruhe, Germany) medium supplemented with 10% (*v*/*v*) heat-inactivated fetal bovine serum (FBS, Merck Millipor, Billerica, USA) and 100 U/mL penicillin and 100 μg/mL streptomycin (Gibco, Thermo Fisher, Karlsruhe, Germany) in a humidified, 37 °C, 5% $CO_2$ incubator. A total of $1 \times 10^6$ cells were plated in $35 \times 10$ mm tissue culture-treated plates (Nunclon™ Delta Surface, Thermo Scientific, Karlsruhe, Germany) in the presence of 15 ng/mL phorbol 12-myristiate-12 acetate (PMA, Sigma-Aldrich, Munich, Germany) for 3 days, followed by incubation in a PMA-free medium for 24 h prior to stimulation with serum samples or LPS. Differentiated THP-1 macrophages were then pretreated with 5 µM Stattic, a STAT3 (Tyr705) inhibitor, for 1 h, followed by incubation with serum samples, which were diluted with an RPMI complete medium in a ratio of 1:2 (T0 = pre-operative, T1 = upon arrival at intensive care unit, T2 = 6 h post-operatively, T3 = 24 h post-operatively) or 0.05 µg/mL LPS for 1 to 4 h. Stattic is well characterized and established in the scientific literature. Stattic specifically inhibits the JAK-STAT3 pathway by inhibiting the STAT3 SH2 domain and therefore STAT3 activation and nuclear translocation. It has been shown to be a reliable inhibitor of short-term cell stimulation with high specificity for inhibiting STAT3 without significantly affecting other cell signaling pathways, which allows for the investigation of the role of STAT3 [51].

### 4.3. Protein Isolation and Western Blot Analysis

For an intracellular protein analysis, cells were harvested after a 1 h stimulation and centrifuged at 6000× $g$ for 10 min and cell pellets were lysed in an RIPA buffer supplemented with protease and phosphatase inhibitors (1:100, Sigma-Aldrich, Munich, Germany). Protein concentration was assessed via a Pierce BCA Protein Assay (Thermo Scientific, Karlsruhe, Germany). Intracellular protein samples were incubated with a Pierce Lane Marker Reducing Sample Buffer (Thermo Scientific, Karlsruhe, Germany) at 95 °C for 5 min and loaded onto a 12% SDS polyacrylamide gel for electrophoresis. Afterwards, proteins were transferred onto an Immobilon® -P polyvinylidene fluoride membrane (Merck Millipore Ltd., Cork, Ireland) overnight at 30 V using a tank-blotting procedure (Bio-Rad Laboratories, Munich, Gernamy). Membranes were then blocked for 1 h at room temperature with 5% BSA (Pierce™, Rockford, IL, USA) or 3% nonfat dried milk (PanReac AppliedChem, Darmstadt, Germany) in TBS + 0.1% Tween 20. Primary antibodies for β-Actin (1:20,000, Cell Signaling, Boston, MA, USA, Cat# 4967), phospho-NF-κB p65 (Ser536) (1:500, Cell Signaling, Boston, USA, Cat# 3033), NF-κB p65 (1:1000, Cell Signaling, Boston, USA, Cat# 3034), phosphor-Stat3 (Tyr705) (1:500, Cell Signaling, Boston, USA, Cat# 9145), and Stat3 (1:1000, Cell Signaling, Boston, USA, Cat# 9139) were diluted in a blocking solution and incubated overnight at 4 °C. Secondary antibodies anti-rabbit IgG-HRP (1:20,000, Cell Signaling, Boston, USA, Cat# 7074) or anti-mouse IgG-HRP (1:10,000, Santa Cruz, Heidelberg, Germany, Cat# sc-516102)) were incubated for 1 h at room temperature. SuperSignal™ West Dura Extended Duration Substrate (Thermo Fisher Scientific, Karlsruhe, Germany) was used to visualize protein expression, captured using a Molecular Imager® ChemiDoc™ XRS System, and Image Lab™ Software V6.1.0 (Bio-Rad, Feldkirchen, Germany) was used for a densitometry analysis.

### 4.4. RNA Isolation and RT-PCR

Total RNA from THP-1 cells was isolated via acidic phenol/chloroform extraction using ROTI®Zol RNA (Carl Roth, Karlsruhe, Germany) according to the manufacturer's instructions. RNA concentration and purity were assessed by spectrophotometric measurements at 260 nm and 280 nm with a Nanodrop 2000 (Nanodrop, Thermo Fisher Scientific, Karlsruhe, Germany) and agarose gel electrophoresis. Reverse transcription was performed using 500 ng total RNA via a qScriber™ cDNA Synthesis Kit (highQu, Kraichtal, Germany) in a thermal cycler (PTC200, MJ Research, St. Bruno, Canada) according to the manufacturer's instructions. The expression of target genes and GAPDH as a reference gene control was analyzed by real-time qPCR using TaqMan Gene Expression Assays (summarized in Table 2) and ORA™ SEE qPCR Probe ROX H Mix (highQu, Kraichtal, Germany) according to manufacturers' recommendations. We assessed the relative quantification of gene expression normalized to GAPDH as a reference gene via the $2^{-\Delta\Delta Ct}$ method and results are depicted as fold changes.

**Table 2.** List of TaqMan Real-Time PCR Assays for RT-qPCR analysis.

| Gene | Assay ID |
|---|---|
| IL-1β | Hs01555410_m1 |
| IL-10 | Hs00961622_m1 |
| TNFα | Hs00174128_m1 |
| ICAM | Hs00164932_m1 |
| VEGF | Hs00900055_m1 |
| SOCS3 | Hs02330328_s1 |
| GAPDH | Hs02786624_g1 |

### *4.5. Statistical Analysis*

A power analysis showed that a sample size of 51 participants will reach a power of 80% at a two-sided level of significance of 5% to detect a difference in TNFα with a Cohen's d effect size of 0.4 (mean difference of −0.58 and standard deviation (SD) of 1.42 based on our preliminary data). The sample size calculation based on the paired *t*-test was used despite the intended analysis with a one-way repeated ANOVA. Regarding a post hoc test, it will lead to the sample size based on the paired *t*-test. Sample size calculation was performed by using program R version 3.6.2 with a package "pwr" and built-in command "pwr.t.test" (https://CRAN.R-project.org/package=pwr, accessed on 31 March 2023).

Western blot and RT-qPCR data were analyzed and illustrated using GraphPad Prism 10 (GraphPad Software, Inc., La Jolla, CA, USA). RM one-way ANOVA with Dunnett's multiple comparisons post-test was relative to an individual pre-operative control (T0), a paired t-test was relative to an untreated sample of the same time point, or an unpaired t-test was relative to a control; * $p \leq 0.05$ was considered significant. For the Western blot analysis, data from at least 3 independent experiments are presented as the mean ± SD and $p$ values < 0.05 were considered significant. For a patient serum sample analysis, the relative quantification of gene expression normalized to GAPDH as a reference gene via the $2^{-\Delta\Delta Ct}$ method was conducted, and results are depicted as fold changes relative to T0. $p$ values < 0.05 were considered significant.

## 5. Conclusions

In summary, we provide evidence that the inflammatory response following cardiac surgery involving CPB is driven in part by the activation of the STAT3 signaling pathway in macrophages. CPB induced an increased expression of cytokines and mediators of vascular permeability in macrophages via the activation of STAT3 by IL-6, IL-8, and IL-10 in the serum samples obtained post-operatively from CHD patients. Pretreatment with Stattic attenuates IL-1β, IL-10, and VEGF but promotes TNFα and SOCS3 expressions in macrophages. However, the effect of Stattic on ICAM expression was dependent on the duration of CPB. Our findings support the hypothesis that STAT3 is involved in the systemic inflammatory response observed in CHD patients following cardiac surgery involving CPB, resulting in both pro- and anti-inflammatory effects. Moreover, the patient population with congenital heart disease (CHD) includes individuals of a diverse age range and weight. We hypothesize that the factors contributing to post-operative inflammation, particularly the complexity of the CHD, surgical duration, and aortic clamping time, which is associated with ischemia–reperfusion-induced injury, have a greater impact on outcomes than age per se. Nonetheless, the limitations of our study to investigate the systemic inflammatory response induced by CPB lie in the heterogeneity of our cohort as well as the use of a monoculture in vitro model.

**Supplementary Materials:** The following supporting information can be downloaded at: https://www.mdpi.com/article/10.3390/ijms252212398/s1.

**Author Contributions:** J.L., K.R.L.S. and G.T.; Data curation, C.P., A.-L.H. and G.T.; Formal analysis, J.L., C.P. and G.T.; Funding acquisition, J.L. and K.R.L.S.; Investigation, C.P. and G.T.; Methodology, J.L., C.P., A.-L.H., L.B. and G.T.; Project administration, J.L., F.B., K.R.L.S. and G.T.; Resources, J.L., R.S., A.-L.H., L.B., F.B. and K.R.L.S.; Supervision, J.L., K.R.L.S. and G.T.; Validation, J.L., C.P., R.S. and G.T.; Writing—original draft, C.P. and G.T.; Writing—review and editing, J.L., C.P., K.R.L.S. and G.T. All authors have read and agreed to the published version of the manuscript.

**Funding:** This research was funded by the Stiftung KinderHerz Deutschland gGmbH, Essen, Germany, grant number 2511-3-19-020, and the Deutsche Stiftung für Herzforschung, grant number F/37/20. We acknowledge support from the Robert Enke Foundation and the German Research Foundation (DFG) for the Open Access Publication Fund of the Charité—Universitätsmedizin Berlin. Dr. Luecht is a participant in the BIH-Charité Junior Clinician Scientist Program funded by the Charité—Universitätsmedizin Berlin and the Berlin Institute of Health at Charité—Universitätsmedizin Berlin. Camila Pauli was supported by a doctoral scholarship from the Deutsches Zentrum für Herz-Kreislauf-Forschung E.V. (DZHK). We acknowledge the Berlin Institute of Health (BIH) @ Charité, BIH Metabolomics Platform, for providing the MagPix® Instrument.

**Institutional Review Board Statement:** The prospective observational feasibility study was conducted at the German Heart Center Berlin after approval by the Ethics Committee of Charité—Universitätsmedizin Berlin, Germany (decision EA2/180/19). The study was registered with the German register for clinical studies before patient recruitment (registration number: DRKS00020885).

**Informed Consent Statement:** Written consent was obtained from the parents of each patient before study inclusion. Patients younger than 18 years undergoing a corrective or palliative cardiac surgery at our center were enrolled. Exclusion criteria were as follows: gestational age $\leq$ 37 weeks, known maternal alcohol or substance abuse during pregnancy, immunodeficiencies or immunosuppressive medication, syndromic diseases (e.g., Trisomy 21 and 18), and congenital kidney disease.

**Data Availability Statement:** The data presented in this study are available on request from the corresponding author. The data are not publicly available due to privacy or ethical restrictions.

**Acknowledgments:** We thank Johanna Fross for her contribution in the development of the THP-1 in vitro model and Pimrapat Gebert, MPH for her counseling in the statistical analysis of the experimental data.

**Conflicts of Interest:** The authors declare no conflicts of interest. The funders had no role in the design of the study; in the collection, analyses, or interpretation of data; in the writing of the manuscript; or in the decision to publish the results.

## References

1. Schwedler, G.; Lindinger, A.; Lange, P.E.; Sax, U.; Olchvary, J.; Peters, B.; Bauer, U.; Hense, H.W. Frequency and spectrum of congenital heart defects among live births in Germany: A study of the Competence Network for Congenital Heart Defects. *Clin. Res. Cardiol.* **2011**, *100*, 1111–1117. [CrossRef] [PubMed]
2. Warren, O.J.; Smith, A.J.; Alexiou, C.; Rogers, P.L.; Jawad, N.; Vincent, C.; Darzi, A.W.; Athanasiou, T. The inflammatory response to cardiopulmonary bypass: Part 1—mechanisms of pathogenesis. *J. Cardiothorac. Vasc. Anesth.* **2009**, *23*, 223–231. [CrossRef] [PubMed]
3. Paparella, D.; Yau, T.M.; Young, E. Cardiopulmonary bypass induced inflammation: Pathophysiology and treatment. *Update. Eur. J. Cardiothorac. Surg.* **2002**, *21*, 232–244. [CrossRef] [PubMed]
4. Seghaye, M.C.; Grabitz, R.G.; Duchateau, J.; Busse, S.; Dabritz, S.; Koch, D.; Alzen, G.; Hornchen, H.; Messmer, B.J.; Von Bernuth, G. Inflammatory reaction and capillary leak syndrome related to cardiopulmonary bypass in neonates undergoing cardiac operations. *J. Thorac. Cardiovasc. Surg.* **1996**, *112*, 687–697. [CrossRef] [PubMed]
5. Hirleman, E.; Larson, D.F. Cardiopulmonary bypass and edema: Physiology and pathophysiology. *Perfusion* **2008**, *23*, 311–322. [CrossRef] [PubMed]
6. Kozik, D.J.; Tweddell, J.S. Characterizing the inflammatory response to cardiopulmonary bypass in children. *Ann. Thorac. Surg.* **2006**, *81*, S2347–S2354. [CrossRef]
7. Brix-Christensen, V. The systemic inflammatory response after cardiac surgery with cardiopulmonary bypass in children. *Acta Anaesthesiol. Scand.* **2001**, *45*, 671–679. [CrossRef]
8. Ungerleider, R.M.; Shen, I. Optimizing response of the neonate and infant to cardiopulmonary bypass. *Semin. Thorac. Cardiovasc. Surg. Pediatr. Card. Surg. Annu.* **2003**, *6*, 140–146. [CrossRef]
9. Güvener, M.; Korun, O.; Demirtürk, O.S. Risk factors for systemic inflammatory response after congenital cardiac surgery. *J. Card. Surg.* **2015**, *30*, 92–96. [CrossRef]

10. Boehne, M.; Sasse, M.; Karch, A.; Dziuba, F.; Horke, A.; Kaussen, T.; Mikolajczyk, R.; Beerbaum, P.; Jack, T. Systemic inflammatory response syndrome after pediatric congenital heart surgery: Incidence, risk factors, and clinical outcome. *J. Card. Surg.* **2017**, *32*, 116–125. [CrossRef]
11. Baumgartner, H.; Bonhoeffer, P.; De Groot, N.M.; de Haan, F.; Deanfield, J.E.; Galie, N.; Gatzoulis, M.A.; Gohlke-Baerwolf, C.; Kaemmerer, H.; Kilner, P.; et al. ESC Guidelines for the management of grown-up congenital heart disease (new version 2010). *Eur. Heart J.* **2010**, *31*, 2915–2957. [CrossRef] [PubMed]
12. Warnes, C.A.; Liberthson, R.; Danielson, G.K.; Dore, A.; Harris, L.; Hoffman, J.I.; Somerville, J.; Williams, R.G.; Webb, G.D. Task force 1: The changing profile of congenital heart disease in adult life. *J. Am. Coll. Cardiol.* **2001**, *37*, 1170–1175. [CrossRef]
13. Durandy, Y. Minimizing systemic inflammation during cardiopulmonary bypass in the pediatric population. *Artif. Organs* **2014**, *38*, 11–18. [CrossRef]
14. Palanzo, D.A.; Wise, R.K.; Woitas, K.R.; Undar, A.; Clark, J.B.; Myers, J.L. Safety and utility of modified ultrafiltration in pediatric cardiac surgery. *Perfusion* **2023**, *38*, 150–155. [CrossRef]
15. Bojan, M. Recent achievements and future developments in neonatal cardiopulmonary bypass. *Paediatr. Anaesth.* **2019**, *29*, 414–425. [CrossRef] [PubMed]
16. Deptula, J.; Glogowski, K.; Merrigan, K.; Hanson, K.; Felix, D.; Hammel, J.; Duncan, K. Evaluation of biocompatible cardiopulmonary bypass circuit use during pediatric open heart surgery. *J. Extra Corpor. Technol.* **2006**, *38*, 22–26. [CrossRef] [PubMed]
17. Golab, H.D.; Bogers, J.J. Small, smaller, smallest. Steps towards bloodless neonatal and infant cardiopulmonary bypass. *Perfusion* **2009**, *24*, 239–242. [CrossRef]
18. Scrascia, G.; Rotunno, C.; Guida, P.; Amorese, L.; Polieri, D.; Codazzi, D.; Paparella, D. Perioperative steroids administration in pediatric cardiac surgery: A meta-analysis of randomized controlled trials. *Pediatr. Crit. Care Med.* **2014**, *15*, 435–442. [CrossRef]
19. Lomivorotov, V.; Kornilov, I.; Boboshko, V.; Shmyrev, V.; Bondarenko, I.; Soynov, I.; Voytov, A.; Polyanskih, S.; Strunin, O.; Bogachev-Prokophiev, A.; et al. Effect of Intraoperative Dexamethasone on Major Complications and Mortality Among Infants Undergoing Cardiac Surgery: The DECISION Randomized Clinical Trial. *JAMA* **2020**, *323*, 2485–2492. [CrossRef]
20. Graham, E.M.; Atz, A.M.; Butts, R.J.; Baker, N.L.; Zyblewski, S.C.; Deardorff, R.L.; DeSantis, S.M.; Reeves, S.T.; Bradley, S.M.; Spinale, F.G. Standardized preoperative corticosteroid treatment in neonates undergoing cardiac surgery: Results from a randomized trial. *J. Thorac. Cardiovasc. Surg.* **2011**, *142*, 1523–1529. [CrossRef]
21. Graham, E.M.; Atz, A.M.; McHugh, K.E.; Butts, R.J.; Baker, N.L.; Stroud, R.E.; Reeves, S.T.; Bradley, S.M.; McGowan, F.X., Jr.; Spinale, F.G. Preoperative steroid treatment does not improve markers of inflammation after cardiac surgery in neonates: Results from a randomized trial. *J. Thorac. Cardiovasc. Surg.* **2014**, *147*, 902–908. [CrossRef] [PubMed]
22. Ponomarev, D.; Boboshko, V.; Shmyrev, V.; Kornilov, I.; Bondarenko, I.; Soynov, I.; Voytov, A.; Polyanskih, S.; Strunin, O.; Bogachev, A.; et al. Dexamethasone in pEdiatric Cardiac Surgery (DECiSion): Rationale and design of a randomized controlled trial. *Contemp. Clin. Trials* **2018**, *72*, 16–19. [CrossRef] [PubMed]
23. Nakamori, Y.; Koh, T.; Ogura, H.; Tanaka, H.; Fujimi, S.; Kasai, K.; Hosotubo, H.; Shimazu, T.; Sugimoto, H. Enhanced expression of intranuclear NF-kappa B in primed polymorphonuclear leukocytes in systemic inflammatory response syndrome patients. *J. Trauma* **2003**, *54*, 253–260. [CrossRef] [PubMed]
24. Nakamori, Y.; Ogura, H.; Koh, T.; Fujita, K.; Tanaka, H.; Sumi, Y.; Hosotsubo, H.; Yoshiya, K.; Irisawa, T.; Kuwagata, Y.; et al. The balance between expression of intranuclear NF-kappaB and glucocorticoid receptor in polymorphonuclear leukocytes in SIRS patients. *J. Trauma* **2005**, *59*, 308–314. [CrossRef] [PubMed]
25. Hirano, T. IL-6 in inflammation, autoimmunity and cancer. *Int. Immunol.* **2021**, *33*, 127–148. [CrossRef]
26. Hu, X.; Li, J.; Fu, M.; Zhao, X.; Wang, W. The JAK/STAT signaling pathway: From bench to clinic. *Signal Transduct. Target. Ther.* **2021**, *6*, 402. [CrossRef]
27. Cai, B.; Cai, J.P.; Luo, Y.L.; Chen, C.; Zhang, S. The Specific Roles of JAK/STAT Signaling Pathway in Sepsis. *Inflammation* **2015**, *38*, 1599–1608. [CrossRef]
28. Lee, H.; Herrmann, A.; Deng, J.H.; Kujawski, M.; Niu, G.; Li, Z.; Forman, S.; Jove, R.; Pardoll, D.M.; Yu, H. Persistently activated Stat3 maintains constitutive NF-kappaB activity in tumors. *Cancer Cell* **2009**, *15*, 283–293. [CrossRef]
29. Kishimoto, T. The biology of interleukin-6. *Blood* **1989**, *74*, 1–10. [CrossRef]
30. Tanaka, T.; Kishimoto, T. The biology and medical implications of interleukin-6. *Cancer Immunol. Res.* **2014**, *2*, 288–294. [CrossRef]
31. Sharma, D.; Farahbakhsh, N.; Shastri, S.; Sharma, P. Biomarkers for diagnosis of neonatal sepsis: A literature review. *J. Matern. Fetal Neonatal Med.* **2018**, *31*, 1646–1659. [CrossRef] [PubMed]
32. Reddy, A.S.S.; Rao, S.S.; Shenoy, V.D.; Shetty, S. Role of Nuclear Factor-Kappa B Activation and Inflammatory Biomarkers in Critically Ill Children. *Indian J. Pediatr.* **2023**, *91*, 1075–1077. [CrossRef] [PubMed]
33. Jenkins, K.J.; Gauvreau, K.; Newburger, J.W.; Spray, T.L.; Moller, J.H.; Iezzoni, L.I. Consensus-based method for risk adjustment for surgery for congenital heart disease. *J. Thorac. Cardiovasc. Surg.* **2002**, *123*, 110–118. [CrossRef] [PubMed]
34. Lücht, J.; Seiler, R.; Herre, A.L.; Brankova, L.; Fritsche-Guenther, R.; Kirwan, J.; Huscher, D.; Münzfeld, H.; Berger, F.; Photiadis, J.; et al. Promising results of a clinical feasibility study: CIRBP as a potential biomarker in pediatric cardiac surgery. *Front. Cardiovasc. Med.* **2024**, *11*, 1247472. [CrossRef]

35. de Jong, P.R.; Schadenberg, A.W.; van den Broek, T.; Beekman, J.M.; van Wijk, F.; Coffer, P.J.; Prakken, B.J.; Jansen, N.J. STAT3 regulates monocyte TNF-alpha production in systemic inflammation caused by cardiac surgery with cardiopulmonary bypass. *PLoS ONE* **2012**, *7*, e35070. [CrossRef] [PubMed]
36. Halter, J.; Steinberg, J.; Fink, G.; Lutz, C.; Picone, A.; Maybury, R.; Fedors, N.; DiRocco, J.; Lee, H.M.; Nieman, G. Evidence of systemic cytokine release in patients undergoing cardiopulmonary bypass. *J. Extra Corpor. Technol.* **2005**, *37*, 272–277. [CrossRef]
37. Sablotzki, A.; Friedrich, I.; Mühling, J.; Dehne, M.G.; Spillner, J.; Silber, R.E.; Czeslik, E. The systemic inflammatory response syndrome following cardiac surgery: Different expression of proinflammatory cytokines and procalcitonin in patients with and without multiorgan dysfunctions. *Perfusion* **2002**, *17*, 103–109. [CrossRef]
38. Sandler, N.; Kaczmarek, E.; Itagaki, K.; Zheng, Y.; Otterbein, L.; Khabbaz, K.; Liu, D.; Senthilnathan, V.; Gruen, R.L.; Hauser, C.J. Mitochondrial DAMPs Are Released During Cardiopulmonary Bypass Surgery and Are Associated with Postoperative Atrial Fibrillation. *Heart Lung Circ.* **2018**, *27*, 122–129. [CrossRef]
39. Takeda, K.; Clausen, B.E.; Kaisho, T.; Tsujimura, T.; Terada, N.; Förster, I.; Akira, S. Enhanced Th1 activity and development of chronic enterocolitis in mice devoid of Stat3 in macrophages and neutrophils. *Immunity* **1999**, *10*, 39–49. [CrossRef]
40. Lang, R.; Patel, D.; Morris, J.J.; Rutschman, R.L.; Murray, P.J. Shaping gene expression in activated and resting primary macrophages by IL-10. *J. Immunol.* **2002**, *169*, 2253–2263. [CrossRef]
41. Murray, P.J. The primary mechanism of the IL-10-regulated antiinflammatory response is to selectively inhibit transcription. *Proc. Natl. Acad. Sci. USA* **2005**, *102*, 8686–8691. [CrossRef] [PubMed]
42. Wang, L.; Astone, M.; Alam, S.K.; Zhu, Z.; Pei, W.; Frank, D.A.; Burgess, S.M.; Hoeppner, L.H. Suppressing STAT3 activity protects the endothelial barrier from VEGF-mediated vascular permeability. *Dis. Model. Mech.* **2021**, *14*, dmm049029. [CrossRef] [PubMed]
43. Bartoli, M.; Gu, X.; Tsai, N.T.; Venema, R.C.; Brooks, S.E.; Marrero, M.B.; Caldwell, R.B. Vascular endothelial growth factor activates STAT proteins in aortic endothelial cells. *J. Biol. Chem.* **2000**, *275*, 33189–33192. [CrossRef] [PubMed]
44. Simons, M.; Gordon, E.; Claesson-Welsh, L. Mechanisms and regulation of endothelial VEGF receptor signalling. *Nat. Rev. Mol. Cell Biol.* **2016**, *17*, 611–625. [CrossRef] [PubMed]
45. Tu, Y.; Guo, Y.; Sun, H.; Zhang, Y.; Wang, Q.; Xu, Y.; Xie, L.; Zhu, M. Tocilizumab attenuates choroidal neovascularization by regulating macrophage polarization through the IL-6R/STAT3/VEGF pathway. *Heliyon* **2024**, *10*, e27893. [CrossRef]
46. Liu, X.; Zhang, A.; Xiang, J.; Lv, Y.; Zhang, X. miR-451 acts as a suppressor of angiogenesis in hepatocellular carcinoma by targeting the IL-6R-STAT3 pathway. *Oncol. Rep.* **2016**, *36*, 1385–1392. [CrossRef] [PubMed]
47. Han, X.; Wang, Y.; Chen, H.; Zhang, J.; Xu, C.; Li, J.; Li, M. Enhancement of ICAM-1 via the JAK2/STAT3 signaling pathway in a rat model of severe acute pancreatitis-associated lung injury. *Exp. Ther. Med.* **2016**, *11*, 788–796. [CrossRef] [PubMed]
48. Schmitt, K.R.; Fedarava, K.; Justus, G.; Redlin, M.; Böttcher, W.; Delmo Walter, E.M.; Hetzer, R.; Berger, F.; Miera, O. Hypothermia During Cardiopulmonary Bypass Increases Need for Inotropic Support but Does Not Impact Inflammation in Children Undergoing Surgical Ventricular Septal Defect Closure. *Artif. Organs.* **2016**, *40*, 470–479. [CrossRef] [PubMed]
49. Chanput, W.; Mes, J.J.; Wichers, H.J. THP-1 cell line: An in vitro cell model for immune modulation approach. *Int. Immunopharmacol.* **2014**, *23*, 37–45. [CrossRef]
50. Daigneault, M.; Preston, J.A.; Marriott, H.M.; Whyte, M.K.; Dockrell, D.H. The identification of markers of macrophage differentiation in PMA-stimulated THP-1 cells and monocyte-derived macrophages. *PLoS ONE* **2010**, *5*, e8668. [CrossRef]
51. Schust, J.; Sperl, B.; Hollis, A.; Mayer, T.U.; Berg, T. Stattic: A small-molecule inhibitor of STAT3 activation and dimerization. *Chem. Biol.* **2006**, *13*, 1235–1242. [CrossRef] [PubMed]

International Journal of
*Molecular Sciences*

MDPI

*Article*

# Synthetic Flavonoid 3,7-Dihydroxy-Isoflav-3-Ene (DHIF) Reduces In-Stent Restenosis in an Atherosclerotic Watanabe Heritable Hyperlipidemic Rabbit Stent Model

**Jarkko P. Hytönen [1,2,†], Olli Leppänen [1,†], Jouni Taavitsainen [1,2] and Seppo Ylä-Herttuala [1,2,*]**

1 A.I. Virtanen Institute, University of Eastern Finland, 70210 Kuopio, Finland; jarkko.hytonen@uef.fi (J.P.H.)
2 Heart Center, Kuopio University Hospital, 70200 Kuopio, Finland
* Correspondence: seppo.ylaherttuala@uef.fi
† These authors contributed equally to this work.

**Abstract:** Inflammation is a major component of the pathogenesis of atherosclerosis and the formation of in-stent restenosis (ISR). A novel flavonoid, DHIF, attenuates reactive oxygen species and nf-κB signaling and has potential to limit ISR via antioxidant action. While current drug eluting stents (DESs) perform well in clinical practice, new therapies to prevent ISR without dependance on cytotoxic drugs are warranted. Our objective was to test whether DHIF reduces ISR in a hyperlipidemic rabbit aorta model of ISR via attenuated inflammatory responses. WHHL rabbit aortas ($n$ = 24) were denuded. Six weeks after injury, stents were implanted into the denuded aortas. DHIF was dissolved in carboxymethyl cellulose (CMC) and administered orally with two doses. CMC served as a control. The animals were sacrificed six weeks after stenting. ISR was evaluated from stent histomorphometry and immunohistology was used to assess the inflammatory and antiproliferative effects of the treatment. ISR was reduced from 20.9 ± 3.0% in controls to 15.2 ± 2.4% ($p$ = 0.0009) and 16.4 ± 2.1% ($p$ = 0.004) in the low- and high-dose groups, respectively. The neointimal area covered by macrophages was 32 ± 9.3% in the controls, 17.2 ± 5.9% ($p$ = 0.005) in the low-dose group and 19.4 ± 7.9% ($p$ = 0.008) in the high-dose group. DHIF significantly reduces ISR and local inflammation in stented arterial regions and could be used to reduce ISR when bare metal stents are used. Targeting local inflammation in the arterial wall may provide a way to reduce ISR in a clinical setting and further studies are warranted.

**Keywords:** antioxidant: redox; superoxide; NF-kappaB; vessel injury; phytochemical; phytoestrogen; restenosis; rabbit model

## 1. Introduction

Coronary artery stenting is a standard treatment for stabile and acute obstructive coronary artery disease [1,2]. The cytostatic drug eluting stents (DESs) reduce in-stent restenosis (ISR) through the inhibition of smooth muscle cell (SMC) proliferation and matrix secretion; nevertheless [3,4], earlier generations of DES have also been shown to prolong endothelial dysfunction, induce neoatherosclerosis and increase the risk of late stent thrombosis after coronary intervention [5,6]. Stent thrombosis is usually prevented with the use of dual antiplatelet therapy (DAPT) following treatment with a DES. Many patients are, however, at a high risk of bleeding while on DAPT and alternative therapeutic approaches are warranted. Although newer generations of DES have demonstrated improved safety and completely bioabsorbable devices have entered the market, it is vital to further investigate the mechanism of restenosis as well as exploring alternative solutions for the prevention and treatment of restenosis [5].

A controlled level of reactive oxygen species (ROS) and a healthy endothelium are essential for vascular homeostasis [6–9]. A percutaneous coronary intervention procedure with a stent causes injury to the endothelium and the deeper layers of the vessel wall, and this injury induces abundant ROS production [3,7,10]. An excess of ROS and the dysfunction of the endothelium contribute significantly to the progression of ISR. Both the restoration of antioxidant balance and the induction of re-endothelialization reduce ISR [8,11,12].

Flavonoids are a group of plant-derived chemicals, or phytochemicals. Due to their phenyl structure, flavonoids possess substantial antioxidant capacity in vitro. The ability of flavonoids to decrease oxidative stress in vivo, however, is suggested to result from the regulation of intracellular signaling and gene expression [13]. Isoflavones, a subgroup of flavonoids, resemble 17-estradiol in structure, and therefore may conduct estrogen agonist or antagonist action within the body, depending on concentration [14]. A novel synthetic flavonoid, 3,7-dihydroxy-isoflav-3-ene (DHIF), has been shown to attenuate NF-κB, prevent elevation in the ROS level, reduce proliferation and induce apoptosis in the neointima after arterial injury [15,16]. All these features are central in reducing ISR. Nevertheless, the efficacy of DHIF for the treatment of ISR has not been previously studied.

In this study, our objective was to examine the effects of oral treatment with DHIF on ISR in an atherosclerotic WHHL rabbit aorta after stent placement. The effects of DHIF administration are examined by determining the inflammatory response post stenting and the development of the neointima six weeks after stent deployment. We hypothesized that an attenuated inflammatory response would, in turn, be associated with reduced neointima formation.

## 2. Results

### 2.1. In-Stent Restenosis

After the 42-day follow-up, histological restenosis in the stented artery was 20.9 ± 3.0% in the control animals. DHIF-treated animals showed a significantly lower ISR of 15.2 ± 2.4% and 16.4 ± 2.1% (low dose, $p = 0.0009$, and high dose, $p = 0.004$, respectively). There were no differences in the injury scores between the groups with scores of 1.2 ± 0.2, 1.1 ± 0.1 and 1.0 ± 0.2 (control, low dose and high dose, respectively; low dose $p = 0.66$; high dose $p = 0.13$) (Figure 1). Descriptive statistics and effect size estimates are shown in Table 1.

**Table 1.** Descriptive statistics and effect size estimates.

| **In-Stent Restenosis** | | | | **Neointimal Apoptosis** | | | |
|---|---|---|---|---|---|---|---|
| | Control | Low Dose | High Dose | | Control | Low Dose | High Dose |
| Mean | 20.9 | 15.2 | 16.4 | Mean | 3.8 | 9.1 | 14.2 |
| Median | 20.3 | 15.2 | 15.8 | Median | 2.4 | 9.1 | 14.2 |
| Range | 8.5 | 6.7 | 5.5 | Range | 7.4 | 11.2 | 15.2 |
| Cohen's d | | 1.6 | 1.3 | Cohen's d | | 1.4 | 2.0 |
| **Endothelialization** | | | | **Pharmacokinetics** | | | |
| | Control | Low Dose | High Dose | | Control | Low Dose | High Dose |
| Mean | 87.1 | 75.4 | 86.1 | Mean | 0.0 | 271.1 | 630.8 |
| Median | 88.2 | 76.0 | 90.0 | Median | 0.0 | 266.5 | 573.7 |
| Range | 42.2 | 20.4 | 53.8 | Range | 0.0 | 259.1 | 891.5 |
| Cohen's d | | 0.7 | 0.1 | Cohen's d | | 4.5 | 3.9 |
| **Neointimal Macrophages** | | | | **Injury Score** | | | |
| | Control | Low Dose | High Dose | | Control | Low Dose | High Dose |
| Mean | 32.0 | 17.2 | 19.4 | Mean | 1.2 | 1.1 | 1.0 |
| Median | 32.9 | 18.1 | 18.1 | Median | 1.1 | 1.1 | 1.0 |
| Range | 30.0 | 16.7 | 22.7 | Range | 0.5 | 0.3 | 0.6 |
| Cohen's d | | 1.5 | 1.2 | Cohen's d | | 0.3 | 0.7 |

**Table 1.** *Cont.*

| **Neointimal Proliferation** | | | | **Inflammation Score** | | | |
|---|---|---|---|---|---|---|---|
| | Control | Low Dose | High Dose | | Control | Low Dose | High Dose |
| Mean | 6.7 | 2.3 | 2.3 | Mean | 1.4 | 1.3 | 1.5 |
| Median | 5.4 | 1.8 | 2.5 | Median | 1.4 | 1.2 | 1.4 |
| Range | 12.6 | 3.8 | 2.8 | Range | 0.6 | 0.4 | 0.5 |
| Cohen's d | | 0.9 | 0.8 | Cohen's d | | 0.7 | 0.0 |

Summary of descriptive statistics for the evaluated parameters and effect size evaluation with Cohen's d values.

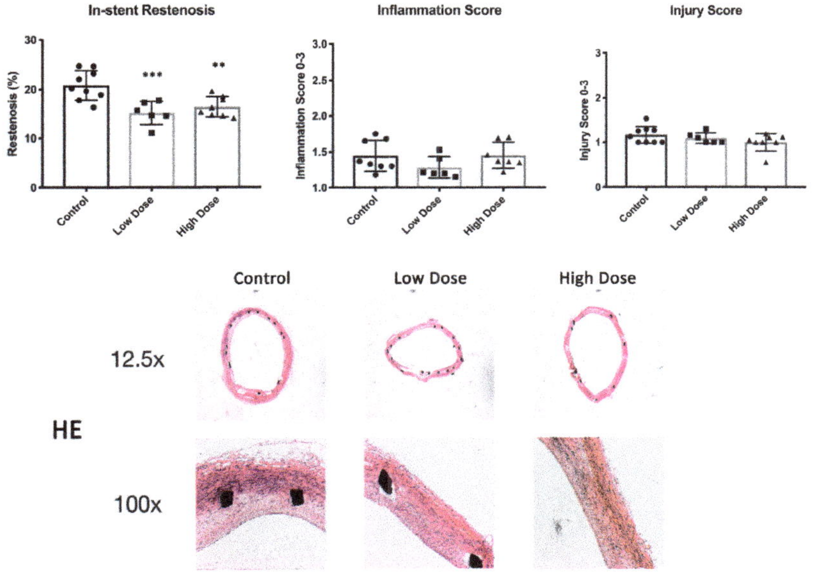


**Figure 1.** In-stent restenosis, inflammation and injury scores. In-stent restenosis was significantly reduced in the treatment groups compared to the controls (top left graph and HE histology). There were no differences in injury or inflammation scores at d42 determined from HE histology (top middle and right graphs). Data in graphs are shown as mean ± SD. Significance in graphs: ** $p < 0.01$; *** $p < 0.001$.

## 2.2. *Inflammation, Proliferation and Apoptosis*

The neointimal area covered densely with macrophages was 32 ± 9.3% in the controls, 17.2 ± 5.9% ($p = 0.005$) in the low-dose group and 19.4 ± 7.9% ($p = 0.008$) in the high-dose group (Figure 2). There were no differences in the inflammation scores of 1.5 ± 0.2, 1.3 ± 0.2 and 1.5 ± 0.2 (control, low dose and high dose, respectively; low dose $p = 0.22$; high dose $p = 0.99$) (Figure 1).

In the neointima, there were on average 7.2 ± 5.2, 2.3 ± 1.7 ($p = 0.03$) and 2.7 ± 1.1 ($p = 0.03$) proliferating cells per square millimeter in the control, low-dose and high-dose groups, respectively. The number of apoptotic cells in the control CMC-treated animals was 3.8 ± 2.5 per $mm^2$ of neointima. For the treatment groups, the numbers of apoptoses were 9.1 ± 3.9 ($p = 0.05$) and 14.2 ± 6.3 ($p = 0.0004$) cells/$mm^2$ (low dose and high dose, respectively) (Figure 3).

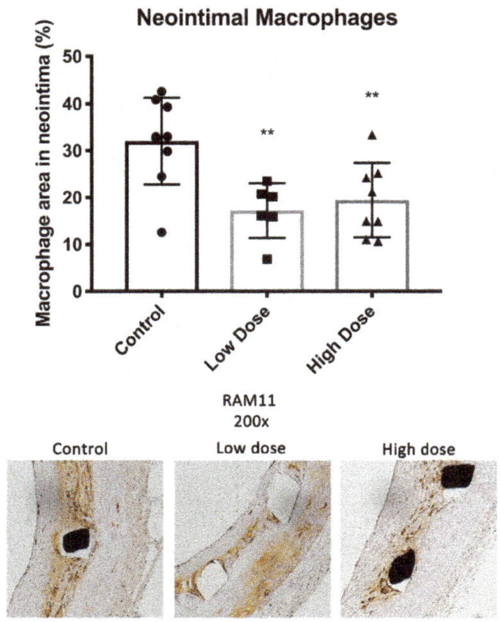


**Figure 2.** Neointimal macrophages. Macrophage areas measured in the neointimas of the treatment group stents (RAM-11 staining) were significantly smaller compared to the controls. Data are shown as mean ± SD, significance ** $p < 0.01$.

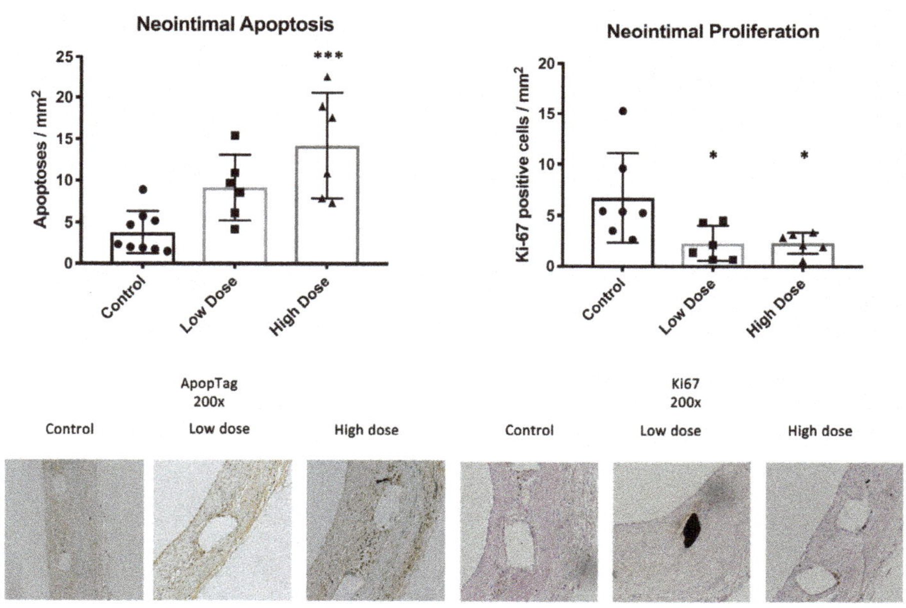


**Figure 3.** Apoptosis and proliferation. The high-dose treatment group showed increased apoptosis in the neointimal tissue at d42. Proliferation was attenuated in the neointimas of both treatment groups compared to the controls. Stainings from ApopTag kit for apoptosis and Ki-67 for proliferation. Data are shown as mean ± SD; significance: * $p < 0.05$; *** $p < 0.001$.

### 2.3. Endothelium

There were no significant differences in the endothelial coverage between the study groups. The endothelial coverage of the total lumen perimeter was 87.1 ± 15.4% in the controls, 75.4 ± 7.9% in the low-dose group and 86.1 ± 18.1% in the high-dose group (low dose $p = 0.32$; high dose $p = 0.98$) (Figure 4).

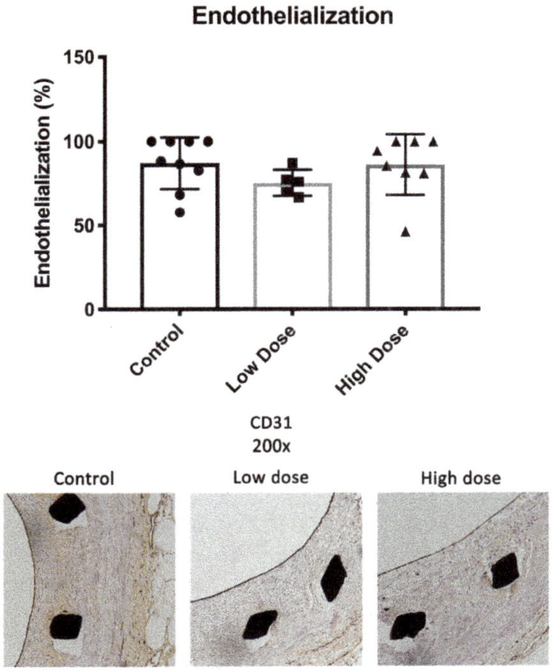


**Figure 4.** Endothelialization. No differences were seen in the rate of endothelial recovery by d42 between the study groups. Complete re-endothelialization was not observed in any of the groups. Data are shown as mean ± SD.

### 2.4. Clinical Chemistry and Safety Tissue Histology

There were no significant differences between the groups in clinical chemistry analyses. Safety tissue histology presented no adverse effects related to the therapy (Table 2).

**Table 2.** Clinical chemistry.

| | Control (CMC) | | Low-Dose DHIF | | High-Dose DHIF | |
|---|---|---|---|---|---|---|
| | d0 | d42 | d0 | d42 | d0 | d42 |
| ALT | 22.3 ± 10.0 | 23.0 ± 5.9 | 31.3 ± 10.5 | 23.2 ± 3.9 | 26.6 ± 19.5 | 26.9 ± 18.5 |
| ALP | 41.3 ± 20.7 | 37.8 ± 13.1 | 34.0 ± 7.5 | 28.5 ± 5.5 | 36.8 ± 11.7 | 32.3 ± 8.7 |
| AST | 15.3 ± 4.5 | 17.3 ± 7.3 | 27.0 ± 13.7 | 18.3 ± 3.6 | 21.0 ± 9.8 | 22.9 ± 9.8 |
| GT | 2.5 ± 2.1 | 2.7 ± 0.9 | 5.2 ± 0.7 | 5.6 ± 1.1 | 7.6 ± 7.1 | 6.6 ± 3.6 |
| Bilirubin | 0.7 ± 0.3 | 0.5 ± 0.3 | 0.4 ± 0.2 | 0.6 ± 0.3 | 0.4 ± 0.4 | 0.8 ± 0.3 |
| Creatinine | 59.0 ± 11.3 | 53.7 ± 5.8 | 66.7 ± 17.5 | 64.2 ± 17.9 | 83.6 ± 15.3 | 87.6 ± 25.1 |
| Cholesterol | 15.1 ± 4.9 | 15.8 ± 6.0 | 12.1 ± 9.7 | 12.9 ± 9.5 | 17.3 ± 2.6 | 16.1 ± 4.7 |
| Triglycerides | 8.6 ± 8.4 | 9.8 ± 6.7 | 6.6 ± 0.7 | 5.3 ± 1.7 | 9.8 ± 3.5 | 7.4 ± 4.7 |
| hS-CRP | 0.12 ± 0.03 | 0.12 ± 0.02 | 0.10 ± 0.02 | 0.12 ± 0.02 | 0.11 ± 0.01 | 0.10 ± 0.02 |

No significant changes were observed in standard laboratory tests between the study groups or between d0 and d42. Serum levels of ALT (Alanine Aminotransferase), ALP (Alkaline Phosphatase), AST (Aspartate Aminotransferase), GT (Gamma-glutamyltransferase), bilirubin, creatinine, total cholesterol, triglycerides and High Sensitivity C-Reactive Protein (hS-CRP) were determined.

*2.5. Pharmacokinetics*

One hour after drug administration, the serum concentrations of DHIF were 271.1 ± 84.7 ng/mL ($p$ = 0.018) and 630.8 ± 230.9 ng/mL ($p$ < 0.0001) in the low- and high-dose groups, respectively. No drug was detected in the control animals (Figure 5).

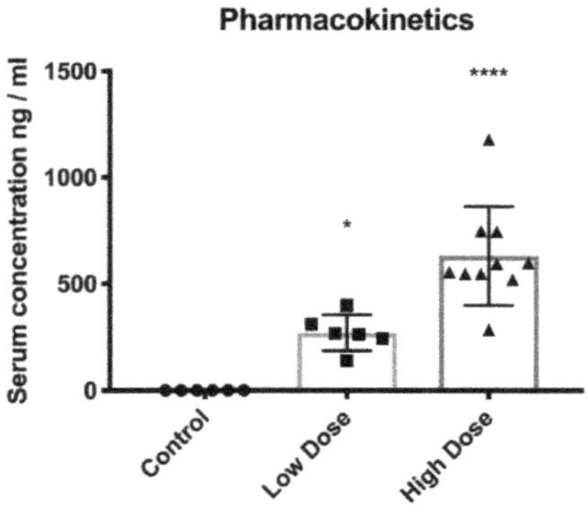


**Figure 5.** Pharmacokinetics. Levels of the study compound were measured one hour after the drug administration. Levels show a dose-dependent plasma concentration. Data are shown as mean ± SD; significance: * $p < 0.05$; **** $p < 0.0001$.

## 3. Discussion

In this study, a synthetic isoflavone, DHIF, significantly reduced ISR six weeks after stent deployment in a clinically relevant atherosclerotic WHHL rabbit model. In addition, treatment with DHIF in both dosage groups reduced the macrophage coverage of the neointima to approximately half of the coverage observed in the controls. This is highly important as inflammation is a key pathogenic factor in atherosclerosis and ISR [17–19]. Neointimal macrophages induce SMC proliferation and medial SMC migration into the neointima, both of which increase the volume of the neointima. The complicated atherosclerotic process is still centered on the properties and function of macrophages and smooth muscle cells when in contact with cholesterol [20–23]. This interplay of SMCs and macrophages in the presence of cholesterol led us to choose this highly relevant atherosclerotic model to conduct the present study in. Neointimal macrophages also promote the development of atherosclerotic vulnerable plaques, which are prone to rupture [3]. Furthermore, DHIF treatment induced apoptosis in the neointimal tissue while reducing proliferation.

Vascular injury causes ROS production to exceed compensatory antioxidant mechanisms and distorts the delicate ROS balance in the artery wall. This distortion results in inflammation and SMC proliferation [10]. Hence, redox regulation may explain the anti-inflammatory effects of DHIF observed in this study.

DHIF has been shown to attenuate NF-κB signaling in injured carotid artery. NF-κB family is a major regulator in injury, inflammation, cell proliferation and cell death [16,24]. All of these processes have a crucial role in the progression of ISR. Flavonoids' ability to decrease oxidative stress in vivo is proposed to result from the regulation of intracellular signaling and gene expression [25]. There is evidence that isoflavones upregulate antioxidant gene expression through the modulation of NF-κB and Nrf2 [13]. An NF-κB decoy eluting stent reduced ISR in hypercholesterolemic rabbits [16].

The endothelial coverage was greater than 75% in our three study groups with no significant differences between the groups. The finding suggests that DHIF treatment causes no delay in endothelial healing six weeks after injury. This successful healing is important as healthy endothelium restricts leukocyte infiltration and SMC proliferation and thus arrests the processes of ISR and neoatherosclerosis. A healthy endothelium also protects from potential in-stent thrombosis. Overall, the extent of re-endothelialization observed in this study is comparable to what we have observed in previous experiments with the WHHL-ISR model [26].

There were no significant differences in ISR or macrophage infiltration of the neointima between the two daily doses of 25 mg/kg and 50 mg/kg of DHIF. In addition, both dosages were equally effective in reducing proliferation in the neointima. Thus, a daily dose higher than 25 mg/kg had very little impact on the treatment outcome. While we observed more apoptotic cells in the neointimas of the high-dose animals, this did not translate into better treatment outcomes with regard to ISR formation.

A limitation of the study is the lack of a DES group as an additional control. The majority of stents implanted today have a drug coating and this will affect the microenvironment within the neointima. A DES study with a prolonged follow-up is warranted to study the effects of this therapy on neoatherosclerosis in DES.

## 4. Materials and Methods

### 4.1. Denudation and Stent Deployment

Adult Watanabe heritable hyperlipidemic (WHHL) rabbits ($n$ = 24) weighing 3.1 to 3.6 kg were kept on a standard diet [27]. A denudation injury of the entire aorta was performed with a 3F Fogarty embolectomy catheter (Edwards Lifesciences, Irvine, CA, USA) introduced by femoral artery cut-down [26]. Six weeks after the injury, a bare metal stent (Trimax, Abbott Vascular, IL, USA) introduced by a carotid artery cut-down was deployed into an infrarenal abdominal aortic segment free of side branches, documented by angiography. The rabbits received 40 mg aspirin in drinking water starting two days before stenting. Clopidogrel was administered p.o., starting with a 30 mg loading dose on the day of stenting, followed by 15 mg daily. Enoxaparin was administered at a dose of 1 mg/kg s.c. daily from stenting [28]. Animals were sacrificed 42 days after stent deployment. Aortic stented segments and safety tissues from the heart, lung, liver, spleen, kidney and semimembranosus muscle as well as segments of the proximal and distal aorta in relation to the stent were collected and further processed for analysis. Instrumentations and euthanasia were performed under medetomidin (Domitor, 0.3 mg/kg, Orion, Espoo, Finland) and ketamine (Ketalar, 20 mg/kg, Pfizer, NY, USA) anesthesia. All animal experiments were authorized by the national Animal Experiment Board in Finland (ELLA).

### 4.2. Drug Administration

DHIF was purified by silica gel chromatography, followed by confirmation of the purity by mass spectrometry–liquid chromatography. DHIF was dissolved in 1% carboxymethyl cellulose (CMC) solution for administration. Animals were randomized into control group receiving CMC only, and two study groups receiving 25 mg/kg or 50 mg/kg (low dose and high dose, respectively) of DHIF daily. The daily dose was divided into two portions equal in size administered every twelve hours with an orogastric tube. Animals were weighed once a week for accurate dosing. The first dose was administered two days before stent implantation and administrations were continued until sacrifice.

### 4.3. Blood Sampling

Arterial blood samples of six milliliters were collected from each rabbit before the first drug administration, before and after stent implantation and at sacrifice. Serum total cholesterol, LDL, HDL, creatinine, liver enzymes and high sensitivity-C Reactive Protein (hsCRP) were determined from the samples. Clinical chemistry analysis was performed at the Eastern Finland laboratory center. In addition, arterial blood samples were collected two weeks after stent implantation, one hour after drug administration, to evaluate DHIF drug serum levels.

### 4.4. Tissue Processing

The stents were paraformaldehyde-fixed and embedded in methylmethacrylate plastic resin (Technovit® 9100, Heraeus Kulzer GmbH, Division Technique, Wehrheim,

Germany) [12]. Safety organ tissues were embedded into paraffin. Embedded stent and safety samples were prepared into 5–7 μm thick sections and stained. The histological stent sections represent the middle section of the stents.

### 4.5. Immunohistochemistry

Immunohistochemistry was performed with monoclonal antibodies (mAbs) against endothelium (CD31, 1:50, Agilent Dako, CA, USA), SMCs (HHF35, 1:50, Enzo, NY, USA), macrophages (RAM-11, 1:200, Dako) and proliferating cells (Ki-67, 1:100, Dako, with citrate buffer antigen retrieval). Apoptosis analyses were performed from sections with the ApopTag-kit (Millipore, MA, USA). Images were captured using an AX70 microscope (Olympus Optical, Japan) and further processed for publication with Photoshop (Adobe, CA, USA), adjusting brightness and contrast evenly over each image.

### 4.6. Histomorphometry

The proportional ISR was determined as the ratio of neointimal area to the area of the internal elastic lamina {[1 − (Luminal area/IEL area) × 100]} from HE-stained sections of the cross-sections of the stented aortas at ×25 magnification. The proportion of intact endothelium of artery lumen perimeter was measured from CD31 immunostained sections of the stented aortas at ×40 magnification. The area covered with dense immunostaining with RAM11 per neointimal area was quantified from sections at ×200 magnification covering the entire intima. The number of proliferating cells per area was quantified from Ki67-immunostained histological sections at ×400 magnification. All histomorphometric analyses were performed using analysis software (Soft Imaging system, version 1) in a blinded manner. Pathological changes in safety tissues were assessed from HE-stained histological sections.

### 4.7. Statistical Analysis

All data represent mean ± standard deviation (SD). The statistical significance of the results is conveyed by values of $p < 0.05$ (*), $p < 0.005$ (**), $p < 0.001$ (***) and $p < 0.0001$ (****). Statistical significance was evaluated using GraphPad Prism software (version 10, GraphPad, MA, USA). Data normality was assessed with Shapiro–Wilk test and the data further analyzed with one-way ANOVA followed by Dunnett's post hoc test. Cohen's d values were determined for effect size analysis.

## 5. Conclusions

In this study, a novel synthetic isoflavone, DHIF, significantly reduced ISR six weeks after stent deployment in a clinically relevant atherosclerotic WHHL rabbit model. The administration of DHIF significantly reduced inflammation and proliferation in the restenotic lesions and did not delay endothelial recovery after the stenting injury. In the days of modern DESs, new concerns, especially regarding the need for dual antiplatelet therapy for high-bleeding-risk patients, have arisen. Therefore, new therapies without cytotoxic agents that impair endothelial recovery are in demand. Safety and the ease of oral administration make DHIF a potent drug candidate to treat restenosis.

**Author Contributions:** Conceptualization, J.P.H. and S.Y.-H.; methodology, J.P.H. and O.L.; validation, J.P.H. and J.T.; formal analysis, J.P.H. and J.T.; investigation, J.P.H. and J.T.; resources, S.Y.-H.; writing—original draft preparation, J.P.H.; writing—review and editing, J.P.H., O.L. and S.Y.-H.; visualization, J.P.H. and J.T.; supervision, J.P.H. and O.L.; project administration, J.P.H.; funding acquisition, S.Y.-H. All authors have read and agreed to the published version of the manuscript.

**Funding:** This study was supported by the Research Council of Finland GeneCellNano Flagship program (37210).

**Institutional Review Board Statement:** All animal experiments were authorized by the national Animal Experiment Board in Finland (license number ESAVI-1289-04.10.03-2011).

**Informed Consent Statement:** Not applicable.

**Data Availability Statement:** The original contributions presented in the study are included in the article, and further inquiries can be directed to the corresponding author.

**Acknowledgments:** The authors thank Johannes Laitinen, MD, for assistance with the experimental work.

**Conflicts of Interest:** The authors declare no conflicts of interest.

## References

1. Rodriguez, A.E.; Santaera, O.; Larribau, M.; Fernandez, M.; Sarmiento, R.; Baliño, N.P.; Newell, J.B.; Roubin, G.S.; Palacios, I.F. Coronary Stenting Decreases Restenosis in Lesions with Early Loss in Luminal Diameter 24 h After Successful {PTCA}. *Circulation* **1995**, *91*, 1397–1402. [CrossRef] [PubMed]
2. Ahn, J.M.; Park, D.W.; Lee, C.W.; Chang, M.; Cavalcante, R.; Sotomi, Y.; Onuma, Y.; Tenekecioglu, E.; Han, M.; Lee, P.H.; et al. Comparison of Stenting Versus Bypass Surgery According to the Completeness of Revascularization in Severe Coronary Artery Disease: Patient-Level Pooled Analysis of the SYNTAX, PRECOMBAT, and BEST Trials. *JACC Cardiovasc. Interv.* **2017**, *10*, 1415–1424. [CrossRef] [PubMed]
3. Ylä-Herttuala, S.; Bentzon, J.F.; Daemen, M.; Falk, E.; Garcia-Garcia, H.M.; Herrmann, J.; Hoefer, I.; Jauhiainen, S.; Jukema, J.W.; Krams, R.; et al. Stabilization of atherosclerotic plaques: An update. *Eur. Heart J.* **2013**, *34*, 3251–3258. [CrossRef] [PubMed]
4. Otsuka, F.; Finn, A.V.; Yazdani, S.K.; Nakano, M.; Kolodgie, F.D.; Virmani, R. The importance of the endothelium in atherothrombosis and coronary stenting. *Nat. Rev. Cardiol.* **2012**, *9*, 439–453. [CrossRef]
5. Cockerill, I.; See, C.W.; Young, M.L.; Wang, Y.; Zhu, D. Designing Better Cardiovascular Stent Materials—A Learning Curve. *Adv. Funct. Mater.* **2021**, *31*, 2005361. [CrossRef]
6. Park, S.J.; Kang, S.J.; Virmani, R.; Nakano, M.; Ueda, Y. In-stent neoatherosclerosis: A final common pathway of late stent failure. *J. Am. Coll. Cardiol.* **2012**, *59*, 2051–2057. [CrossRef]
7. Nakazawa, G.; Finn, A.V.; Joner, M.; Ladich, E.; Kutys, R.; Mont, E.K.; Gold, H.K.; Burke, A.P.; Kolodgie, F.D.; Virmani, R. Delayed arterial healing and increased late stent thrombosis at culprit sites after drug-eluting stent placement for acute myocardial infarction patients: An autopsy study. *Circulation* **2008**, *118*, 1138–1145. [CrossRef]
8. Glagov, S.; Stankunavicius, R.; Zarins, C.K.; Weisenberg, E.; Kolettis, G.J. Compensatory Enlargement of Human Atherosclerotic Coronary Arteries. *N. Engl. J. Med.* **2010**, *316*, 1371–1375. [CrossRef]
9. Levonen, A.L.; Vähäkangas, E.; Koponen, J.K.; Ylä-Herttuala, S. Antioxidant gene therapy for cardiovascular disease: Current status and future perspectives. *Circulation* **2008**, *117*, 2142–2150. [CrossRef]
10. Juni, R.P.; Duckers, H.J.; Vanhoutte, P.M.; Virmani, R.; Moens, A.L. Oxidative stress and pathological changes after coronary artery interventions. *J. Am. Coll. Cardiol.* **2013**, *61*, 1471–1481. [CrossRef]
11. Clarkson, T.B.; Prichard, R.W.; Morgan, T.M.; Petrick, G.S.; Klein, K.P. Remodeling of Coronary Arteries in Human and Nonhuman Primates. *JAMA J. Am. Med. Assoc.* **1994**, *271*, 289–294. [CrossRef]
12. Ribichini, F.; Joner, M.; Ferrero, V.; Finn, A.V.; Crimins, J.; Nakazawa, G.; Acampado, E.; Kolodgie, F.D.; Vassanelli, C.; Virmani, R. Effects of oral prednisone after stenting in a rabbit model of established atherosclerosis. *J. Am. Coll. Cardiol.* **2007**, *50*, 176–185. [CrossRef] [PubMed]
13. Levonen, A.L.; Inkala, M.; Heikura, T.; Jauhiainen, S.; Jyrkkänen, H.K.; Kansanen, E.; Määttä, K.; Romppanen, E.; Turunen, P.; Rutanen, J.; et al. Nrf2 gene transfer induces antioxidant enzymes and suppresses smooth muscle cell growth in vitro and reduces oxidative stress in rabbit aorta in vivo. *Arterioscler. Thromb. Vasc. Biol.* **2007**, *27*, 741–747. [CrossRef]
14. Siow, R.C.M.; Li, F.Y.L.; Rowlands, D.J.; de Winter, P.; Mann, G.E. Cardiovascular targets for estrogens and phytoestrogens: Transcriptional regulation of nitric oxide synthase and antioxidant defense genes. *Free Radic. Biol. Med.* **2007**, *42*, 909–925. [CrossRef]
15. Karin, M.; Yamamoto, Y.; Wang, Q.M. The IKK NF-κB system: A treasure trove for drug development. *Nat. Rev. Drug Discov.* **2004**, *3*, 17–26. [CrossRef]
16. Ohtani, K.; Egashira, K.; Nakano, K.; Zhao, G.; Funakoshi, K.; Ihara, Y.; Kimura, S.; Tominaga, R.; Morishita, R.; Sunagawa, K. Stent-based local delivery of nuclear factor-κB decoy attenuates in-stent restenosis in hypercholesterolemic rabbits. *Circulation* **2006**, *114*, 2773–2779. [CrossRef] [PubMed]
17. Costa, M.A.; Simon, D.I. Molecular basis of restenosis and drug-eluting stents. *Circulation* **2005**, *111*, 2257–2273. [CrossRef] [PubMed]
18. Libby, P.; Ridker, P.M.; Hansson, G.K. Inflammation in Atherosclerosis. From Pathophysiology to Practice. *J. Am. Coll. Cardiol.* **2009**, *54*, 2129–2138. [CrossRef]
19. Hansson, G.K.; Libby, P. The immune response in atherosclerosis: A double-edged sword. *Nat. Rev. Immunol.* **2006**, *6*, 508–519. [CrossRef]
20. Park, S.H. Regulation of Macrophage Activation and Differentiation in Atherosclerosis. *J. Lipid Atheroscler.* **2021**, *10*, 251–267. [CrossRef]
21. Khallou-Laschet, J.; Varthaman, A.; Fornasa, G.; Compain, C.; Gaston, A.T.; Clement, M.; Dussiot, M.; Levillain, O.; Graff-Dubois, S.; Nicoletti, A.; et al. Macrophage plasticity in experimental atherosclerosis. *PLoS ONE* **2010**, *5*, e8852. [CrossRef] [PubMed]

22. Wang, Y.; Dubland, J.A.; Allahverdian, S.; Asonye, E.; Sahin, B.; Jaw, J.E.; Sin, D.D.; Seidman, M.A.; Leeper, N.J.; Francis, G.A. Smooth Muscle Cells Contribute the Majority of Foam Cells in ApoE (Apolipoprotein E)-Deficient Mouse Atherosclerosis. *Arterioscler. Thromb. Vasc. Biol.* **2019**, *39*, 876–887. [CrossRef] [PubMed]
23. Janoudi, A.; Shamoun, F.E.; Kalavakunta, J.K.; Abela, G.S. Cholesterol crystal induced arterial inflammation and destabilization of atherosclerotic plaque. *Eur. Heart J.* **2016**, *37*, 1959–1967. [CrossRef]
24. Perkins, N.D. Integrating cell-signalling pathways with NF-κB and IKK function. *Nat. Rev. Mol. Cell Biol.* **2007**, *8*, 49–62. [CrossRef]
25. Chen, L.; Teng, H.; Jia, Z.; Battino, M.; Miron, A.; Yu, Z.; Cao, H.; Xiao, J. Intracellular signaling pathways of inflammation modulated by dietary flavonoids: The most recent evidence. *Crit. Rev. Food Sci. Nutr.* **2018**, *58*, 2908–2924. [CrossRef] [PubMed]
26. Hytönen, J.P.; Taavitsainen, J.; Laitinen, J.T.T.; Partanen, A.; Alitalo, K.; Leppänen, O.; Ylä-Herttuala, S. Local adventitial anti-angiogenic gene therapy reduces growth of vasa-vasorum and in-stent restenosis in WHHL rabbits. *J. Mol. Cell Cardiol.* **2018**, *121*, 145–154. [CrossRef]
27. Watanabe, Y. Serial inbreeding of rabbits with hereditary hyperlipidemia (WHHL-rabbit). *Atherosclerosis* **1980**, *36*, 261–268. [CrossRef]
28. Hytönen, J.; Leppänen, O.; Braesen, J.H.; Schunck, W.-H.; Mueller, D.; Jung, F.; Mrowietz, C.; Jastroch, M.; Von Bergwelt-Baildon, M.; Kappert, K.; et al. Activation of Peroxisome Proliferator-Activated Receptor-δ as Novel Therapeutic Strategy to Prevent In-Stent Restenosis and Stent Thrombosis. *Arterioscler. Thromb. Vasc. Biol.* **2016**, *36*, 1534–1548. [CrossRef]

International Journal of
*Molecular Sciences*

MDPI

*Article*

# A Novel Strategy for the Treatment of Aneurysms: Inhibition of MMP-9 Activity through the Delivery of TIMP-1 Encoding Synthetic mRNA into Arteries

**Sonia Golombek †, Isabelle Doll †, Louisa Kaufmann, Mario Lescan, Christian Schlensak and Meltem Avci-Adali ***

Department of Thoracic and Cardiovascular Surgery, University Hospital Tuebingen, Calwerstraße 7/1, 72076 Tuebingen, Germany
* Correspondence: meltem.avci-adali@uni-tuebingen.de
† These authors contributed equally to this work.

**Abstract:** Aneurysms pose life-threatening risks due to the dilatation of the arteries and carry a high risk of rupture. Despite continuous research efforts, there are still no satisfactory or clinically effective pharmaceutical treatments for this condition. Accelerated inflammatory processes during aneurysm development lead to increased levels of matrix metalloproteinases (MMPs) and destabilization of the vessel wall through the degradation of the structural components of the extracellular matrix (ECM), mainly collagen and elastin. Tissue inhibitors of metalloproteinases (TIMPs) directly regulate MMP activity and consequently inhibit ECM proteolysis. In this work, the synthesis of TIMP-1 protein was increased by the exogenous delivery of synthetic TIMP-1 encoding mRNA into aortic vessel tissue in an attempt to inhibit MMP-9. In vitro, TIMP-1 mRNA transfection resulted in significantly increased TIMP-1 protein expression in various cells. The functionality of the expressed protein was evaluated in an appropriate ex vivo aortic vessel model. Decreased MMP-9 activity was detected using in situ zymography 24 h and 48 h post microinjection of 5 μg TIMP-1 mRNA into the aortic vessel wall. These results suggest that TIMP-1 mRNA administration is a promising approach for the treatment of aneurysms.

**Keywords:** aneurysm; synthetic mRNA; cardiovascular diseases; TIMP-1; protein replacement therapies; cardiovascular interventions

## 1. Introduction

Cardiovascular diseases (CVDs) are the leading cause of death worldwide and cause more than 1.8 million deaths each year, representing 37% of all deaths in the European Union (EU) [1] and placing a significant burden on medical healthcare systems. A subfamily of CVDs are aneurysms, which are caused by progressive focal abnormal dilatation of the arteries due to local weakening of the blood vessel wall. They can develop in various locations throughout the body, including the aorta (abdominal or thoracic aortic aneurysm), blood vessels supplying the brain (cranial aneurysm), and other parts of the body (peripheral aneurysm).

Aortic aneurysms affect the body's main artery, the aorta, and if left untreated, can lead to catastrophic dissections and ruptures with massive internal bleeding, with an overall mortality rate of 90% [2,3], corresponding to more than 200,000 deaths per year worldwide [4]. Abdominal aortic aneurysms (AAA) are the most commonly identified aortic aneurysms and have the highest prevalence in the over-65 age group (approximately 4–7% of men and 1–2% of women) [5].

Current therapeutic options for the treatment of aneurysms include medical stabilization, monitoring of progression, and surgical treatment for late-stage AAA patients. Medical stabilization relies on regular and systemic administration of antihypertensive drugs to reduce the shear forces on the vessel wall to slow the growth of the aneurysm without being able to restore the fragile vessel. Several drugs, such as β-blockers, calcium antagonists, and angiotensin-converting enzyme (ACE) inhibitors, have been tested in this regard, with some beneficial effects reported but not consistently confirmed in larger and controlled studies. Thus, there is currently no effective and specific drug therapy to stabilize or heal aneurysms. Consequently, the monitoring of aneurysm progression and surgical treatment of patients involving open aneurysm repair (OAR) or endovascular aneurysm repair (EVAR) in late aneurysm stages are the only options to mitigate the risk of fatal rupture [6].

Several factors may play a role in the development of AAA, including advanced age, male gender, smoking, family history, hypertension, and atherosclerosis. At the tissue level, extracellular matrix (ECM) degradation leads to aortic wall weakening and dilatation, which may subsequently result in aneurysm rupture with high mortality and morbidity.

The wall of the arteries consists of three layers: The innermost layer is the tunica intima, which is in contact with the blood and consists of an endothelial cell layer. The second layer, the tunica media, contains smooth muscle cells (SMCs) and the ECM composed of elastic and collagen fibers, which provide stability as well as elasticity. The tunica adventitia is the outer layer composed of fibroblasts, vasa vasorum, lymphatic vessels, nerves, and immune and stem cells [7]. In pathological conditions, such as aortic aneurysms, the vessel wall is destabilized due to defective or deficient structural components. Inflammation and upregulation of proteolytic processes lead to the degradation of the major structural proteins, elastin and collagen, in the aortic vessel wall. The ECM protein elastin is the load-bearing component and is responsible for the elasticity and resilience of large arteries, whereas collagen is responsible for the structural integrity and tensile strength bearing load at high pressures or when elastin fails. Over time, elastin degradation, cyclic strain, and increased wall tension lead to progressive aortic dilatation [8].

High levels of elastin degradation products have been found in serum samples from AAA patients [9]. These products might lead to disease progression by increasing chemotaxis and activation of mononuclear cells. Oxidative stress, inflammation, and accelerated SMC apoptosis further promote aortic wall destabilization [8–10]. Moreover, inflammation in the vessel wall caused by chemokines and elastin degradation products, among others, leads to the recruitment of leukocytes to the aortic vessel wall and activation of macrophages, which in turn produce pro-inflammatory molecules and matrix metalloproteinases (MMPs) [11–13].

MMPs are a group of enzymes involved in the physiological turnover of the ECM during wound healing or tissue remodeling processes by degrading structural components, such as elastin or collagen. MMP-2, -9, and -12 play a major role in the degradation of the ECM in degenerative AAAs [12]. In particular, MMP-9, also called gelatinase B, is responsible for the degradation of elastin and collagen type I and IV. Increased MMP-9 plasma levels as well as mRNA expression levels could be detected in AAA compared to healthy aortic vessels [14]. Interestingly, an AAA size dependency of the MMP-9 content was observed in patients with AAA [15]. MMP-9 mRNA expression levels were significantly higher in moderate diameter sized (5- to 6.9 cm) compared to small (<4.0 cm) and large (>7.0 cm) AAAs, indicating a crucial role of MMP-9 in the progression of the AAA process. The activity of MMPs is regulated by their endogenous inhibitors, the family of tissue inhibitors of metalloproteinases (TIMPs), and thus, they also play an important role in ECM remodeling and disease development. Therefore, an imbalance between MMPs and TIMPs has been associated with the pathophysiology and progression of various diseases, such as CVDs [16]. The stoichiometric ratio between the non-covalent binding of TIMP and MMP has been reported to be 1:1 [17]. Thus, maintaining the MMP:TIMP ratio is the principal basis for the development of several new drugs. TIMP-1 is a soluble protein secreted by cells into the ECM and is the endogenous inhibitor of MMP-9 [18]. Moreover, TIMP-1 has a high affinity for almost all active MMPs and can bind to pro-MMP-9, slowing its activation [18]. Decreased TIMP-1 levels have been detected in aortic aneurysms, providing a favorable environment for ECM degradation and further aneurysm expansion [19].

A promising new approach for tissue regeneration by reproducing missing or damaged proteins is the use of synthetic messenger RNA (mRNA) [20]. Synthetic mRNA offers the possibility of protein synthesis under physiological conditions. There are already numerous in vivo studies of synthetic mRNA-based drugs for protein replacement therapy, immunotherapy, and vaccine production. The breakthrough came with mRNA-based SARS-CoV-2 vaccines developed during the COVID-19 pandemic, as the first-ever approved mRNA therapeutics [21,22]. For more than two decades, mRNA technology has been optimized. Special attention has been paid to the optimization of mRNA to reduce unwanted immune-activating side effects and to increase mRNA stability and protein expression efficiency [23]. Both can be achieved by incorporating modified nucleotides [24], integrating a 3′poly(A)-tail and a 5′-end cap structure [25], as well as optimization of the codon sequences and untranslated regions [26], and optimizing the codon sequences (increasing GC-rich codons and reducing AU-rich codons) can be used instead of modified nucleotides [27]. Moreover, compared to viral vectors and recombinant proteins, in vitro transcribed mRNA has several advantages, e.g., there is no risk of insertional mutagenesis [28,29], and the protein of interest is only transiently expressed, avoiding potential protein-dependent side effects [28]. The local overexpression of TIMP-1 protein prevented elastin depletion and aneurysm formation as well as rupture in rats [30].

In this study, TIMP-1 protein expression was analyzed after the in vitro TIMP-1 mRNA transfection of cells. Afterward, a microinjection-based method was developed to deliver the TIMP-1 mRNA into the vessel wall. The successful expression of TIMP-1 after the delivery of synthetic TIMP-1 mRNA in ex vivo porcine and human aortic vessel walls was demonstrated and the functionality of the produced TIMP-1 protein was evaluated using in situ zymography.

## 2. Results

### *2.1. Transfection of Cells with Synthetic TIMP-1 mRNA*

To analyze TIMP-1 protein production, $3 \times 10^5$ EA.hy926 cells, NUFFs (newborn foreskin fibroblasts), or HUVECs (human umbilical vein endothelial cells) were transfected with 0.5, 1.0, or 1.5 µg synthetic TIMP-1 mRNA. Using ELISA, significantly increased TIMP-1 levels were detected in the supernatants of all cell types 24 h after TIMP-1 mRNA transfection (Figure 1A). Except for EA.hy926 cells, increasing the TIMP-1 mRNA concentration to 1.5 µg resulted in a significant increase in TIMP-1 protein expression. Higher levels of TIMP-1 were detected in the supernatants of the cells than in the cell lysates (Figure 1B), indicating the secretion of TIMP-1 protein into the extracellular space after translation. In EA.hy926 cells and NUFFs, only transfection of 1.5 µg TIMP-1 mRNA resulted in significantly higher TIMP-1 levels in cells compared with cells transfected with transfection reagent (L2000) alone (Figure 1B). In HUVECs, transfection of 0.5 µg TIMP-1 mRNA already resulted in significantly increased TIMP-1 levels in cells.

### *2.2. Delivery of Synthetic mRNA into the Aortic Vessel Wall by Microinjection*

To investigate the delivery of synthetic mRNA into the blood vessel wall, 1 µg of Cy3-labeled hGLuc mRNA was injected ex vivo using hollow microneedles from the intraluminal side into porcine and human aortic vessel walls. In paraffin cross-sections, the localization of Cy3-labeled hGLuc mRNA was analyzed using fluorescence microscopy. In these histologic sections, the individual compartments of the vessels, tunica intima, tunica media, and tunica adventitia, were readily recognizable. The cell nuclei were stained with DAPI (blue). Cy3-labeled hGLuc mRNA (red) was detected mainly in the tunica adventitia of the porcine aortic wall (Figure 2A) and the tunica media of the human aortic wall (Figure 2B).

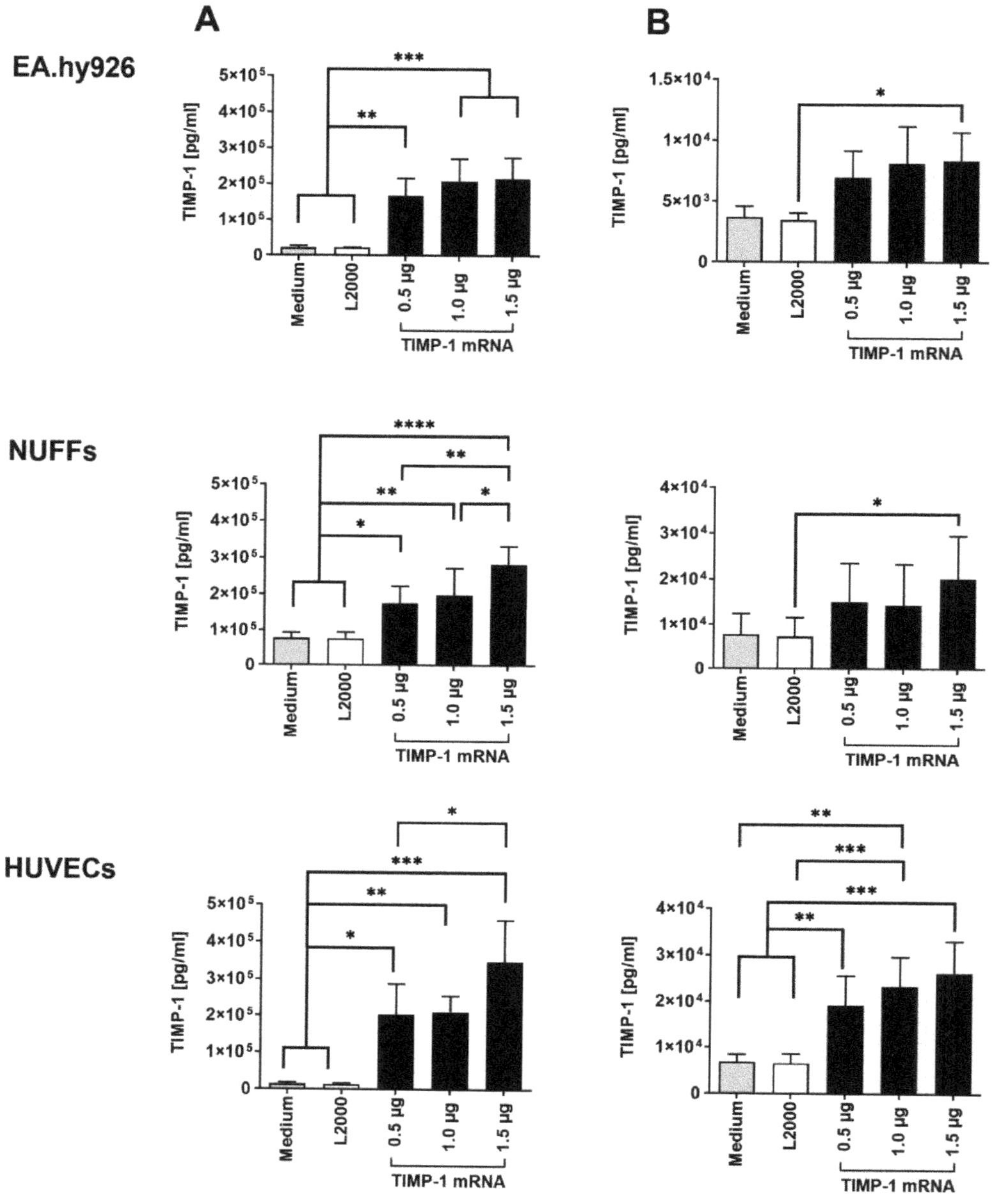


**Figure 1.** Analysis of TIMP-1 protein expression after the transfection of different cells with synthetic TIMP-1 mRNA. During the analysis, $3 \times 10^5$ EA.hy926 cells, NUFFs, or HUVECs were transfected with 0.5, 1.0, and 1.5 µg TIMP-1 mRNA complexed 1:1 with Lipofectamine 2000 (L2000). After 4 h incubation at 37 °C and 5% $CO_2$, the transfection complexes were replaced with the appropriate corresponding cell culture medium. After additional incubation for 24 h at 37 °C and 5% $CO_2$, the TIMP-1 protein concentration in the (**A**) supernatants and (**B**) cell lysates was determined using human TIMP-1 specific ELISA. Cells treated with medium or medium containing L2000 served as controls. The results are shown as the mean + SD (n = 3). Statistical differences were determined using one-way ANOVA followed by Tukey's multiple comparisons test. (* $p < 0.05$, ** $p < 0.01$, *** $p < 0.001$, and **** $p < 0.0001$).

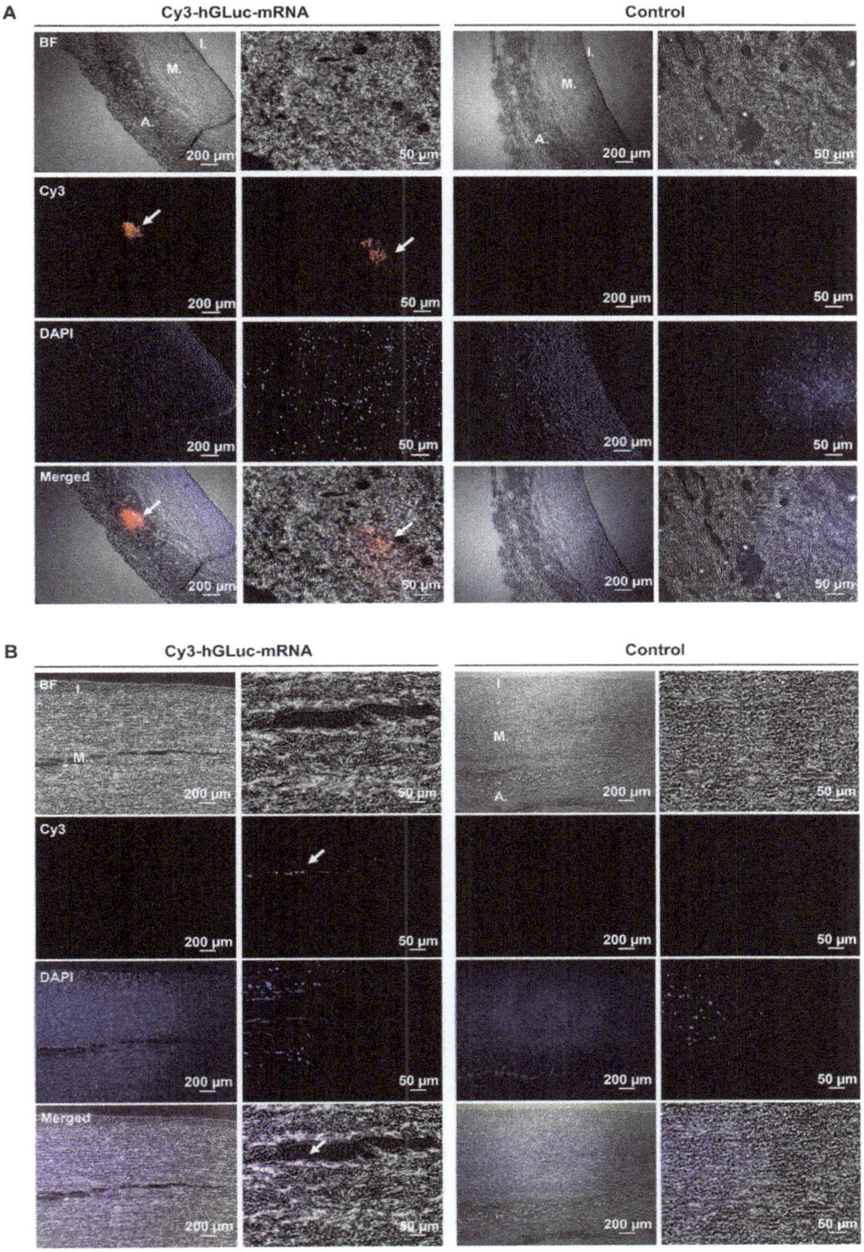


**Figure 2.** Application of synthetic hGLuc mRNA into ex vivo porcine and human aortic vessel walls. 1 µg of Cy3-labeled hGLuc mRNA complexed with 1 µL L2000 in 50 µL Opti-MEM was injected into porcine (**A**) and human (**B**) aortic vessel walls from the intraluminal side. After microinjection, the samples were fixed in 4% PFA, and paraffin sections were prepared (2.5 µm). Cell nuclei were stained with DAPI. White arrows indicate the Cy3-labeled hGLuc mRNA in the tunica media and adventitia. DAPI (blue), Cy3-labeled hGLuc mRNA (red). I.: tunica intima, M.: tunica media, and A.: tunica adventitia. Left column: 5× magnification, right column: 40× magnification. (porcine n = 3; human n = 4).

To determine whether the administered hGLuc mRNA could be translated into protein, the supernatants of ex vivo cultivated aortic vessels were collected at 24 and 48 h post-injection. Compared with control groups injected with medium alone or medium with L2000, microinjection of hGLuc mRNA in both human as well as porcine aortic vessels led to significantly increased luciferase activity after 24 h of incubation (Figure 3). After 48 h of incubation, similar luciferase activity was measured in the supernatant of porcine vessels as after 24 h, indicating that most of the injected mRNA was translated in the first 24 h of incubation (Figure 3, left). In human aortic vessels, slightly higher luciferase activity was detected after 48 h compared to 24 h. Although cell activity might be impaired in human pathogenic vessels derived from aortic aneurysms compared with healthy vessels, a significant amount of expressed luciferase protein was detected after 24 and 48 h of incubation after mRNA injection (Figure 3, right).

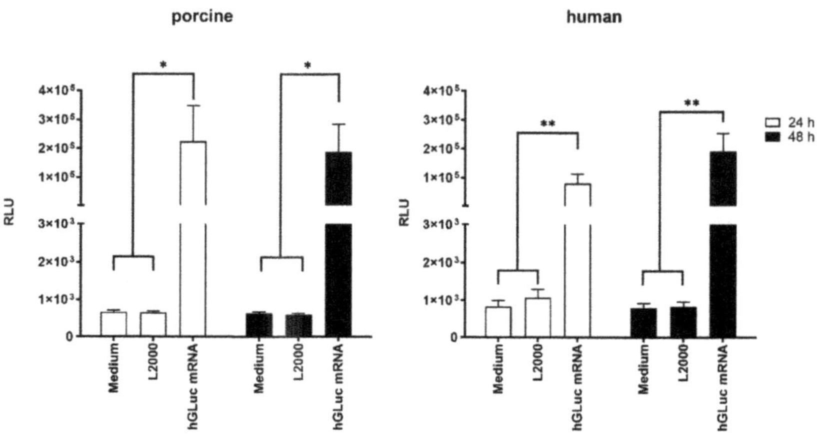


**Figure 3.** Protein expression after the injection of synthetic hGLuc mRNA into ex vivo porcine and human aortic vessel walls. Luciferase activity was detected after ex vivo microinjection of 1 μg synthetic hGLuc mRNA complexed with 1 μL L2000 in the porcine aorta (**left**) and human aorta (**right**). Cell supernatants were analyzed using luciferase assay after incubation at 37 °C and 5% $CO_2$ for 24 and 48 h, respectively. Aortic vessels injected with medium or medium containing 1 μL L2000 were used as the control. For statistical analysis, two-way ANOVA test with subsequent Tukey's multiple comparisons test was performed (* $p < 0.05$; ** $p < 0.01$). The results are shown as means + SEM (porcine n = 4; human n = 5).

### 2.3. *Production of TIMP-1 Protein after the Delivery of Synthetic TIMP-1 mRNA in the Porcine Aortic Vessel Wall*

To determine whether the administered TIMP-1 mRNA leads to increased TIMP-1 protein expression, the supernatants of ex vivo cultivated porcine aortic vessels were collected at 24 and 48 h post injection with 3 μg TIMP-1 mRNA. Significantly increased TIMP-1 levels were detected after 48 h of incubation using ELISA (Figure 4).

### 2.4. *Detection of MMP-9 Inhibition after the Application of Synthetic TIMP-1 mRNA in Porcine and Human Aortic Vessel Walls*

After demonstrating that ex vivo aortic tissue can express the desired protein up to 48 h post injection of synthetic mRNA using hGLuc encoding mRNA, the same ex vivo model was used to evaluate the functionality of expressed TIMP-1 protein. In situ zymography can visualize and assess the proteolytic activity of MMP-9 in histological sections using the fluorescently labeled MMP substrate DQ-gelatin. Enzymatic cleavage of DQ-gelatin by active MMPs results in a green fluorescence signal [31]. Since the activity of MMP-9 is inhibited by EDTA, which binds catalytically necessary $Zn^{2+}$ ions, the tissue sections incubated with EDTA served as negative controls for the in situ zymography. TIMP-1 binds covalently to MMP-9 and inhibits DQ-gelatin cleavage. Thus, the ability of TIMP-1 protein, which is produced after the microinjection of TIMP-1 mRNA into the vessel wall, to inhibit MMP-9 was investigated in both porcine and human aortic vessels. For this purpose, 5 µg of human synthetic TIMP-1 mRNA was injected ex vivo into porcine and human aortic vessels from the side of the intima and incubated for 24 or 48 h. After fixation, MMP-9 activity was assessed in paraffin sections using in situ zymography, MMP-9-dependent DQ-gelatin cleavage efficiency was determined, and TIMP-1 mRNA-treated versus untreated groups were compared. Both 24 and 48 h after injection of TIMP-1 mRNA, a significantly lower signal of DQ-gelatin was detected in the porcine vessel wall than in vessels treated with L2000 alone (Figure 5). As expected, the lowest MMP-9 activity was detected in EDTA-treated vessel sections. The synthetic mRNA could be also microinjected from the side of the adventitia in porcine aortic samples (Figure S1). Here, the results were similar to those obtained with mRNA injections from the side of the intima.

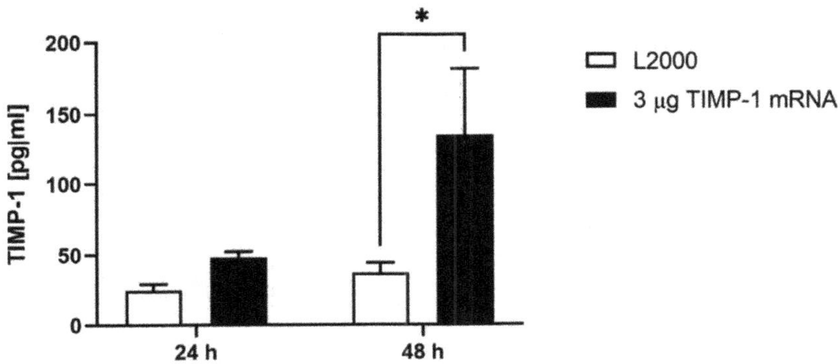


**Figure 4.** Detection of TIMP-1 levels in the supernatant of porcine aorta after the microinjection of 3 µg TIMP-1 mRNA. 3 µg of TIMP-1 mRNA was complexed with L2000 in 50 µL Opti-MEM and injected into the porcine aortic vessel wall. TIMP-1 levels in the supernatant were detected 24 and 48 h post injection using TIMP-1 ELISA. The results are shown as means + SEM (n = 3). Statistical differences were determined using two-way ANOVA followed by Tukey's multiple comparisons test. (* $p < 0.05$).

The ECM structures of the porcine (Figure S2) and human (Figure S3) aorta samples were also analyzed after hematoxylin and eosin, elastin, and collagen staining of the histological sections. No differences between the control and treated groups could be observed.

Similar results were obtained in human aortic tissue after the injection of TIMP-1 mRNA and incubation for 24 and 48 h (Figure 6). Here, a strongly reduced DQ-gelatin signal was observed both 24 and 48 h after the injection of synthetic TIMP-1 mRNA compared with the L2000 control groups. Thereby, the inhibitory function of the produced TIMP-1 protein was confirmed. The proteolytic MMP-9 activity was successfully inhibited and decreased in mRNA-treated tissues.

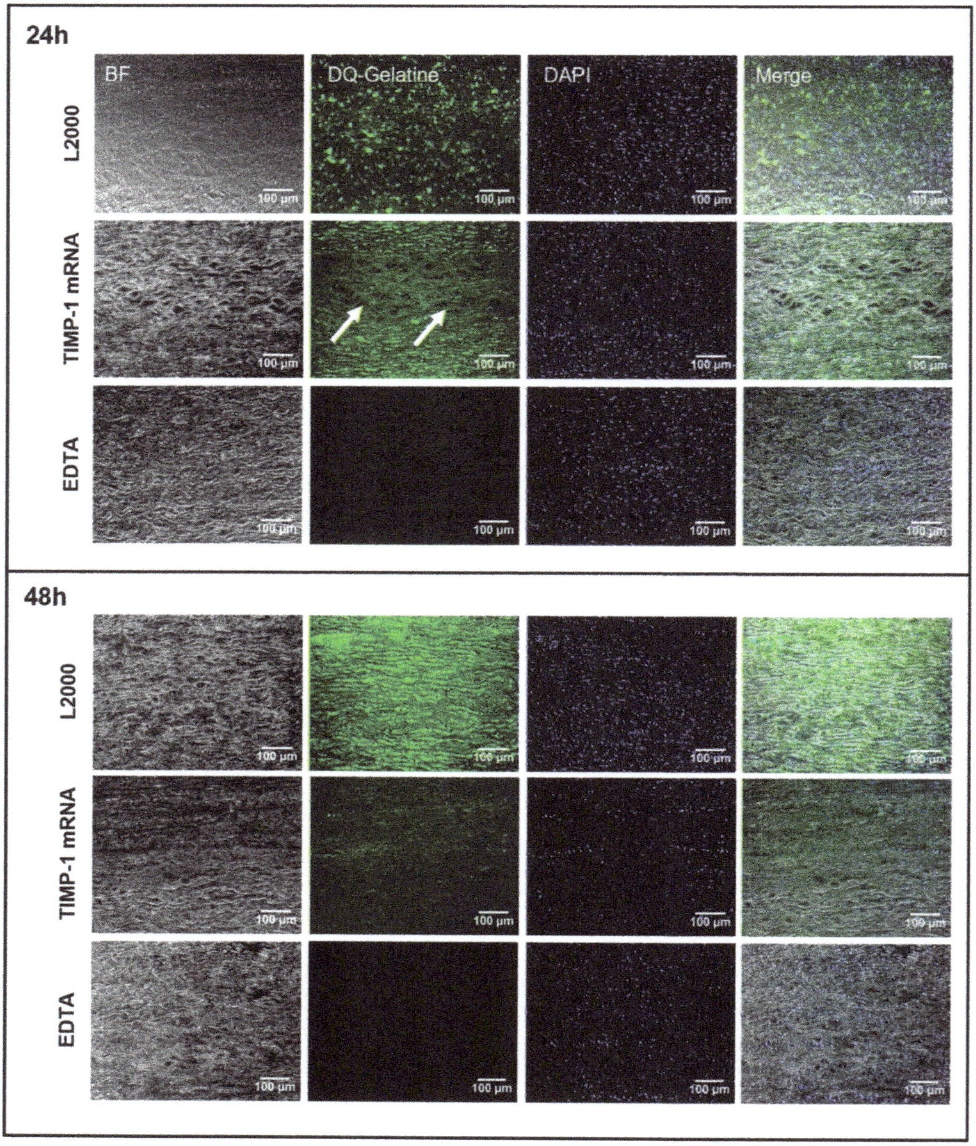


**Figure 5.** Analysis of MMP activity using in situ zymography in porcine aorta after the microinjection of human TIMP-1 mRNA and incubation for 24 and 48 h. During the analysis, 5 µg of TIMP-1 mRNA complexed 1:1 with L2000 in 50 µL Opti-MEM was injected ex vivo from the side of the intima into the porcine aorta. Tissues were incubated for 24 and 48 h at 37 °C and 5% $CO_2$. Injection of L2000 only in Opti-MEM served as a control. Post incubation, the tissues were fixed, and 5 µm paraffin sections were prepared. MMP-9 activity was visualized by using DQ-gelatin-based in situ zymography. To verify enzymatic gelatin cleavage by MMPs, control sections were inhibited with EDTA before the addition of DQ substrate. Cell nuclei were stained using a DAPI-containing mounting medium. DQ gelatin: green; nuclei: blue. 20× magnification; (n = 3). Arrows indicate areas with decreased DQ-gelatine fluorescence signal.

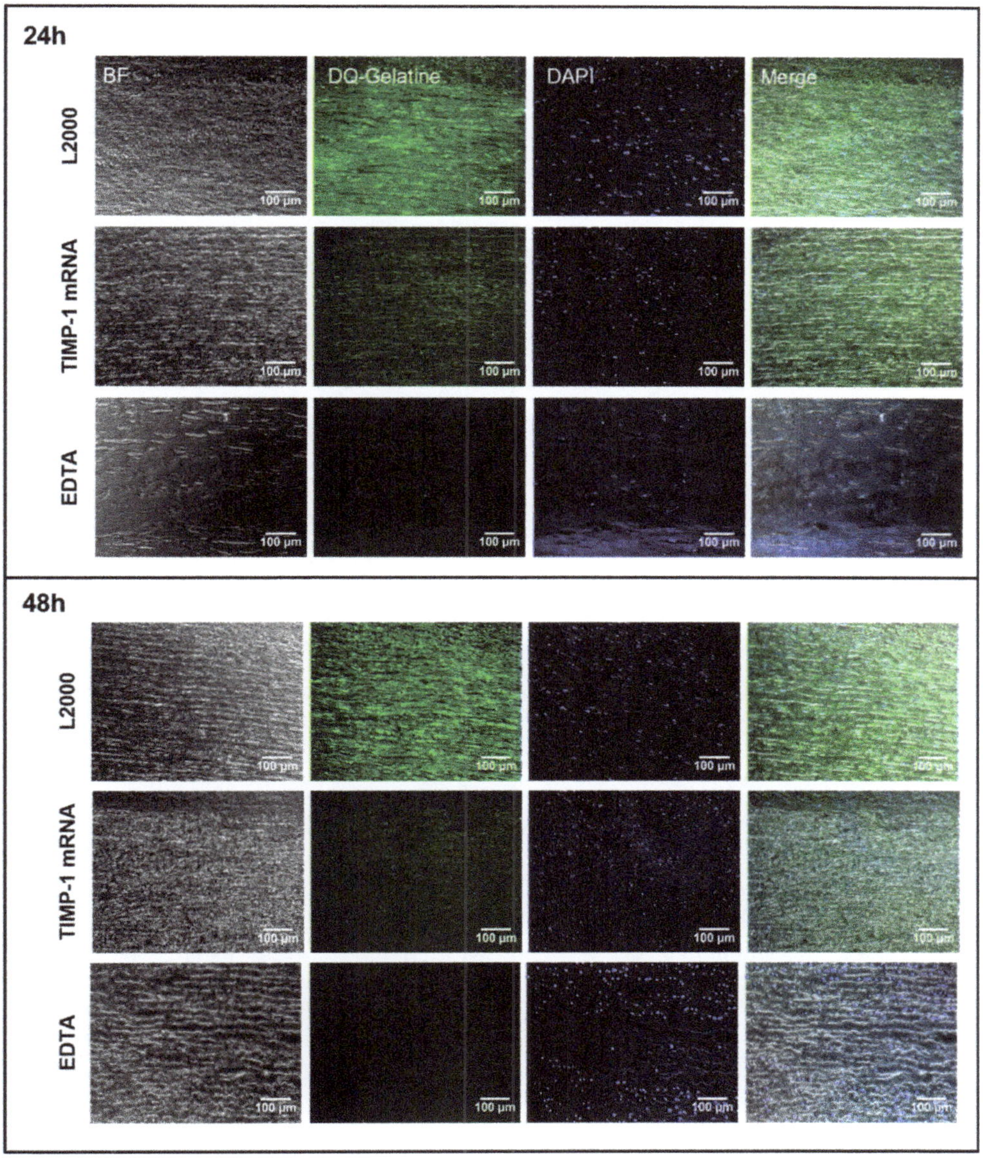


**Figure 6.** Analysis of MMP activity using in situ zymography in the human aorta after microinjection of human TIMP-1 mRNA and incubation for 24 and 48 h. During the analysis, 5 µg of TIMP-1 mRNA complexed 1:1 with L2000 in 50 µL Opti-MEM was injected ex vivo from the side of the intima into the human aorta. Tissues were incubated for 24 and 48 h at 37 °C and 5% $CO_2$. Injection of L2000 only in medium served as a control. Post-incubation tissues were fixed, 5 µm paraffin sections were prepared, and MMP-9 activity was visualized by using DQ-gelatin-based in situ zymography. To verify enzymatic gelatin cleavage by MMPs, control sections were inhibited with EDTA before the addition of DQ substrate. Cell nuclei were stained using a DAPI-containing mounting medium. DQ gelatine: green; nuclei: blue. 20× magnification; (n = 4).

*2.5. Analysis of TIMP-1 mRNA Translation Efficiency by Variation of Nucleotide Modifications*

To improve the protein expression efficiency of the mRNA, different nucleotide modifications were incorporated into the mRNA and tested. In the prior analysis, the nucleotide modifications Ψ/5mC were used. Ψ was replaced by $me^1Ψ$, and TIMP-1 mRNA was modified with either $me^1Ψ$/5mC or $me^1Ψ$ only. Next, 1.5 µg of each mRNA variant was complexed with L2000, and $3 \times 10^5$ EA.hy926 cells were transfected. After 24 h, protein translation efficiency was analyzed via TIMP-1-specific ELISA in supernatants (Figure 7). Through the incorporation of $me^1Ψ$ only, the protein expression efficiency was significantly enhanced. The additional modification with 5mC did not improve translation.

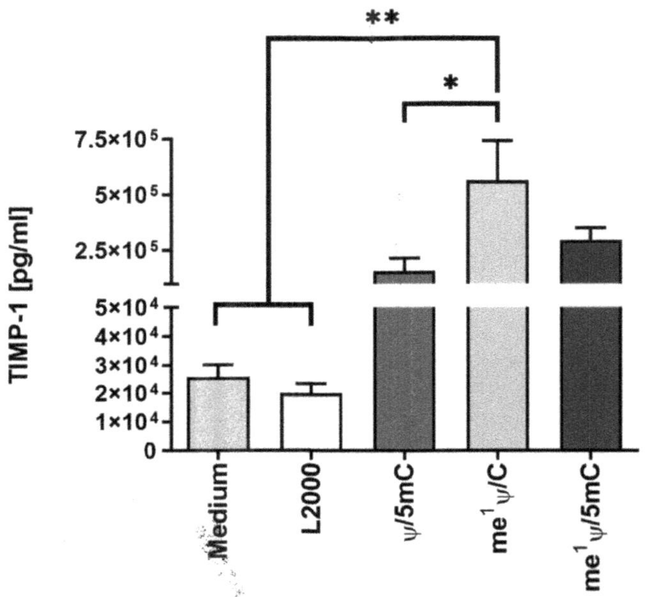


**Figure 7.** Impact of nucleotide modification on synthetic TIMP-1 mRNA expression efficiency in EA.hy926 cells. First, $3 \times 10^5$ EA.hy926 cells were seeded and transfected with 1.5 µg of TIMP-1 mRNA complexed with 1.5 µL of L2000 in Opti-MEM for 4 h at 37 °C and 5% $CO_2$. Thereafter, the transfection complexes were replaced by cell culture medium, and the cells were incubated at 37 °C and 5% $CO_2$. Cells treated with Opti-MEM or Opti-MEM plus L2000 served as controls. After 24 h, the TIMP-1 concentration was determined in collected supernatants via human TIMP-1-specific ELISA. The results are shown as the mean + SEM (n = 3). Statistical differences between the mRNA-treated groups were determined using one-way ANOVA following Tukey's multiple comparison test (* $p < 0.05$, ** $p < 0.01$).

## 3. Discussion

Increased MMP activity is known to play a significant role in various diseases, including CVDs as well as other diseases like cancer [16]. In cardiovascular diseases, MMPs are involved in processes such as plaque destabilization, vascular remodeling, and aneurysm formation [32]. In cancer, MMPs facilitate tumor invasion, metastasis, and angiogenesis [33].

AAAs are characterized by arterial dilatation, degeneration of arterial architecture, disruption of the ECM, inflammatory infiltration, oxidative stress, and, in particular the presence of matrix-degrading enzymes like MMPs, necessitating multifactorial potential treatments. ECM degradation, particularly mediated by matrix MMPs, is considered crucial for AAA occurrence and progression. Due to its multifactorial pathogenesis, potential AAA treatments have included anti-inflammatory agents, endogenous proteinase inhibitors such as TIMPs, and genetic and pharmacological inhibition of MMPs [34]. Early attempts to develop MMP inhibitors were based on hydroxamate-based inhibitors, which were based on the structure of collagen and contained a group that inactivates the catalytic zinc ion via chelation [35]. However, this approach led to intolerable side effects like joint stiffening [36]. Later, beyond others, antibody-based therapeutics were developed for selective inhibition of MMPs [37–39]. Multiple studies have proven that the administration of doxycycline suppresses aortic dilation, inhibits MMP-9 expression and activity, and reduces the incidence of AAA formation [40]. However, currently, the antibiotic substance doxycycline is the only MMP inhibitor approved by the Food and Drug Administration (FDA) for use in humans [41] and is used to treat periodontitis [42]. Thus, there is a great need to develop selective and locally effective MMP inhibitors, which are also well tolerated. A very interesting approach is using antibody-conjugated nanoparticles encapsulating different MMP inhibitors such as batimastat, doxycycline, or pentagalloylglucose in AAA models showing positive effects on aneurysms [43–45]. In addition, the nanoparticles can specifically target the damaged tissue area following systemic application by specific binding of the conjugated antibody to the damaged elastic fibers present in the affected vessel area.

TIMPs could form the basis for another new class of MMP inhibitors. Studies using MMP- and TIMP-knockout mice have definitively shown the importance of MMP/TIMP imbalances in the onset and progression of AAA [11,46], making the endogenous MMP inhibitor TIMP-1 a relevant candidate. Allaire et al. used retroviral vectors to introduce TIMP-1 cDNA into SMCs, which they then injected into the aortic lumen of rats with aneurysms [30]. Local TIMP-1 overexpression prevented aneurysmal degeneration and rupture. Since MMPs also lead to unstable plaques in atherosclerosis, Rouis et al. injected TIMP-1 DNA via an adenoviral vector venously into an atherosclerotic mouse model [47], which resulted in high hepatic expression of TIMP-1. The increased TIMP-1 concentration in the plasma led to a significantly reduced number of atherosclerotic lesions in atherosclerosis-susceptible hypercholesterolemic apoE−/− mice. While this is a promising strategy, the further effects of a systemic increase in TIMP-1 on the organism still need to be evaluated. Giraud et al. showed that gingival fibroblasts (GFs) with increased secretion of TIMP-1 deposited on the adventitia of AAAs decreased inflammation and ECM destruction and prevented aneurysm progression and rupture [48].

mRNA-based therapeutics represent an innovative approach for producing effective proteins directly within the targeted cells and in affected areas. Unlike DNA, synthetic mRNA circumvents the need to enter the cell nucleus, thus mitigating the risk of insertional mutations. Moreover, it also offers the advantage of fast, uncomplicated, and cost-effective production. Additionally, its transient activity and subsequent elimination via physiological degradation pathways curtail possible adverse effects. Notably, existing mRNA therapeutics, such as vaccines, have demonstrated their adaptability to swiftly respond to viral mutations, exemplifying the versatility of this approach [28]. Despite the most common use of synthetic mRNA as vaccines, mRNA-based protein replacement therapeutics have already entered the clinical stage, offering tremendous potential as new therapy options for several diseases [49,50]. Nevertheless, tissue- and cell-specific targeted delivery still remains one of the biggest hurdles.

To our knowledge, we have demonstrated for the first time the applicability of synthetic TIMP-1 mRNA as a new approach for the inhibition of MMP activity in blood vessels. Following these promising results, we plan to conduct in vivo studies in aneurysm models. These will allow a more comprehensive evaluation of the TIMP-1 mRNA effect in a physiologically relevant context. However, the current ex vivo data provide essential insights that will guide our in vivo research.

In our study, the intraluminal injection of the synthetic TIMP-1 mRNA using microneedles resulted in successful delivery to the tissue and the expression of the target protein, serving as a suitable delivery method for the used ex vivo model. Further possible application methods for therapeutic use are mRNA-coated stents, direct injection into the aortic wall via a catheter, or systemic targeted delivery systems such as nanoparticles. The intraluminal route using a catheter infusion was used in a recent study for the local delivery of pentagalloylglucose to AAAs in pigs. A first-in-human pilot study is being conducted to assess the novel localized treatment for stabilizing small- to medium-sized infrarenal abdominal aortic aneurysms [51].

In summary, the use of synthetic TIMP-1 mRNA represents a promising new strategy for the treatment of cardiovascular diseases characterized by the loss of elastic structural proteins or increased inflammation, such as aneurysms, dissections, and atherosclerosis, by decreasing MMP-9 activity. It is therefore an innovative and novel strategy that has the potential to prevent the progression of aneurysms in the future.

## 4. Materials and Methods

### 4.1. Cultivation of Cells

EA.hy926 cells (ATCC) and human fibroblasts (NUFFs, newborn foreskin fibroblasts, AMS Biotechnology (Europe) Ltd., Abingdon, UK) were cultivated in Dulbecco's modified Eagle medium (DMEM) with high glucose and L-glutamine with 10% heat-inactivated fetal bovine serum (FBS) at 37 °C and 5% $CO_2$. Cells were washed once with Dulbecco's phosphate-buffered saline (DPBS) and detached with 0.05% trypsin-EDTA. All cell culture reagents were obtained from Thermo Fisher Scientific (Waltham, MA, USA). Human umbilical vein endothelial cells (HUVECs, (PromoCell, Heidelberg, Germany)) were cultivated in VascuLife EnGS endothelial cell culture medium (without hydrocortisone) (Lifeline Cell Technology, Frederick, MD, USA) at 37 °C and 5% $CO_2$. The cells were washed once with DPBS and detached using 0.04% trypsin/0.03% EDTA and trypsin neutralizing solution (TNS) 0.05% in 0.1% BSA (both Promocell, Heidelberg, Germany). Cell culture flasks were coated with 0.1% gelatin (Fluka, Morristown, NJ, USA) in PBS for 15 min at room temperature (RT) before seeding the cells. The medium was changed every 3–4 days, and the cells were passaged upon reaching a confluency of 80–90%.

### 4.2. mRNA Synthesis

Synthetic mRNA was produced via in vitro transcription (IVT) as previously described [52]. Plasmids containing either secretable humanized Gaussian luciferase (hGLuc) or human TIMP-1 encoding sequences were ordered from Eurofins Genetics (Ebersberg, Germany). These plasmids were used to generate the DNA templates containing the coding sequence of hGLuc or TIMP-1 via PCR. Therefore, 50–100 ng of plasmid and the HotStar HiFidelity Polymerase Kit (Qiagen, Hilden, Germany) were used according to the manufacturer's instructions. Forward (5′-TTGGACCCTCGTACAGAAGCTAATACG-3′) and reverse primers (5′-T120-CTTCCTACTCAGGCTTTATTCAAAGACCA-3) were purchased from ELLA Biotech (Martinsried, Germany). PCR was performed using the following PCR cycling conditions: 94 °C for 3 min, 30 cycles of denaturation to generate single strands at 94 °C for 45 s, hybridization of primers at 60 °C for 1 min, elongation at 72 °C for 1 min, and final elongation at 72 °C for 5 min. The amplified PCR products were purified using the QIAquick PCR purification kit (Qiagen, Hilden, Germany) according to the manufacturer's instructions. The concentration and purity of the DNA were analyzed photometrically, and the size of the amplified DNA fragments was controlled via 1% agarose gel electrophoresis.

IVT was performed using a T7 MEGAscript Kit (Life Technologies, Darmstadt, Germany) according to the manufacturer's instructions. Therefore, 1.5 µg of the hGLuc or TIMP-1 template DNA was used. During IVT, a 2.5 mM 3′-O-Me-m7G(5′)ppp(5′)G RNA cap structure analog (ARCA, New England Biolabs, Frankfurt am Main, Germany) and nucleotides (1.875 mM GTP, 7.5 mM ATP, 7.5 mM 5-methyl-CTP (5mC), 7.5 mM Ψ). were incorporated. This mRNA modification was named Ψ/5mC. To analyze the impact of nucleotide modification on TIMP-1 protein expression, TIMP-1 mRNA was also produced with 7.5 mM N1-methylpseudouridine ($me^1Ψ$) instead of Ψ and 7.5 mM 5mC (named $me^1Ψ$/5mC) or 7.5 mM CTP (named $me^1Ψ$/C). ATP, GTP, and CTP from the MEGAscript T7 Kit were used, and the others were from TriLink (BioTechnologies, San Diego, CA, USA).

To each IVT reaction, 40 U RiboLock RNase inhibitor (Thermo Scientific, Waltham, MA, USA) was added, and the IVT reaction was incubated for 4 h at 37 °C. For the removal of the DNA template, 1 µL of TurboDNase was added and incubated for a further 15 min before performing mRNA purification with the RNeasy Mini Kit (Qiagen, Hilden, Germany) according to the manufacturer's instructions. Subsequently, mRNA was dephosphorylated at 37 °C for 30 min with 15 U Antarctic phosphatase (New England Biolabs, Frankfurt am Main, Germany) and purified as described above. The concentration and purity of mRNAs were analyzed photometrically, and the size of the mRNAs was confirmed via 1% agarose gel electrophoresis and staining with 1x GelRed (Biotium, Fremont, CA, USA) in 1x Tris-borate-EDTA (TBE) buffer for 1 h at RT.

### *4.3. Cy3 Labeling of hGLuc mRNA*

To evaluate the localization of the mRNA in the blood vessel walls, hGluc mRNA was labeled with the fluorophore Cy3. Therefore, Cy3 was linked to the mRNA using a copper-free click chemistry reaction. During IVT, 5-azido-C3-UTP (Jena Bioscience, Jena, Germany) molecules were integrated into the mRNA and subsequently labeled with Cy3. For this purpose, 25% 5-azido-C3-UTP (1.875 mM) and 75% Ψ (5.625 mM) were used for IVT as described in our previous study [53]. The azido mRNA was purified and coupled with DBCO-sulfo-Cy3 (Jena Bioscience) in a 5:1 ratio in a total volume of 40 µL nuclease-free water for 1 h at 37 °C. The labeled mRNA was purified using the RNeasy Mini Kit (Qiagen, Hilden, Germany) and evaluated via 1% agarose gel electrophoresis.

### *4.4. Transfection of Cells with TIMP-1 mRNA*

During this stage, $3 \times 10^5$ HUVECs, NuFFs, or EA.hy926 cells were seeded per well of a 6-well plate and cultivated overnight at 37 °C and 5% $CO_2$ in the respective cell culture medium. Before the seeding of HUVECs, cell culture plates were coated with 0.1% gelatin in PBS for 15 min. The cells were transfected with 0.5, 1, or 1.5 µg synthetic mRNA complexed with Lipofectamine™ 2000 (L2000) in a 1:1 ratio in 1 mL Opti-MEM (except for 0.5 µg mRNA, where 1 µL L2000 was used). Opti-MEM alone (medium) or Opti-MEM with 1.5 µL L2000 (L2000) were used as controls. For the analysis of modified nucleotides on TIMP-1 mRNA expression efficiency, 1.5 µg of each modified TIMP-1 mRNA variant was complexed with 1.5 µL L2000 and transfected into EA.hy926 cells. Transfection mixtures were incubated for 20 min at RT. Before adding the transfection mixtures, the cells were washed once with DPBS, and the cells were incubated for 4 h at 37 °C and 5% $CO_2$ with the transfection complexes. Then, the transfection mixtures were replaced with 2 mL of cell culture medium, and the cells were incubated for another 24 h at 37 °C and 5% $CO_2$. Cell culture supernatants were collected, snap-frozen in liquid nitrogen, and stored at −80 °C until performing TIMP-1 protein-specific ELISA. Furthermore, the cells were lysed to detect intracellular TIMP-1 protein amounts via ELISA.

### *4.5. Lysis of Cells after TIMP-1 mRNA Transfection*

The cells were washed 2x with 1 mL cold DPBS, overlaid with 300 µL cold 1x RIPA buffer with 1:100 Halt™ Protease Inhibitor Cocktail (both from Thermo Fisher Scientific, Waltham, MA, USA), and incubated on ice for 5 min. The cells were then lysed via repeated pipetting, transferred to a reaction tube, and treated in an ultrasonic bath for $3 \times 10$ s with a 10 s pause in between. Subsequently, the cell lysates were centrifuged at 13,000 rpm, 4 °C for 25 min. The supernatant was collected, snap-frozen in liquid nitrogen, and stored at −80 °C until performing the TIMP-1 protein-specific ELISA.

### *4.6. Application of Synthetic mRNA into Blood Vessel Walls*

#### 4.6.1. Tissue Preparation

Porcine and human aortic tissues were transferred to 0.9% NaCl solution immediately after collection and used for microinjections on the same day. Before microinjection, the tissue was cut into $0.5 \times 0.5$ cm pieces and incubated for 30 min in an antibiotic solution composed of 250 mg/mL gentamicin (Sigma-Aldrich, St. Louis, MO, USA) and 1.25 mg/mL amphotericin B (PromoCell, Heidelberg, Germany) in DMEM followed by washing with DPBS (both from Thermo Fisher Scientific (Waltham, MA, USA)).

4.6.2. Injection of Synthetic hGLuc or TIMP-1 mRNA

To analyze synthetic mRNA-mediated protein expression in the aortic vessel wall, synthetic mRNA was delivered ex vivo to porcine and human aortic vessel walls by microinjection from the side of the intima using MicronJet600™ hollow microneedles from NanoPass Technologies (Nes Ziona, Israel). Therefore, 1 µg of hGLuc mRNA or 3 or 5 µg TIMP-1 mRNA was complexed in a 1:1 ratio (µg:µL) with L2000 and incubated in a total volume of 50 µL Opti-MEM for 20 min at RT. The prepared mRNA solutions were drawn into a 1 mL syringe (Luer-LokTM, Becton, Dickinson and Company, Franklin Lakes, NJ, (USA)) using a cannula, which was replaced by the microinjection needle for the injection from the intraluminal side. After injection, samples were incubated for 5 min at RT, washed once in DPBS, and incubated at 37 °C in 5% $CO_2$ in 1 mL EC medium in a 12-well plate for 24 to 48 h to analyze the luciferase activity, TIMP-1 amount by ELISA and the activity of exogenously produced TIMP-1 protein in the vessel wall using in situ zymography. Microinjection of 50 µL Opti-MEM and the corresponding amount of L2000 into the tissue served as controls. For the in-situ zymography, the tissue samples were incubated for 48 and 72 h at 4 °C in 10x zinc fixative (Formalin free, BD Pharmingen™, Becton, Dickinson and Company, Franklin Lakes, NJ, USA).

To evaluate the localization of the mRNA after microinjection into the blood vessel wall, 1 µg Cy3 labeled hGLuc mRNA was complexed with 1 µL L2000 in 50 µL Opti-MEM for 20 min at RT. Tissue treated with Opti-MEM and the corresponding amount of L2000 served as the control. After the microinjection, the tissues were washed 1x in DPBS and fixed overnight in 4% paraformaldehyde (PFA) at 4 °C for histological analyses.

4.6.3. Detection of Luciferase Activity

After the delivery of synthetic hGLuc mRNA into the aortic vessel wall, the expression of hGLuc in tissue supernatants was determined using a luciferase assay. Therefore, 40 µL of each supernatant was transferred in triplicate into a 96-well plate (Nunc Maxisorp, Thermo Fisher Scientific (Waltham, MA, USA)), 100 µL of 20 µg/mL coelenterazine (Carl Roth, Karlsruhe, Germany) in DPBS (w/o $Ca^{2+}/Mg^{2+}$) was injected automatically into each well, and luminescence was detected in Relative Light Units (RLU) using the microplate reader Mithras LB 940 (Berthold Technologies, Bad Wildbad, Germany).

*4.7. Histological Analysis*

4.7.1. Detection of Microinjected mRNA in the Aortic Vessel Wall

After fixation, the samples were washed twice in 2 mL DPBS, transferred to embedding cassettes, and coated with 100% ethanol. Subsequently, automated dehydration and embedding in paraffin were performed by the Institute of Pathology at the University Hospital of Tübingen. Subsequently, tissue sections with a thickness of 2.5 µm were prepared. Afterward, the sections were deparaffinized 4× for 5 min in xylene rehydrated stepwise in an ethanol series: 2× 99% for 1 min, 2× 96% for 1 min, and 2× 70% for 1 min, washed twice in nuclease-free water, and then boiled for 2 min in TBE buffer (pH 9). The slides were cooled under running water, washed three times with DPBS, and, in the case of Cy3 mRNA injection, embedded in Fluoroshield Mounting Medium with DAPI (Vector Laboratories, Burlingame, CA, USA). Xylene and ethanol were both provided by AnalaR NORMAPUR (VWR, Darmstadt, Germany). To detect Cy3-labeled mRNA, fluorescence images were taken using an Axiovert135 microscope (Carl Zeiss, Oberkochen, Germany) and analyzed with AxioVision Rel 4.8 software.

4.7.2. In Situ Zymography

After fixation, the tissue was transferred to embedding cassettes, dehydrated, and embedded using a tissue embedding device and an automated tissue processor at the Institute of Pathology at the University Hospital of Tübingen. The tissue was then poured into paraffin blocks and 5 μm thick sections were prepared using the Microtome Microm HM 355S (Thermo Fisher Scientific (Waltham, MA, USA)) and mounted on SuperFrost® microscope slides (R. Langenbrinck, Emmendingen, Germany). The tissue sections were deparaffinized in xylene (100% xylene, mixture of isomers, AnalaR NORMAPUR, VWR, Darmstadt, Germany) twice for 2 min and rehydrated using descending ethanol (AnalaR NORMAPUR, VWR, Darmstadt, Germany) series (100%, 80%, 70%, and 60%). The slides were washed in double-distilled water (dd$H_2O$) and stained in a humidity chamber using an EnzChek™ Gelatinase/Collagenase Assay Kit (Thermo Fisher Scientific (Waltham, MA, USA)) by adding 250 μL substrate solution consisting of 1 mg/mL of the fluorescence-labeled DQ-gelatine diluted 1:50 in reaction buffer ((150 mM NaCl (NORMAPUR®, VWR International, LLC., Radnor, PA, USA), 5 mM $CaCl_2$, 50 mM Tris-HCl, 0.2 mM sodium azide (all from Sigma-Aldrich, Darmstadt, Germany); pH 7.6)) to each section and incubation at 37 °C for 2 h. As a control, tissue sections were treated for 1 h with 20 mM EDTA to inhibit enzymatic metalloproteinase activity. These control slides were then incubated at 37 °C for 2 h with the substrate solution also containing 20 mM EDTA. Afterward, the sections were washed 3× for 1 min in dd$H_2O$ and fixed in 250 μL 4% PFA for 10 min in the dark, followed by 2× washing for 5 min in DPBS. Finally, the tissue sections were covered with Fluoroshield mounting medium containing DAPI (Vector Laboratories, Newark, NJ, USA) to stain the cell nuclei. Images were taken using the Axiovert135 fluorescence microscope (Carl Zeiss, Oberkochen, Germany) and analyzed using AxioVision Rel 4.8 program software.

*4.8. Human TIMP-1 ELISA*

TIMP-1 amount was detected in the cell lysates and cell culture supernatants after TIMP-1 mRNA transfection and the microinjection of mRNA into the aortic tissue. The amount of TIMP-1 was detected using human TIMP-1 DuoSet ELISA (R&D Systems, Minneapolis, MN, USA) according to the manufacturer's instructions. Cell supernatants and lysates were diluted (cell supernatants: 1:150–1:250; cell lysates 1:50) in 1% BSA in DPBS before ELISA. Supernatants of aortic vessel tissue cultivation were not diluted. The absorbance of the samples was measured using a microplate reader (Eon Synergy 2, BioTek Instruments, Winooski, VT, USA) at 450 nm with a correction wavelength of 540 nm.

*4.9. Statistical Analysis*

The data are shown as the mean + standard deviation (SD) or standard error of the mean (SEM). One- or two-way analysis of variance (ANOVA) was performed followed by Tukey's multiple comparisons test. All of the analyses were performed using Origin Pro 2024b (v10.15) and GraphPad Prism version 9.3.1. Differences of $p < 0.05$ were considered statistically significant.

## 5. Conclusions

In this study, the synthesis of TIMP-1 protein was induced by the exogenous delivery of synthetic TIMP-1 encoding mRNA into the aortic vessel to inhibit MMP-9. In vitro experiments demonstrated a significant increase in TIMP-1 protein expression in various cells following TIMP-1 mRNA transfection. Additionally, TIMP-1 protein expression was further increased through nucleotide modifications and the replacement of Ψ/m5C with $me^1\Psi$. The functionality of the expressed protein was assessed using an appropriate ex vivo aortic vessel model, revealing a decrease in MMP-9 activity detected using in situ zymography 24 and 48 h after the microinjection of 5 μg TIMP-1 mRNA into the aortic vessel wall. These results indicate that administering TIMP-1 mRNA holds potential as a treatment strategy for aneurysms.

**Supplementary Materials:** The supporting information can be downloaded at: https://www.mdpi.com/article/10.3390/ijms25126599/s1.

**Author Contributions:** S.G. and M.A.-A. conceived and designed the experiments. S.G., I.D., and L.K. performed the experiments and analyzed the data. M.L. and C.S. contributed the reagents/materials/analysis tools. S.G. and I.D. wrote the paper. M.A.-A. corrected the paper and supervised the project. All authors have read and agreed to the published version of the manuscript.

**Funding:** We acknowledge the support provided by the Open Access Publishing Fund of the University of Tübingen.

**Institutional Review Board Statement:** The porcine abdominal aorta samples used in the experiments were collected from already euthanized, dead pigs (German Landrace pigs (70 ± 10 kg, female, 4–6 months old)) that were euthanized in the animal operating room of the Department of Exper-imental Surgery at the University Hospital of Tübingen, Germany, within the scope of the organ harvesting application of 22. December 2017. The human aortic samples were obtained from aneurysm excisions of patients who were surgically treated at the Department of Thoracic and Cardiovascular Surgery of the University Hospital Tübingen. The study was approved by the Ethical Committee of the Faculty of Medicine of the University of Tübingen (298/2023BO2).

**Informed Consent Statement:** Informed consent was obtained from all subjects involved in the study.

**Data Availability Statement:** The original contributions presented in the study are included in the article/Supplementary Material, further inquiries can be directed to the corresponding author.

**Conflicts of Interest:** The authors declare no conflict of interest.

## References

1. European Cardiovascular Disease Statistics 2017. Available online: https://ehnheart.org/cvd-statistics.html (accessed on 12 September 2023).
2. Wei, L.; Bu, X.; Wang, X.; Liu, J.; Ma, A.; Wang, T. Global Burden of Aortic Aneurysm and Attributable Risk Factors from 1990 to 2017. *Glob. Heart* **2021**, *16*, 35. [CrossRef] [PubMed]
3. Kent, K.C. Abdominal aortic aneurysms. *N. Engl. J. Med.* **2014**, *371*, 2101–2108. [CrossRef] [PubMed]
4. Knypl, K. The abdominal aortic artery aneurysm and cardiovascular risk factors. *e-J. Cardiol. Pract.* **2020**, *18*, 28.
5. Robinson, D.; Mees, B.; Verhagen, H.; Chuen, J. Aortic aneurysms: Screening, surveillance and referral. *Aust. Fam. Physician* **2013**, *42*, 364–369.
6. Lu, J.; Egger, J.; Wimmer, A.; Grosskopf, S.; Freisleben, B. Detection and visualization of endoleaks in CT data for monitoring of thoracic and abdominal aortic aneurysm stents. *Med. Imaging 2008 Vis. Image-Guid. Proced. Model.* **2008**, *6918*, 492–498.
7. Milutinović, A.; Šuput, D.; Zorc-Pleskovič, R. Pathogenesis of atherosclerosis in the tunica intima, media, and adventitia of coronary arteries: An updated review. *Bosn. J. Basic Med. Sci.* **2020**, *20*, 21. [CrossRef] [PubMed]
8. Joviliano, E.E.; Ribeiro, M.S.; Tenorio, E.J.R. MicroRNAs and Current Concepts on the Pathogenesis of Abdominal Aortic Aneurysm. *Braz. J. Cardiovasc. Surg.* **2017**, *32*, 215–224. [CrossRef] [PubMed]
9. Lindholt, J.S. Activators of plasminogen and the progression of small abdominal aortic aneurysms. *Ann. N. Y. Acad. Sci.* **2006**, *1085*, 139–150. [CrossRef] [PubMed]
10. Miller, F.J., Jr.; Sharp, W.J.; Fang, X.; Oberley, L.W.; Oberley, T.D.; Weintraub, N.L. Oxidative stress in human abdominal aortic aneurysms: A potential mediator of aneurysmal remodeling. *Arterioscler. Thromb. Vasc. Biol.* **2002**, *22*, 560–565. [CrossRef] [PubMed]
11. Longo, G.M.; Xiong, W.; Greiner, T.C.; Zhao, Y.; Fiotti, N.; Baxter, B.T. Matrix metalloproteinases 2 and 9 work in concert to produce aortic aneurysms. *J. Clin. Investig.* **2002**, *110*, 625–632. [CrossRef] [PubMed]
12. Rabkin, S.W. The Role Matrix Metalloproteinases in the Production of Aortic Aneurysm. *Prog. Mol. Biol. Transl. Sci.* **2017**, *147*, 239–265. [PubMed]
13. Ramella, M.; Boccafoschi, F.; Bellofatto, K.; Follenzi, A.; Fusaro, L.; Boldorini, R.; Casella, F.; Porta, C.; Settembrini, P.; Cannas, M. Endothelial MMP-9 drives the inflammatory response in abdominal aortic aneurysm (AAA). *Am. J. Transl. Res.* **2017**, *9*, 5485–5495. [PubMed]
14. McMillan, W.D.; Pearce, W.H. Increased plasma levels of metalloproteinase-9 are associated with abdominal aortic aneurysms. *J. Vasc. Surg.* **1999**, *29*, 122–129. [CrossRef] [PubMed]
15. McMillan, W.D.; Tamarina, N.A.; Cipollone, M.; Johnson, D.A.; Parker, M.A.; Pearce, W.H. Size matters: The relationship between MMP-9 expression and aortic diameter. *Circulation* **1997**, *96*, 2228–2232. [CrossRef] [PubMed]
16. Cabral-Pacheco, G.A.; Garza-Veloz, I.; Castruita-De la Rosa, C.; Ramirez-Acuna, J.M.; Perez-Romero, B.A.; Guerrero-Rodriguez, J.F.; Martinez-Avila, N.; Martinez-Fierro, M.L. The Roles of Matrix Metalloproteinases and Their Inhibitors in Human Diseases. *Int. J. Mol. Sci.* **2020**, *21*, 9739. [CrossRef] [PubMed]

17. Gomis-Rüth, F.X.; Maskos, K.; Betz, M.; Bergner, A.; Huber, R.; Suzuki, K.; Yoshida, N.; Nagase, H.; Brew, K.; Bourenkov, G.P.; et al. Mechanism of inhibition of the human matrix metalloproteinase stromelysin-1 by TIMP-1. *Nature* **1997**, *389*, 77–81. [CrossRef] [PubMed]
18. Baker, A.H.D.R. Edwards, and G. Murphy, Metalloproteinase inhibitors: Biological actions and therapeutic opportunities. *J. Cell Sci.* **2002**, *115*, 3719–3727. [CrossRef] [PubMed]
19. Wilson, W.R.; Schwalbe, E.C.; Jones, J.L.; Bell, P.R.; Thompson, M.M. Matrix metalloproteinase 8 (neutrophil collagenase) in the pathogenesis of abdominal aortic aneurysm. *Br. J. Surg.* **2005**, *92*, 828–833. [CrossRef] [PubMed]
20. Steinle, H.; Weber, J.; Stoppelkamp, S.; Grosse-Berkenbusch, K.; Golombek, S.; Weber, M.; Canak-Ipek, T.; Trenz, S.M.; Schlensak, C.; Avci-Adali, M. Delivery of synthetic mRNAs for tissue regeneration. *Adv. Drug Deliv. Rev.* **2021**, *179*, 114007. [CrossRef]
21. Baden, L.R.; El Sahly, H.M.; Essink, B.; Kotloff, K.; Frey, S.; Novak, R.; Diemert, D.; Spector, S.A.; Rouphael, N.; Creech, C.B.; et al. Efficacy and Safety of the mRNA-1273 SARS-CoV-2 Vaccine. *N. Engl. J. Med.* **2020**, *384*, 403–416. [CrossRef] [PubMed]
22. Polack, F.P.; Thomas, S.J.; Kitchin, N.; Absalon, J.; Gurtman, A.; Lockhart, S.; Perez, J.L.; Perez Marc, G.; Moreira, E.D.; Zerbini, C.; et al. Safety and Efficacy of the BNT162b2 mRNA Covid-19 Vaccine. *N. Engl. J. Med.* **2020**, *383*, 2603–2615. [CrossRef] [PubMed]
23. Jia, L.; Qian, S.B. Therapeutic mRNA Engineering from Head to Tail. *Acc. Chem. Res.* **2021**, *54*, 4272–4282. [CrossRef] [PubMed]
24. Kwon, H.; Kim, M.; Seo, Y.; Moon, Y.S.; Lee, H.J.; Lee, K.; Lee, H. Emergence of synthetic mRNA: In vitro synthesis of mRNA and its applications in regenerative medicine. *Biomaterials* **2018**, *156*, 172–193. [CrossRef] [PubMed]
25. Patel, S.; Athirasala, A.; Menezes, P.P.; Ashwanikumar, N.; Zou, T.; Sahay, G.; Bertassoni, L.E. Messenger RNA Delivery for Tissue Engineering and Regenerative Medicine Applications. *Tissue Eng. Part A* **2019**, *25*, 91–112. [CrossRef] [PubMed]
26. Sultana, N.; Hadas, Y.; Sharkar, M.T.K.; Kaur, K.; Magadum, A.; Kurian, A.A.; Hossain, N.; Alburquerque, B.; Ahmed, S.; Chepurko, E.; et al. Optimization of 5′ Untranslated Region of Modified mRNA for Use in Cardiac or Hepatic Ischemic Injury. *Mol. Ther. Methods Clin. Dev.* **2020**, *17*, 622–633. [CrossRef] [PubMed]
27. Zhong, Z.; Mc Cafferty, S.; Combes, F.; Huysmans, H.; De Temmerman, J.; Gitsels, A.; Vanrompay, D.; Catani, J.P.; Sanders, N.N. mRNA therapeutics deliver a hopeful message. *Nano Today* **2018**, *23*, 16–39. [CrossRef]
28. Sahin, U.; Karikó, K.; Türeci, Ö. mRNA-based therapeutics–developing a new class of drugs. *Nat. Rev. Drug Discov.* **2014**, *13*, 759–780. [CrossRef] [PubMed]
29. Yang, E.; van Nimwegen, E.; Zavolan, M.; Rajewsky, N.; Schroeder, M.; Magnasco, M.; Darnell, J.E., Jr. Decay rates of human mRNAs: Correlation with functional characteristics and sequence attributes. *Genome Res.* **2003**, *13*, 1863–1872. [CrossRef] [PubMed]
30. Allaire, E.; Forough, R.; Clowes, M.; Starcher, B.; Clowes, A.W. Local overexpression of TIMP-1 prevents aortic aneurysm degeneration and rupture in a rat model. *J. Clin. Investig.* **1998**, *102*, 1413–1420. [CrossRef] [PubMed]
31. Hadler-Olsen, E.; Kanapathippillai, P.; Berg, E.; Svineng, G.; Winberg, J.O.; Uhlin-Hansen, L. Gelatin in situ zymography on fixed, paraffin-embedded tissue: Zinc and ethanol fixation preserve enzyme activity. *J. Histochem. Cytochem.* **2010**, *58*, 29–39. [CrossRef]
32. Holm Nielsen, S.; Jonasson, L.; Kalogeropoulos, K.; Karsdal, M.; Reese-Petersen, A.L.; Auf dem Keller, U.; Genovese, F.; Nilsson, J.; Goncalves, I. Exploring the role of extracellular matrix proteins to develop biomarkers of plaque vulnerability and outcome. *J. Intern. Med.* **2020**, *287*, 493–513. [CrossRef] [PubMed]
33. Wang, X.; Khalil, R.A. Matrix metalloproteinases, vascular remodeling, and vascular disease. *Adv. Pharmacol.* **2018**, *81*, 241–330. [PubMed]
34. Li, Y.; Wang, W.; Li, L.; Khalil, R.A. MMPs and ADAMs/ADAMTS inhibition therapy of abdominal aortic aneurysm. *Life Sci.* **2020**, *253*, 117659. [CrossRef] [PubMed]
35. Reich, R.; Thompson, E.W.; Iwamoto, Y.; Martin, G.R.; Deason, J.R.; Fuller, G.C.; Miskin, R. Effects of inhibitors of plasminogen activator, serine proteinases, and collagenase IV on the invasion of basement membranes by metastatic cells. *Cancer Res.* **1988**, *48*, 3307–3312. [PubMed]
36. Pavlaki, M.; Zucker, S. Matrix metalloproteinase inhibitors (MMPIs): The beginning of phase I or the termination of phase III clinical trials. *Cancer Metastasis Rev.* **2003**, *22*, 177–203. [CrossRef] [PubMed]
37. Fischer, T.; Riedl, R. Inhibitory antibodies designed for matrix metalloproteinase modulation. *Molecules* **2019**, *24*, 2265. [CrossRef] [PubMed]
38. Naito, S.; Takahashi, T.; Onoda, J.; Yamauchi, A.; Kawai, T.; Kishino, J.; Yamane, S.; Fujii, I.; Fukui, N.; Numata, Y. Development of a neutralizing antibody specific for the active form of matrix metalloproteinase-13. *Biochemistry* **2012**, *51*, 8877–8884. [CrossRef] [PubMed]
39. Kaimal, R.; Aljumaily, R.; Tressel, S.L.; Pradhan, R.V.; Covic, L.; Kuliopulos, A.; Zarwan, C.; Kim, Y.B.; Sharifi, S.; Agarwal, A. Selective blockade of matrix metalloprotease-14 with a monoclonal antibody abrogates invasion, angiogenesis, and tumor growth in ovarian cancer. *Cancer Res.* **2013**, *73*, 2457–2467. [CrossRef] [PubMed]
40. Atkinson, G.; Bianco, R.; Di Gregoli, K.; Johnson, J.L. The contribution of matrix metalloproteinases and their inhibitors to the development, progression, and rupture of abdominal aortic aneurysms. *Front. Cardiovasc. Med.* **2023**, *10*, 1248561. [CrossRef] [PubMed]
41. Liu, J.; Khalil, R.A. Matrix metalloproteinase inhibitors as investigational and therapeutic tools in unrestrained tissue remodeling and pathological disorders. *Prog. Mol. Biol. Transl. Sci.* **2017**, *148*, 355–420. [PubMed]

42. Sorsa, T.; Tjäderhane, L.; Konttinen, Y.T.; Lauhio, A.; Salo, T.; Lee, H.M.; Golub, L.M.; Brown, D.L.; Mäntylä, P. Matrix metalloproteinases: Contribution to pathogenesis, diagnosis and treatment of periodontal inflammation. *Ann. Med.* **2006**, *38*, 306–321. [CrossRef] [PubMed]
43. Nosoudi, N.; Nahar-Gohad, P.; Sinha, A.; Chowdhury, A.; Gerard, P.; Carsten, C.G.; Gray, B.H.; Vyavahare, N.R. Prevention of Abdominal Aortic Aneurysm Progression by Targeted Inhibition of Matrix Metalloproteinase Activity With Batimastat-Loaded Nanoparticles. *Circ. Res.* **2015**, *117*, e80–e89. [CrossRef] [PubMed]
44. Wang, X.; Parasaram, V.; Dhital, S.; Nosoudi, N.; Hasanain, S.; Lane, B.A.; Lessner, S.M.; Eberth, J.F.; Vyavahare, N.R. Systemic delivery of targeted nanotherapeutic reverses angiotensin II-induced abdominal aortic aneurysms in mice. *Sci. Rep.* **2021**, *11*, 8584. [CrossRef] [PubMed]
45. Dhital, S.; Rice, C.D.; Vyavahare, N.R. Reversal of elastase-induced abdominal aortic aneurysm following the delivery of nanoparticle-based pentagalloyl glucose (PGG) is associated with reduced inflammatory and immune markers. *Eur. J. Pharmacol.* **2021**, *910*, 174487. [CrossRef] [PubMed]
46. Ikonomidis, J.S.; Gibson, W.C.; Butler, J.E.; McClister, D.M.; Sweterlitsch, S.E.; Thompson, R.P.; Mukherjee, R.; Spinale, F.G. Effects of deletion of the tissue inhibitor of matrix metalloproteinases-1 gene on the progression of murine thoracic aortic aneurysms. *Circulation* **2004**, *110* (Suppl. 1), II268–II273. [CrossRef] [PubMed]
47. Rouis, M.; Adamy, C.; Duverger, N.; Lesnik, P.; Horellou, P.; Moreau, M.; Emmanuel, F.; Caillaud, J.; Laplaud, P.; Dachet, C. Adenovirus-mediated overexpression of tissue inhibitor of metalloproteinase-1 reduces atherosclerotic lesions in apolipoprotein E–deficient mice. *Circulation* **1999**, *100*, 533–540. [CrossRef]
48. Giraud, A.; Zeboudj, L.; Vandestienne, M.; Joffre, J.; Esposito, B.; Potteaux, S.; Vilar, J.; Cabuzu, D.; Kluwe, J.; Seguier, S.; et al. Gingival fibroblasts protect against experimental abdominal aortic aneurysm development and rupture through tissue inhibitor of metalloproteinase-1 production. *Cardiovasc. Res.* **2017**, *113*, 1364–1375. [CrossRef] [PubMed]
49. Rohner, E.; Yang, R.; Foo, K.S.; Goedel, A.; Chien, K.R. Unlocking the promise of mRNA therapeutics. *Nat. Biotechnol.* **2022**, *40*, 1586–1600. [CrossRef] [PubMed]
50. Qin, S.; Tang, X.; Chen, Y.; Chen, K.; Fan, N.; Xiao, W.; Zheng, Q.; Li, G.; Teng, Y.; Wu, M.; et al. mRNA-based therapeutics: Powerful and versatile tools to combat diseases. *Signal Transduct. Target. Ther.* **2022**, *7*, 166. [CrossRef] [PubMed]
51. Cheng, S.W.K.; Eagleton, M.; Echeverri, S.; Munoz, J.G.; Holden, A.H.; Hill, A.A.; Krievins, D.; Ramaiah, V. A pilot study to evaluate a novel localized treatment to stabilize small- to medium-sized infrarenal abdominal aortic aneurysms. *J. Vasc. Surg.* **2023**, *78*, 929–935.e1. [CrossRef] [PubMed]
52. Avci-Adali, M.; Behring, A.; Steinle, H.; Keller, T.; Krajeweski, S.; Schlensak, C.; Wendel, H.P. In vitro synthesis of modified mRNA for induction of protein expression in human cells. *JoVE (J. Vis. Exp.)* **2014**, *93*, e51943.
53. Golombek, S.; Pilz, M.; Steinle, H.; Kochba, E.; Levin, Y.; Lunter, D.; Schlensak, C.; Wendel, H.P.; Avci-Adali, M. Intradermal Delivery of Synthetic mRNA Using Hollow Microneedles for Efficient and Rapid Production of Exogenous Proteins in Skin. *Mol. Ther. Nucleic Acids* **2018**, *11*, 382–392. [CrossRef]

International Journal of
*Molecular Sciences*

MDPI

*Article*

# Does Cell-Type-Specific Silencing of Monoamine Oxidase B Interfere with the Development of Right Ventricle (RV) Hypertrophy or Right Ventricle Failure in Pulmonary Hypertension?

**Paulin Brosinsky [1,*], Jacqueline Heger [1], Akylbek Sydykov [2], Astrid Weiss [2], Stephan Klatt [3], Laureen Czech [1], Simone Kraut [2], Ralph Theo Schermuly [2], Klaus-Dieter Schlüter [1] and Rainer Schulz [1]**

[1] Physiologisches Institut, Justus-Liebig-Universität, 35392 Gießen, Germany; jacqueline.heger@physiologie.med.uni-giessen.de (J.H.); laureen.czech@physiologie.med.uni-giessen.de (L.C.); klaus-dieter.schlueter@physiologie.med.uni-giessen.de (K.-D.S.); rainer.schulz@physiologie.med.uni-giessen.de (R.S.)
[2] Excellence Cluster Cardiopulmonary System (ECCPS), Justus-Liebig-Universität, 35392 Gießen, Germany; akylbek.sydykov@innere.med.uni-giessen.de (A.S.); astrid.weiss@innere.med.uni-giessen.de (A.W.); simone.kraut@innere.med.uni-giessen.de (S.K.); ralph.schermuly@innere.med.uni-giessen.de (R.T.S.)
[3] Vascular Research Centre, Goethe Universität, 60590 Frankfurt, Germany; klatt@vrc.uni-frankfurt.de
* Correspondence: paulin.brosinsky@physiologie.med.uni-giessen.de; Tel.: +49-641-99-47211

**Abstract:** Increased mitochondrial reactive oxygen species (ROS) formation is important for the development of right ventricular (RV) hypertrophy (RVH) and failure (RVF) during pulmonary hypertension (PH). ROS molecules are produced in different compartments within the cell, with mitochondria known to produce the strongest ROS signal. Among ROS-forming mitochondrial proteins, outer-mitochondrial-membrane-located monoamine oxidases (MAOs, type A or B) are capable of degrading neurotransmitters, thereby producing large amounts of ROS. In mice, MAO-B is the dominant isoform, which is present in almost all cell types within the heart. We analyzed the effect of an inducible cardiomyocyte-specific knockout of MAO-B (cmMAO-B KO) for the development of RVH and RVF in mice. Right ventricular hypertrophy was induced by pulmonary artery banding (PAB). RV dimensions and function were measured through echocardiography. ROS production (dihydroethidium staining), protein kinase activity (PamStation device), and systemic hemodynamics (in vivo catheterization) were assessed. A significant decrease in ROS formation was measured in cmMAO-B KO mice during PAB compared to Cre-negative littermates, which was associated with reduced activity of protein kinases involved in hypertrophic growth. In contrast to littermates in which the RV was dilated and hypertrophied following PAB, RV dimensions were unaffected in response to PAB in cmMAO-B KO mice, and no decline in RV systolic function otherwise seen in littermates during PAB was measured in cmMAO-B KO mice. In conclusion, cmMAO-B KO mice are protected against RV dilatation, hypertrophy, and dysfunction following RV pressure overload compared to littermates. These results support the hypothesis that cmMAO-B is a key player in causing RV hypertrophy and failure during PH.

**Keywords:** pulmonary hypertension; monoamine oxidase; right heart

## 1. Introduction

Cardiac hypertrophy occurs as a result of a variety of heart diseases, including pulmonary arterial hypertension (PAH). PAH increases right ventricular (RV) afterload, inducing—if prolonged—RV hypertrophy (RVH). With sustained pressure overload, pathological remodeling of the RV occurs, leading to RV dilation and, finally, RV failure (RVF) [1]. While smaller amounts of reactive oxygen species (ROS) act as signaling molecules and contribute to RVH, excessive ROS formation contributes to the transition of adaptive to maladaptive hypertrophy (maladaptive remodeling) and the development of RVF [2,3]. Excessive ROS formation can induce cardiac dysfunction by activating stress protein kinases, through oxidative modification of contractile proteins, by disturbing metabolism and the intracellular ion homeostasis, or by damaging mitochondria and thereby inducing cell death [4].

Apart from a number of cytosolic enzymes, mitochondria also contribute to increased ROS formation [5], and scavenging of mitochondrial ROS attenuates pressure overload- or hypoxia-induced RVH and/or RVF [6]. Mitochondrial ROS are not only derived from the respiratory chain complexes but also from a variety of other mitochondrial proteins. Some of these proposed proteins are uncoupling proteins (UCPs), including p66shc and monoamine oxidases (MAOs) [2,3]. While the importance of UCP, namely UCP2, for ROS formation is debated [3], its importance for the development of RVF during pressure overload has been established [7]. On the contrary, while there is no doubt that p66shc contributes to ROS formation, especially in stress situations [8,9], p66shc has no impact on RVH or RVF during pressure overload [10], which, in part, might be related to its low expression in cardiomyocytes [11].

MAOs are located at the outer mitochondrial membrane where they degrade neurotransmitters to produce ROS. There are two isoforms, MAO-A and MAO-B, presenting 92% of sequence identity [12,13]. MAO-B, which is expressed to a greater extent in the myocardium of mice and humans [14], differs from MAO-A with respect to substrate specificity and inhibitor sensitivity. The two MAO isoforms have common substrates, such as dopamine but also specific substrates. MAO-B can metabolize 1-methyl histamine [15], produced by the histamine-N-methyltransferase [16], while MAO-A metabolizes serotonin (or 5-hydroxytryptamin, 5-HT) and catecholamines (for review, see [17]). MAO requires flavin adenine dinucleotide as a cofactor that is reduced by the reaction and subsequently re-oxidized by oxygen and water, generating hydrogen peroxide [18]. MAO can also form reactive aldehydes, such as 4-hydroxynonenal, as a byproduct of catecholamine metabolism through cardiolipin peroxidation inside mitochondria in primary cardiomyocytes.

While the importance of MAOs in the pathophysiology of left ventricular diseases is well-established [19–22], data on their importance for RV diseases are rare, but involvement in PH has been proposed [23]. In rats, PH secondary to monocrotaline injection [24], sugen5416/hypoxia, or pulmonary artery banding [25] upregulates MAO-A expression in the pulmonary vasculature and the failing RV. The MAO-A inhibitor clorgyline reduces RV afterload and pulmonary vascular remodeling in sugen/hypoxia rats through reduced pulmonary vascular proliferation and oxidative stress, resulting in improved RV stiffness and relaxation and reversed RV hypertrophy. In rats with PAB, clorgyline has no direct effect on the RV [25].

MAOs are expressed in all cardiac cells, including cardiomyocytes, fibroblasts, and endothelial and vascular smooth muscle cells (for review, see [3]). A permanent cardiomyocyte-specific knockout of MAO-A reduces the incidence of catecholamine-induced arrhythmias in mice [26]. Recently, we demonstrated that a lack of cardiomyocyte-specific MAO-B reduces left ventricular infarct size in vitro [27] and in vivo [28]. Information on the importance of cardiomyocyte-specific MAOs in RV disease is absent. We therefore investigated the effects of MAO-B-mediated ROS formation on RV dilatation and function in response to pulmonary artery banding in the inducible cardiomyocyte-specific MAO-B knockout mouse model [27].

## 2. Results

To determine the role of MAO-B during RVH, Myh6-MCreM_x_MAOB$^{fl/fl}$ mice were used to induce a cardiomyocyte-specific knockout of MAO-B (cmMAO-B KO) and compared to Cre-negative littermates (MAO-B$^{fl/fl}$). Pulmonary artery banding (PAB) was performed to achieve pressure overload, and SHAM-operated animals served as controls.

Body weight and heart rate were similar among the different groups of mice (Table S1). Also, the effect of PAB on left ventricular function was comparable between MAO-B$^{fl/fl}$ and cmMAO-B KO mice (Table S1). Pulmonary artery banding resulted in a similar increase in RV systolic pressure in MAO-B$^{fl/fl}$ and cmMAO-B KO mice measured during RV catheterization (Figure S2).

### 2.1. *Effect of cmMAO-B Knockout on Reactive Oxygen Species Formation*

MAO-B is known to produce ROS during oxidative deamination of neurotransmitters. In certain stress situations, like for left ventricular diseases, the importance of ROS generated by MAOs was already described [22]. To investigate the influence of MAO-B during RV pressure overload, intracellular ROS detection was performed for MAO-B$^{fl/fl}$ and cmMAO-B KO mice. Following three weeks of PAB, ROS formation increased in MAO-B$^{fl/fl}$ mice. In contrast, the effect was completely blunted in cmMAO-B KO mice (Figure 1).

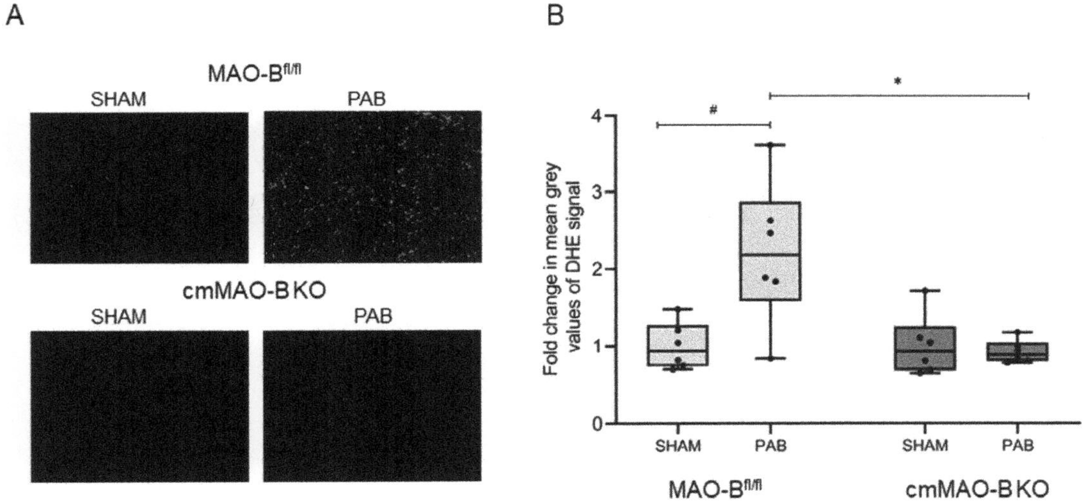


**Figure 1.** Increased ROS production in response to PAB was abrogated in cmMAO-B KO mice. Representative images (20× microscope's magnification) of cryosections of RV tissue stained with dihydroethidium (DHE) dye from PAB- or SHAM-operated MAO-B$^{fl/fl}$ and cmMAO-B KO mice (**A**). Quantification of ROS formation in RV tissue from MAO-B$^{fl/fl}$ (SHAM and PAB n = 6 each) and cmMAO-B KO (SHAM and PAB n = 6 each) mice by measuring mean grey values of the DHE signal. Data are presented as mean fold change compared to MAO-B$^{fl/fl}$ SHAM ± SD, * and #: $p < 0.05$ analyzed through two-side ANOVA (**B**).

In concordance with the proposed substrates of MAO-B, histamine and ethanolamine concentrations were increased in cmMAO-B KO mice following three weeks of PAB (LC-MS/MS analysis) when compared to MAO-B$^{fl/fl}$ mice (Figure S3).

### 2.2. *Effect of cmMAO-B Knockout on Right Ventricular Dimension and Function*

RV pressure overload can induce RV remodeling, finally leading to RVF. It was demonstrated that MAO-B is a key regulator for cardiac structural and functional disarrangement in the left ventricle [29]. The role of cmMAO-B for the development of RVH and/or RVF is unknown.

Echocardiographic analysis revealed that banded MAO-B$^{fl/fl}$ mice showed a significant increase in RV inner diameter (RVID) and a decrease in RV systolic function (TAPSE), both effects being absent in cmMAO-B KO mice (Figure 2).

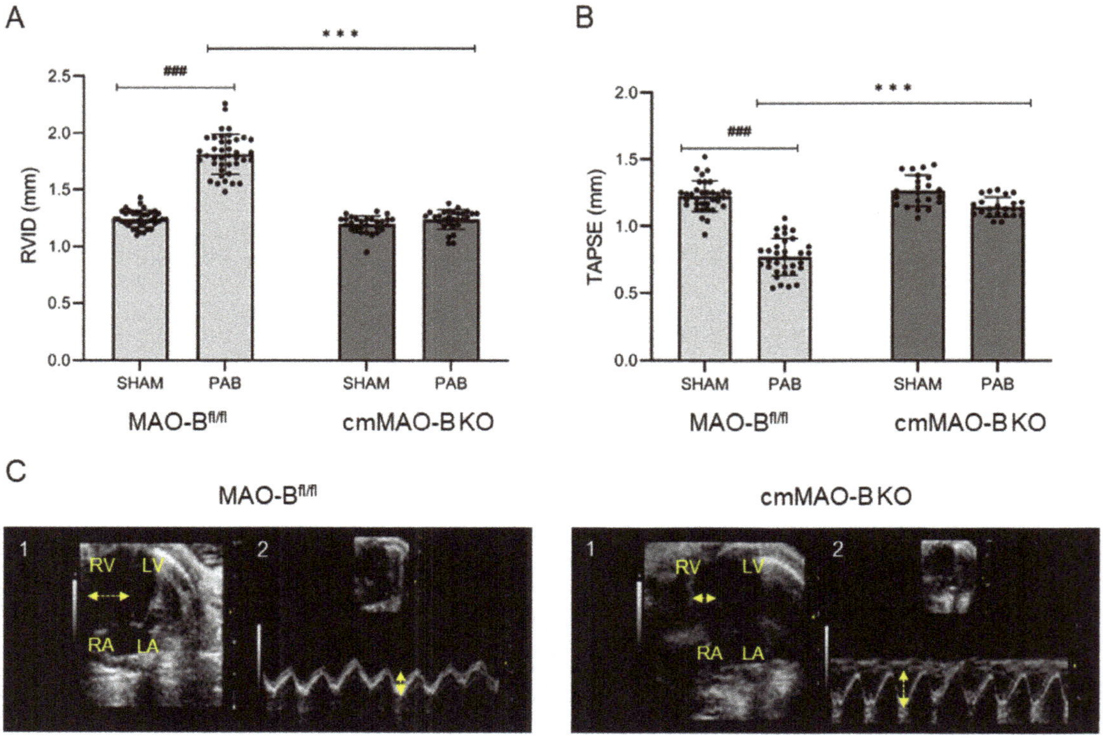


**Figure 2.** PAB-operated cmMAO-B KO mice were unaffected in their RV dimension and systolic function. RV geometry and function of MAO-$B^{fl/fl}$ (SHAM n = 33, PAB n = 33) and cmMAO-B KO (SHAM n = 22, PAB n = 23) mice were measured through echocardiography three weeks after PAB or SHAM surgery. Data are shown for right ventricular inner diameter (RVID in mm) (**A**) and tricuspid annular plane systolic excursion (TAPSE in mm) (**B**). The dots represent the individual data points. Data represent the mean ± SD, ### and ***: $p < 0.005$ analyzed through two-side ANOVA. Images of apical four-chamber view (1) and the systolic excursion of the tricuspid valve (TAPSE) (2) of one representative MAO-$B^{fl/fl}$ and cmMAO-B KO heart (**C**). LA: left atrium, LV: left ventricle, RA: right atrium, RV: right ventricle. Horizontal arrows represent RVID; vertical arrows represent TAPSE.

RV wall thickness (RVWT, echocardiography) was increased in MAO-$B^{fl/fl}$ mice; again, the effect was blunted in cmMAO-B KO mice (Figure 3A). The lack of increased cardiomyocyte hypertrophy in cmMAO-B KO mice compared to MAO-$B^{fl/fl}$ mice during PAB was confirmed on the protein level measuring the hypertrophy marker Myosin Heavy Chain (MYH) 7 in the right part of the septum (The Jess Simple Western system) (Figure 3B).

### 2.3. *Effect of cmMAO-B Knockout on Protein Kinase Activity during PAB*

Because ROS production was increased in MAO-$B^{fl/fl}$ mice during PAB, we assessed the protein kinase activity to identify the de-regulated kinases. Redox-sensitive kinases are known to be involved in cardiac hypertrophy [30,31], so it was assumable that the loss of MAO-B could have an impact. Interestingly, differences in kinase activity could be observed within the types of surgery as well as within the genotypes, as the activities of the kinase were influenced by the cardiomyocyte ablation of MAO-B as well as by PAB (Figure S4). However, in order to assess the role of cmMAO-B during RVH, the focus was mainly on the PAB groups.

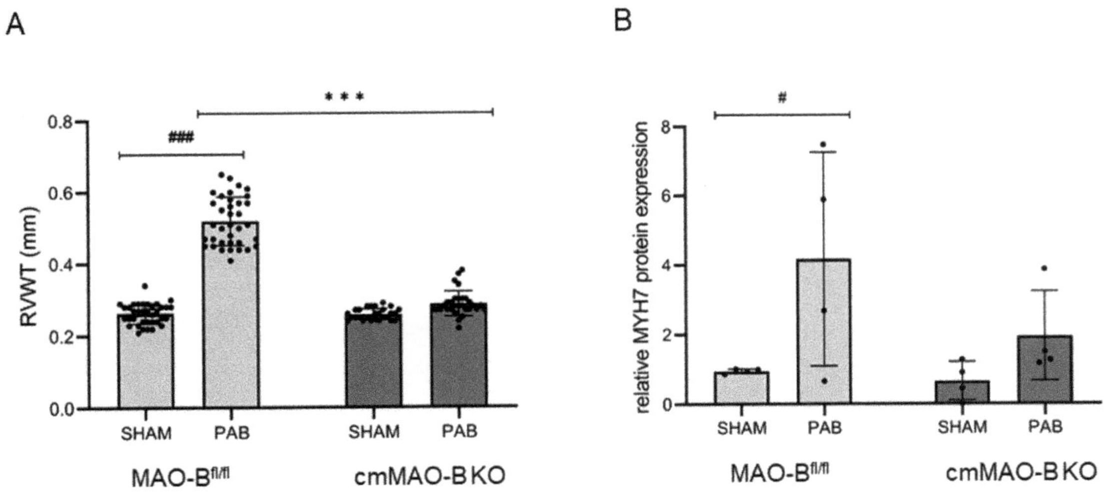


**Figure 3.** Cardiomyocyte-specific deletion of MAO-B protected the hearts from RVH. Three weeks after SHAM/PAB surgery, mice were evaluated for RVH development. Echocardiography was used to determine right ventricular wall thickness RVWT (in mm). Each dot represents the individual data point. Data represent the mean ± SD for MAO-B$^{fl/fl}$ (SHAM n = 36, PAB n = 36) and cmMAO-B KO mice (SHAM n = 28, PAB n = 30) (**A**). The Jess Simple Western system was used to detect the hypertrophy marker Myosin Heavy Chain (MYH) 7 at the protein level in the right part of the septum of MAO-B$^{fl/fl}$ and cmMAO-B KO mice (n = 4 for each of the four conditions) normalized to vinculin. Data represent the mean ± SD relative to the MAO-B$^{fl/fl}$ SHAM group (**B**). Both diagrams were analyzed through two-side ANOVA, #: $p < 0.05$, ### and ***: $p < 0.005$.

The involvement of ROS as indirect secondary messengers during RVH has already been described [32]. The Ras/Raf/MAPK(MEK)/ERK signaling pathway represents an important signal cascade regulating cardiac hypertrophy [33,34], and half of the de-regulated kinases belong to this MAP kinase signaling pathway (Figure 4, Table S2). In particular, RAF kinases (ARAF/BRAF/RAF1) have already been identified as key regulators during cardiac hypertrophy in mice [30,35].

The activity of protein kinases, which are considered to be involved in cardiac hypertrophy, was increased in MAO-B$^{fl/fl}$ mice during PAB when compared to cmMAO-B KO mice. In addition, all three RAF kinases were significantly downregulated after PAB in cmMAO-B KO mice (Figure 4).

### 2.4. *Effect of cmMAO-B Knockout on Cardiomyocyte Function*

Next, we examined the contraction behavior of cardiomyoyctes as RV geometry and function were impaired in MAO-B$^{fl/fl}$ mice after PAB.

Isolated RV cardiomyocytes were analyzed from PAB- and SHAM-operated mice. Cardiomyocyte contraction, relaxation, and shortening velocities were not affected by pressure overload, and no significant differences were detected between MAO-B$^{fl/fl}$ and cmMAO-B KO mice (Table 1).

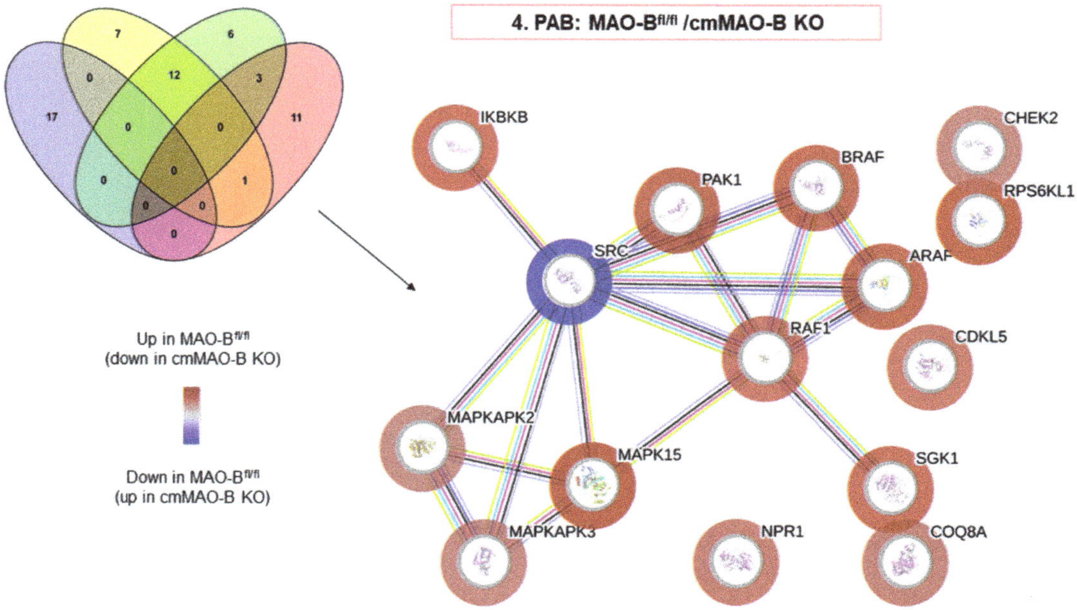


**Figure 4.** The activity of the kinases involved in cardiac hypertrophy was decreased in cmMAO-B KO mice in response to pressure overload. Functional protein association networks of kinases were analyzed in the right part of the septum of mice due pressure overload. MAO-B$^{fl/fl}$ and cmMAO-B KO mice underwent PAB surgery (n = 4, for each group). After three weeks, protein lysates from the right part of the septum were prepared for kinome profiling. Sequential two-group comparisons were performed to identify kinases, which are differentially regulated because of the genetic background and in response to pressure overload. The resulting hits were uploaded into the String protein database webpage to create images representing functional relationships revealing protein, i.e., kinase, association networks. The Venn diagram highlights the number of kinases that are significantly ($p < 0.05$) de-regulated in not only one but multiple two-group comparisons. For the reactome diagram, the outer color reflects the kinase activity highlighted for PAB: MAO-B$^{fl/fl}$ vs. cmMAO-B KO.

**Table 1.** cmMAO-B KO and pressure overload had no influence on contraction behavior of cardiomyocytes. Cell shortening of isolated cardiomyocytes from RV three weeks after PAB (MAO-B$^{fl/fl}$ n = 6, cmMAO-B KO n = 5) or SHAM (MAO-B$^{fl/fl}$ n = 6, cmMAO-B KO n = 6) surgery (n = number of mice, 36 cardiomyocytes were analyzed per animal). Data of diastolic cell length (Ldiast in µm), contraction velocity (Con Vel in µm/s), relaxation velocity (Rel Vel in µm/s), and load-free cell shortening (quantified as percent shortening amplitude normalized to the diastolic cell length of individual cells (dL/L(%)) are shown. Data represent the mean ± SD.

| | RV | Ldiast (µm) | Con Vel (µm/s) | Rel Vel (µm/s) | dL/L (%) |
|---|---|---|---|---|---|
| MAO-B$^{fl/fl}$ | SHAM | 109.93 ± 9.79 | 202.63 ± 51.36 | 166.05 ± 45.05 | 8.53 ± 1.82 |
| | PAB | 111.32 ± 7.23 | 182.99 ± 46.88 | 151.03 ± 45.33 | 7.65 ± 7.65 |
| cmMAO-B KO | SHAM | 112.08 ± 11.66 | 215.58 ± 72.75 | 180.14 ± 59.07 | 9.10 ± 2.44 |
| | PAB | 108.36 ± 11.00 | 183.64 ± 48.88 | 152.43 ± 45.10 | 8.30 ± 2.08 |

## 3. Discussion

### *3.1. Main Findings*

The current study aims to clarify the role of cardiomyocyte MAO-B in the development of RV hypertrophy and failure. The main findings of this study are (1) that RV pressure overload increases myocardial ROS formation in a MAO-B-dependent manner; (2) that functions of isolated RV cardiomyocytes are neither affected by pressure overload nor by MAO-B knockout; and (3) that the development of RV hypertrophy and failure upon pressure overload is abolished by an inducible cardiomyocyte-specific knockout of MAO-B.

### *3.2. MAO and ROS Formation*

In isolated mitochondria from mice hearts, we and others could demonstrate both MAO-A and MAO-B expression, with MAO-B being the dominant isoform [3]. However, cardiomyocyte MAO-A also exerts cardiac effects; cardiomyocyte-specific MAO-A inhibition exerts an anti-arrhythmic effect by enhancing diastolic calcium handling under catecholamine stress. Mechanistically, this is facilitated by a reduction in ROS generation, consequently leading to decreased oxidation of type II protein kinase A and calmodulin kinase II [26]. Thus, MAO-A can also contribute to myocardial ROS formation under certain stress conditions (catecholamines). However, with pressure overload as used in the present study, increased myocardial ROS formation seen in MAO-$B^{fl/fl}$ mice was completely abolished following cardiomyocyte-specific knockout of MAO-B. Therefore, the contribution of MAOs to stress-induced ROS formation depends on the underlying stimulus.

### *3.3. MAO Substrates*

Mice deficient for both MAO-A and MAO-B demonstrate increased tissue levels of serotonin, norepinephrine, dopamine, and phenylethylamine [36], and genetic ablation of MAO-A increases the serotonin concentration in the blood and tissue in rats [37]. Similarly, blockade of MAO by drugs used for other indications (e.g., antidepressants) can alter histamine levels in mice hearts [38]. Here, we demonstrated that cardiomyocyte-specific knockout of MAO-B increased myocardial histamine and ethanolamine levels, confirming the above substrates for MAO-B and their usage for the formation of myocardial ROS. Interestingly, histamine profoundly impacts the pathophysiology of the heart, resulting in hypertension-induced cardiac hypertrophy (for review, see [39]) through activation of histamine-H2 receptors [40] and histamine-H2 receptor polymorphisms, which altered heart failure development in patients [41]. However, despite the increase in histamine levels following knockout of MAO-B in cardiomyocytes, hypertrophic growth induced by pressure overload following pulmonary artery banding was decreased in cmMAO-B KO mice.

### *3.4. MAO and Pulmonary Hypertension*

Data on the importance of MAOs for RV diseases are rare, but involvement in PH has been proposed [23]. In rats, PH secondary to monocrotaline injection [24], sugen5416/hypoxia, or pulmonary artery banding [25] upregulated MAO-A expression in the pulmonary vasculature and the failing RV. The MAO-A inhibitor clorgyline reduced RV afterload and pulmonary vascular remodeling in sugen/hypoxia rats, while clorgyline had no direct effect on the RV following PAB [25]. This is in contrast to the findings of the present study, where inducible cmMAO-B KO preserved RV geometry and function following PAB. Potential explanations for the observed difference could relate to (1) species differences (rats vs. mice), (2) genetic vs. pharmacological approaches, (3) the mode of pressure overload (sugen/hypoxia vs. banding), or (4) as discussed in Section 3.2., substrate availability (catecholamines vs. histamine).

### *3.5. MAO and RV Geometry*

In hearts [42] as well as in isolated cardiomyocytes [43] of spontaneously hypertensive rats, MAO activity is significantly increased even before the development of cardiac hypertrophy [43]. Increased MAO activity might represent an early event in the development of cardiac hypertrophy [42] due to its potential impact on cardiac metabolism [44], as cardiac hypertrophy normally goes along with a metabolic switch to preferential use of carbohydrates rather than fatty acids [45–47]. In the present study, RV hypertrophy during PAB was dependent on ROS produced by MAO-B, as cardiomyocyte-specific MAO-B knockout abolishes both ROS formation and RV hypertrophy (measured with echocardiography and on the protein level). Data on MAO-dependent effects on myocardial hypertrophy are supported by the measured alterations in protein kinase activity; while PAB increased the activity of redox-sensitive protein kinases involved in hypertrophic growth [31,33], this effect of PAB was blunted in the myocardium from mice with the cardiomyocyte-specific knockout of MAO-B.

For instance, recently, it was shown that BRAF was elevated in humans with heart failure. In addition, cardiomyocyte-specific activation of BRAF in mice led to hypertrophy [30]. Furthermore, a cardiomyocyte-specific KO of BRAF resulted in protection against cardiomyocyte hypertrophy, induced by angiotensin II, in mice [35].

Here, we were able to confirm the connection of activated BRAF and hypertrophy, as these kinases were downregulated in myocardium without cmMAO-B compared to their respective littermates during pressure overload.

In mice, transverse aortic constriction or doxorubicin intoxication led to an increase in left ventricular dimensions (end-diastolic, end-systolic left ventricular diameter) [21,29]. Either global knockout of MAO-B [29] or its pharmacological blockade [21] attenuated the stress-induced changes in left ventricular geometry. Similarly, in the present study, cardiomyocyte-specific knockout of MAO-B abolished the pressure-overload-induced increase in RV diameter.

### 3.6. MAO and Cardiomyocyte Function

It is accepted that a functional impact of hypertrophy is an increased contractility, a common feature of patients with PAH [48] and animals with pressure overload [49]. In a previously published study, we indeed described an improved function of wildtype RV cardiomyocytes following three weeks of PAB [50]. However, whether the observed effect in the above study was ROS-dependent or secondary to changes in cardiac metabolism remains unknown. On the contrary, increased ROS formation can directly reduce cardiomyocyte function through oxidative modification of contractile proteins [51,52]. In the present study, although myocardial ROS were increased with PAB and remained unaffected following cardiomyocyte-specific MAO-B knockout, neither intervention altered the function of isolated RV cardiomyocytes. Thus, the amount of ROS formed by MAO-B during PAB appears to be insufficient to directly modify contractile proteins.

In conclusion, cmMAO-B KO mice were protected against RV dilatation, hypertrophy, and dysfunction following RV pressure overload compared to littermates. These results support the hypothesis that cmMAO-B is a key player in causing RV hypertrophy and failure during PH.

## 4. Materials and Methods

### 4.1. Animals and Ethical Concerns

The conditions of the used animals in the present study conform to the Guide for the Care and Use of Laboratory Animals published by the US National Institutes of Health (NIH publication No. 85-23, revised 1996) and were authorized by the "Regierungspräsidium Gießen" (GI 20/1 Nr. G 31/2019). Mice were bred in the animal facility of the Physiological Institute in Gießen, Germany. Generation, breeding, and induction of knockout mice were performed as previously described in detail [27]. Cardiomyocyte-specific knockout was induced by tamoxifen feeding; Myh6-MCreM_x_MAO-$B^{fl/fl}$ mice (cmMAO-B KO) were fed 400 mg/kg tamoxifen citrate for two weeks, followed by ten weeks of standard chow. Cre-negative MAO-$B^{fl/fl}$ littermates that underwent the same protocol were used as control mice. In the present study, 12–18-week-old female and male mice were used, which were kept in a 12 h light/dark cycle and had free access to standard chow and drinking water, unless otherwise indicated. A schematic overview of the experimental design is shown in Figure S1.

### 4.2. Pulmonary Artery Banding (PAB) In Vivo

The operations were performed as previously described in detail [53]. Thirty minutes prior to the surgery, analgesic buprenorphine hydrochloride (Temgesic®, 0.1 mg/kg, Sigma-Aldrich, Steinheim, Germany) was given subcutaneously. To initiate the inhalation anesthesia, 3–4% isoflurane supplemented with 100% oxygen were used and maintained at 1.5–2.5% during the surgery. Before the mice were placed on a heating surface, they were intubated and so mechanically ventilated by using the mouse ventilator MiniVent type 845 (Hugo Sachs Elektronik, March-Hugstetten, Germany). Via the second intercostal space, a left anterolateral thoracotomy was performed. Afterwards, a small titanium clip (Hemoclip®, Edward Weck & Co., Inc., Research Triangle Park, NC, USA) was placed with a modified hemoclip applier around the pulmonary trunk. In this way, a 65–70% construction of the pulmonary artery could be achieved. Subsequently, the chest and skin were closed with 6.0 prolene sutures. The SHAM group underwent the same procedure without the application of the hemoclip.

### 4.3. Echocardiography

Three-week-postoperative transthoracic echocardiography was performed by using the Vevo2100 high-resolution imaging system equipped with a 30 MHz transducer (VisualSonics, Toronto, ON, Canada). Inhalation anesthesia was initiated with 3–4% and maintained with 1.5–2% isoflurane in oxygen. The heart rate was monitored by taping all legs to ECG electrodes, and the temperature of the mice was controlled during the procedure. To evaluate the right heart function in vivo, right ventricular wall thickness, right ventricular internal diameter, and tricuspid annular plane systolic excursion were measured. Calculations were performed offline with the software Vevo LAB (version 5.5.0).

### 4.4. Invasive Hemodynamic Measurement

Three weeks after PAB or SHAM operations, hemodynamic measurements were conducted under anesthesia. The animals were anaesthetized with 3–4% isoflurane in oxygen and ventilated with a rodent ventilator (Harvard Apparatus, Holiston, MA, USA). Maintenance of anesthesia could be achieved with 2–3% isoflurane supplemented with oxygen. The mice were laid supine on a heating platform with three legs taped to electrocardiogram electrodes for monitoring of the heart rate. A rectal thermometer (Indus Instruments, Houston, TX, USA) was used to control the body temperature. The heating pad helped to keep the body temperature at 36.5–37.5 °C. Invasive hemodynamic measurements were performed using a micro pressure catheter (Millar instruments, Houston, TX, USA). Systemic arterial pressure was measured via the right carotid artery.

### 4.5. Isolation and Culture of Adult Mouse Ventricular Cardiomyocytes

As described by Bøtker et al., cardiomyocytes were isolated from the control group and the cmMAO-B KO group [54]. Following to the anesthesia with 4–5% isoflurane and the cervical dislocation, hearts were extracted, rapidly rinsed with 4 °C cold 0.9% NaCl, and attached on a cannula of the Langendorff apparatus. With retrograde perfusion in the Langendorff system containing collagenase and calcium-free buffer with a pH of 7.4 at 37 °C (in mmol/L: 10 Glucose monohydrate $D^+$, 25 Hepes, 2.5 KCl, 1.2 $KH_2PO_4$, 1.2 $MgSO_4 \cdot 7H_2O$, 110 NaCl), hearts were digested. Thereafter, the right ventricular wall, left ventricular wall, and ventricular septum were separated. The ventricular tissues were minced and incubated for another 5 min in the recirculating buffer. The septum was stored and frozen at −80 °C. The two suspensions were filtered through a 200 μm nylon mesh and via centrifugation separated from other cells. Within these washing steps, cells were resuspended in culture medium (in mmol/L: 2.5 $CaCl_2$-dihydrate, 5 Glucose, 10 Hepes, 4.7 KCl, 1.2 $KH_2PO_4$, 0.8 $MgSO_4$, 118 NaCl and 1.9 Na-Pyruvate) by step-wise increasing the concentrations of calcium up to 1 mM. Subsequently, cells were plated on laminin-coated 35 mm culture dishes (Falcon, type 3001). To remove non-attached cardiomyocytes, cells were washed with fresh culture medium after 45 min. The cardiomyocytes were cultured in this medium and ready for analysis; they were kept in an incubator at 37 °C, 5.5% $CO_2$, and a humidity of 95% until then.

*4.6. Determination of Cell Contraction*

Cell contraction behavior of MAO-$B^{fl/fl}$ and cmMAO-B KO animals was analyzed at room temperature by using a cell-edge-detection system. The isolated and cultured cardiomyocytes were stimulated via two AgCl electrodes with biphasic electrical stimuli constructed out of two equal but opposite rectangular 50 V stimuli of 0.5 ms duration. With a voltage of 2 Hz each cell was stimulated and measured four times. For the cell shortening calculation, the mean of these measurements was taken. Additionally, cell lengths were analyzed by using a line camera, recorded at 500 Hz. The results were represented as cell shortening and normalized to diastolic cell length (dL/L(%)).

*4.7. Metabolome Analysis: Sample Extraction*

Heart tissues were removed from the mice and immediately frozen in liquid nitrogen. Next, the frozen tissues were homogenized on dry ice and stored at −80 °C. For metabolite extraction, 10–20× the volume of ice-cold methanol extraction buffer was added to the samples (e.g., 20 mg of tissue equals 200–400 µL of extraction buffer). The methanol extraction buffer contains 100% methanol with 1 mM of TCEP, 1 mM of ascorbic acid, 0.1% formic acid, and 0.1 µM of internal standards. The samples were vortexed and sonicated on ice until completely homogenous. Next, the samples were put on ice for 15 min and subsequently centrifuged at 14,000 rpm for 10 min at 4 °C to remove protein. The supernatant was transferred to another tube. Two times the volume of MilliQ-water was added. The samples were frozen at −80 °C, followed by a sublimation step on a freeze-dryer using the Alpha 3-4 LSCbasic system (Martin Christ, Osterode am Harz, Germany). Next, the dried samples were reconstituted in 80 µL of 200 mM of boric acid buffer, shortly vortexed, and incubated on ice for 15 min. Next, 20 µL of AQC reagent was added (from AccQ-Taq Derivatization Kit, Milford, CT, USA, Waters application note). The samples were shortly vortexed and derivatized for 10 min at 55 °C. After the incubation step, the samples were put back on ice and centrifuged again at 14,000 rpm for 10 min at 4 °C. At last, the supernatant was transferred to MS glass vials.

*4.8. LC-MS/MS Analysis*

Metabolites were identified and quantified by using a 1290 Infinity II Bio LC system coupled to a 6495C QQQ MS in dynamic MRM mode (both Agilent Technologies, Waldbronn, Germany). In detail, metabolites were identified with authentic standards and/or via retention time, elution order from the column, and 1–2 transitions. LC separation of metabolites was performed on an Agilent Zorbax Extend RR HD 1.8 µm (2.1 × 150 mm) column with a solvent system of 0.1% formic acid in water (A) and 0.1% formic acid in acetonitrile (B). The LC gradient was 19 min long, with the following schedule: 0 min 1% B, 2 min 1% B, 9 min 15% B, 14 min 30% B, 16 min 60% B, and 17 min 65% B until 19 min 1% B. The flow-rate was set to 300 µL/min, and the column temperature was set to 30 °C. Data acquisition was performed in positive ionization mode. The gas temperature was set to 290 °C, and the gas flow was set to 20 L/min. The nebulizer was set to 45 psi. The sheath gas flow was set to 11 L/min, with a temperature of 400 °C. The capillary voltage was set to 3800 V with a nozzle voltage of 500 V. The voltages of the High-Pressure RF and Low-Pressure RF were set to 150/60 V, respectively.

*4.9. Detection of Intracellular ROS*

To detect intracellular ROS, cell-permeable fluorescent dye dihydroethidium (DHE) was used. Cryosections of RV from MAO-$B^{fl/fl}$ and cmMAO-B KO were incubated with DHE (dissolved in 1× PBS) for 5 min at 37 °C in the dark. Subsequently, the cryosections were washed with PBS and fixed with Dako Fluorescent Mounting Medium (Dako, North America, Inc., Carpinteria, CA, USA). Slides were then placed under a fluorescence microscope (BZ-X810 Keyence, Neu-Isenburg, Germany) and analyzed with the Quantity One 1-D Analysis Software (BIO-RAD, Version 4.6.6).

### 4.10. Peptide-Based Kinase Activity Assay

Protein isolation and peptide-based kinase activity assays for tyrosine as well as serine/threonine kinases (using the PamStation device, Pamgene, 's Hertogenbosch, The Netherlands) were conducted according to the manufacturer's instructions, as previously described [55–57]. As the right ventricle consists of a small amount of tissue and this tissue had already been used for other analyses, the right parts of the septum were used for the subsequently performed analysis of kinase activity and protein expression. This meant that the same set of mice could be used. It was assumed that tissue from the right ventricle was mainly included on the right side of the septum. For lysis of the right part of the septum, the tissues were mechanically homogenized using the Precellys tissue homogenizers and 1.4 mm ceramic beads (zirconium oxide) at 5500 rpm for 10 s two times, in the presence of 100 µL of M-PER lysis buffer (Thermo Fisher Scientific, Waltham, MA, USA) supplemented with protease and phosphatase inhibitor cocktail (Thermo Fisher Scientific). Afterwards, the homogenates were incubated for 1 h at 4 °C and centrifuged at 13,000 rpm at 4 °C for 15 min, and the supernatant was immediately aliquoted, flash-frozen, and stored at −80 °C until the time point of measurement. The protein concentration was determined using a bicinchoninic acid (BCA) protein assay kit (Thermo Fisher Scientific) according to the manufacturer's instructions. Then, 10 µg of protein lysate was applied on the tyrosine kinase PamChips (2 µg for serine/threonine kinase, respectively) to investigate the kinase-mediated phosphorylation of the substrate peptides. The assay was performed according to the manufacturer's instructions using the Evolve software (PamGene, Evolve 3, version 3.1.0.5). Based on the pattern of phosphorylation for the four experimental conditions, bioinformatic analyses through BioNavigator software (PamGene, BioNavigator 6, version 6.3.67.0) using protein databases allowed for a prediction of kinases upstream of these events.

### 4.11. Jess Simple Western System (ProteinSimple, San Jose, CA, USA)

This is an automated, capillary-based size separation immunoassay. This technology was used to detect the hypertrophy marker Myosin Heavy Chain 7 (MYH7) (Invitrogen, Waltham, MA, USA, NOQ7.5.4D, diluted 1:50) at the protein level in the right part of the septum of mice. MYH7 protein was normalized to Vinculin (Sigma-Aldrich, V9131, diluted 1:125,000). The standard reagent pack containing the components for preparing the samples and the size marker (12–230 kDa) was applied according to the manufacturer's instructions. For each capillary, 2.4 µL of the sample with an average concentration of 5.6 µg/µL (resulting in a total protein amount of 13.4 µg on average) was mixed with 0.6 µL of 5× fluorescence master mix (provided as part of the standard reagent pack). Next, the samples were denatured for 5 min at 95 °C, quickly centrifuged, and stored on ice. The plate was loaded with all components according to the manufacturer's pipetting scheme and afterwards centrifuged for 5 min at 1000× *g*. The plate as well as the cartridge (i.e., capillaries) were placed in the Jess device (ProteinSimple, San Jose, CA, USA), and a fully automated separation together with the immunodetection of proteins took place within three hours. Detection was based on chemiluminescence using horseradish peroxidase-coupled secondary antibodies. The light emission was detected using a CCD camera and then analyzed using the Compass software (version 4.1.0, Protein Simple). The results, i.e., protein expression reflected by band intensities, are displayed in traditional lane view as well as electropherograms, which allow for quantification by defining the respective area under the curve.

### 4.12. Statistical Analysis

Data presented in the figures are expressed in means ± SD and individual data points. Data presented in the tables are expressed as means ± SD. A $p$-value $< 0.05$ is considered to indicate a significant difference. All data were analyzed through two-side ANOVA. The program GraphPad Prism 9.4.1 was used for statistical analysis.

**Supplementary Materials:** The following supporting information can be downloaded at: https://www.mdpi.com/article/10.3390/ijms25116212/s1.

**Author Contributions:** Conceptualization, R.S.; funding acquisition, K.-D.S.; investigation, A.S., A.W., L.C., P.B., S.K. (Stephan Klatt), S.K. (Simone Kraut) and J.H.; methodology, A.S., A.W., S.K. (Stephan Klatt) and K.-D.S.; project administration, J.H. and R.S.; supervision, R.S.; validation, J.H. and R.S.; visualization, A.S. and P.B.; writing—original draft, P.B.; writing—review and editing, K.-D.S., R.T.S. and R.S. All authors have read and agreed to the published version of the manuscript.

**Funding:** This research received no external funding. The study is part of the collaborative research centre 1213 (B05 and CP02).

**Institutional Review Board Statement:** All animal experiments were conformed to the Guide for the Care and Use of Laboratory Animals published by the US National Institutes of Health (NIH publication No. 85-23, revised 1996) and were authorized by the "Regierungspräsidium Gießen" (GI 20/1 Nr. G 31/2019).

**Informed Consent Statement:** Not applicable.

**Data Availability Statement:** The data presented in this study are available on request from the corresponding author.

**Acknowledgments:** The authors thank the following individuals: Alissia Hartmann, Anna Reis, Birgit Störr, Christina Pilz, Julia Baldauf, and Martin Heisler for their excellent technical assistance and Ann-Christin Nickel, Patrizia Ewald, and Sabrina Schick for animal care.

**Conflicts of Interest:** The authors declare no conflicts of interest.

## References

1. Maarman, G.J.; Schulz, R.; Sliwa, K.; Schermuly, R.T.; Lecour, S. Novel putative pharmacological therapies to protect the right ventricle in pulmonary hypertension: A review of current literature. *Br. J. Pharmacol.* **2017**, *174*, 497–511. [CrossRef] [PubMed]
2. Schlüter, K.-D.; Kutsche, H.S.; Hirschhäuser, C.; Schreckenberg, R.; Schulz, R. Review on Chamber-Specific Differences in Right and Left Heart Reactive Oxygen Species Handling. *Front. Physiol.* **2018**, *9*, 1799. [CrossRef] [PubMed]
3. Schulz, R.; Schlüter, K.-D. Importance of Mitochondria in Cardiac Pathologies: Focus on Uncoupling Proteins and Monoamine Oxidases. *Int. J. Mol. Sci.* **2023**, *24*, 6459. [CrossRef]
4. Heusch, P.; Canton, M.; Aker, S.; Van De Sand, A.; Konietzka, I.; Rassaf, T.; Menazza, S.; Brodde, O.; Di Lisa, F.; Heusch, G.; et al. The contribution of reactive oxygen species and p38 mitogen-activated protein kinase to myofilament oxidation and progression of heart failure in rabbits. *Br. J. Pharmacol.* **2010**, *160*, 1408–1416. [CrossRef]
5. Andreadou, I.; Schulz, R.; Papapetropoulos, A.; Turan, B.; Ytrehus, K.; Ferdinandy, P.; Daiber, A.; Di Lisa, F. The role of mitochondrial reactive oxygen species, NO and $H_2S$ in ischaemia/reperfusion injury and cardioprotection. *J. Cell. Mol. Med.* **2020**, *24*, 6510–6522. [CrossRef] [PubMed]
6. Pak, O.; Scheibe, S.; Esfandiary, A.; Gierhardt, M.; Sydykov, A.; Logan, A.; Fysikopoulos, A.; Veit, F.; Hecker, M.; Kroschel, F.; et al. Impact of the mitochondria-targeted antioxidant MitoQ on hypoxia-induced pulmonary hypertension. *Eur. Respir. J.* **2018**, *51*, 1701024. [CrossRef]
7. Kutsche, H.S.; Schreckenberg, R.; Weber, M.; Hirschhäuser, C.; Rohrbach, S.; Li, L.; Niemann, B.; Schulz, R.; Schlüter, K.-D. Alterations in Glucose Metabolism During the Transition to Heart Failure: The Contribution of UCP-2. *Cells* **2020**, *9*, 552. [CrossRef]
8. Wu, J.; Xia, S.; Kalionis, B.; Wan, W.; Sun, T. The Role of Oxidative Stress and Inflammation in Cardiovascular Aging. *BioMed Res. Int.* **2014**, *2014*, 615312. [CrossRef]
9. Boengler, K.; Bornbaum, J.; Schlüter, K.-D.; Schulz, R. P66shc and its role in ischemic cardiovascular diseases. *Basic Res. Cardiol.* **2019**, *114*, 29. [CrossRef]
10. Hirschhäuser, C.; Sydykov, A.; Wolf, A.; Esfandiary, A.; Bornbaum, J.; Kutsche, H.S.; Boengler, K.; Sommer, N.; Schreckenberg, R.; Schlüter, K.-D.; et al. Lack of Contribution of p66shc to Pressure Overload-Induced Right Heart Hypertrophy. *Int. J. Mol. Sci.* **2020**, *21*, 9339. [CrossRef]
11. Obreztchikova, M.; Elouardighi, H.; Ho, M.; Wilson, B.A.; Gertsberg, Z.; Steinberg, S.F. Distinct Signaling Functions for Shc Isoforms in the Heart. *J. Biol. Chem.* **2006**, *281*, 20197–20204. [CrossRef]
12. Addonizio, G.; Alexopoulos, G.S. Drug-induced dystonia in young and elderly patients. *Am. J. Psychiatry* **1988**, *145*, 869–871. [CrossRef] [PubMed]

13. Bach, A.W.; Lan, N.C.; Johnson, D.L.; Abell, C.W.; Bembenek, M.E.; Kwan, S.W.; Seeburg, P.H.; Shih, J.C. cDNA cloning of human liver monoamine oxidase A and B: Molecular basis of differences in enzymatic properties. *Proc. Natl. Acad. Sci. USA* **1988**, *85*, 4934–4938. [CrossRef] [PubMed]
14. Dorris, R.L. A simple method for screening Monoamine Oxidase (MAO) inhibitory drugs for type preference. *J. Pharmacol. Methods* **1982**, *7*, 133–137. [CrossRef] [PubMed]
15. Costiniti, V.; Spera, I.; Menabò, R.; Palmieri, E.M.; Menga, A.; Scarcia, P.; Porcelli, V.; Gissi, R.; Castegna, A.; Canton, M. Monoamine oxidase-dependent histamine catabolism accounts for post-ischemic cardiac redox imbalance and injury. *Biochim. Biophys. Acta Mol. Basis Dis.* **2018**, *1864*, 3050–3059. [CrossRef] [PubMed]
16. Maintz, L.; Novak, N. Histamine and histamine intolerance. *Am. J. Clin. Nutr.* **2007**, *85*, 1185–1196. [CrossRef] [PubMed]
17. Mialet-Perez, J.; Santin, Y.; Parini, A. Monoamine oxidase-A, serotonin and norepinephrine: Synergistic players in cardiac physiology and pathology. *J. Neural Transm.* **2018**, *125*, 1627–1634. [CrossRef] [PubMed]
18. Mialet-Perez, J.; Bianchi, P.; Kunduzova, O.; Parini, A. New insights on receptor-dependent and monoamine oxidase-dependent effects of serotonin in the heart. *J. Neural Transm.* **2007**, *114*, 823–827. [CrossRef] [PubMed]
19. Deshwal, S.; Di Sante, M.; Di Lisa, F.; Kaludercic, N. Emerging role of monoamine oxidase as a therapeutic target for cardiovascular disease. *Curr. Opin. Pharmacol.* **2017**, *33*, 64–69. [CrossRef] [PubMed]
20. Deshwal, S.; Forkink, M.; Hu, C.-H.; Buonincontri, G.; Antonucci, S.; Di Sante, M.; Murphy, M.P.; Paolocci, N.; Mochly-Rosen, D.; Krieg, T.; et al. Monoamine oxidase-dependent endoplasmic reticulum-mitochondria dysfunction and mast cell degranulation lead to adverse cardiac remodeling in diabetes. *Cell Death Differ.* **2018**, *25*, 1671–1685. [CrossRef]
21. Antonucci, S.; Di Sante, M.; Tonolo, F.; Pontarollo, L.; Scalcon, V.; Alanova, P.; Menabò, R.; Carpi, A.; Bindoli, A.; Rigobello, M.P.; et al. The Determining Role of Mitochondrial Reactive Oxygen Species Generation and Monoamine Oxidase Activity in Doxorubicin-Induced Cardiotoxicity. *Antioxid. Redox Signal.* **2021**, *34*, 531–550. [CrossRef] [PubMed]
22. Kaludercic, N.; Arusei, R.J.; Di Lisa, F. Recent advances on the role of monoamine oxidases in cardiac pathophysiology. *Basic Res. Cardiol.* **2023**, *118*, 41. [CrossRef] [PubMed]
23. Sommer, N.; Schulz, R. Mitochondrial Monoamine Oxidase: Another Player in Pulmonary Hypertension? *Am. J. Respir. Cell Mol. Biol.* **2021**, *64*, 277–278. [CrossRef] [PubMed]
24. van Eif, V.W.W.; Bogaards, S.J.P.; van der Laarse, W.J. Intrinsic cardiac adrenergic (ICA) cell density and MAO-A activity in failing rat hearts. *J. Muscle Res. Cell Motil.* **2014**, *35*, 47–53. [CrossRef] [PubMed]
25. Sun, X.-Q.; Peters, E.L.; Schalij, I.; Axelsen, J.B.; Andersen, S.; Kurakula, K.; Gomez-Puerto, M.C.; Szulcek, R.; Pan, X.; da Silva Goncalves Bos, D.; et al. Increased MAO-A Activity Promotes Progression of Pulmonary Arterial Hypertension. *Am. J. Respir. Cell Mol. Biol.* **2021**, *64*, 331–343. [CrossRef] [PubMed]
26. Shi, Q.; Malik, H.; Crawford, R.M.; Streeter, J.; Wang, J.; Huo, R.; Shih, J.C.; Chen, B.; Hall, D.; Abel, E.D.; et al. Cardiac MAO-A inhibition protects against catecholamine-induced ventricular arrhythmias via enhanced diastolic calcium control. *Cardiovasc. Res.* **2024**, *120*, cvae012. [CrossRef] [PubMed]
27. Heger, J.; Hirschhäuser, C.; Bornbaum, J.; Sydykov, A.; Dempfle, A.; Schneider, A.; Braun, T.; Schlüter, K.-D.; Schulz, R. Cardiomyocytes-specific deletion of monoamine oxidase B reduces irreversible myocardial ischemia/reperfusion injury. *Free Radic. Biol. Med.* **2021**, *165*, 14–23. [CrossRef] [PubMed]
28. Heger, J.; Szabados, T.; Brosinsky, P.; Bencsik, P.; Ferdinandy, P.; Schulz, R. Sex Difference in Cardioprotection against Acute Myocardial Infarction in MAO-B Knockout Mice In Vivo. *Int. J. Mol. Sci.* **2023**, *24*, 6443. [CrossRef] [PubMed]
29. Kaludercic, N.; Carpi, A.; Nagayama, T.; Sivakumaran, V.; Zhu, G.; Lai, E.W.; Bedja, D.; De Mario, A.; Chen, K.; Gabrielson, K.L.; et al. Monoamine Oxidase B Prompts Mitochondrial and Cardiac Dysfunction in Pressure Overloaded Hearts. *Antioxid. Redox Signal.* **2014**, *20*, 267–280. [CrossRef]
30. Clerk, A.; Meijles, D.N.; Hardyman, M.A.; Fuller, S.J.; Chothani, S.P.; Cull, J.J.; Cooper, S.T.E.; Alharbi, H.O.; Vanezis, K.; Felkin, L.E.; et al. Cardiomyocyte BRAF and type 1 RAF inhibitors promote cardiomyocyte and cardiac hypertrophy in mice in vivo. *Biochem. J.* **2022**, *479*, 401–424. [CrossRef]
31. Wen, Y.; Liu, R.; Lin, N.; Luo, H.; Tang, J.; Huang, Q.; Sun, H.; Tang, L. NADPH Oxidase Hyperactivity Contributes to Cardiac Dysfunction and Apoptosis in Rats with Severe Experimental Pancreatitis through ROS-Mediated MAPK Signaling Pathway. *Oxid. Med. Cell. Longev.* **2019**, *2019*, 4578175. [CrossRef] [PubMed]
32. Bogaard, H.J.; Abe, K.; Vonk Noordegraaf, A.; Voelkel, N.F. The Right Ventricle Under Pressure. *Chest* **2009**, *135*, 794–804. [CrossRef] [PubMed]
33. Wang, Y. Mitogen-Activated Protein Kinases in Heart Development and Diseases. *Circulation* **2007**, *116*, 1413–1423. [CrossRef] [PubMed]
34. Maulik, S.K.; Kumar, S. Oxidative stress and cardiac hypertrophy: A review. *Toxicol. Mech. Methods* **2012**, *22*, 359–366. [CrossRef]
35. Alharbi, H.O.; Hardyman, M.A.; Cull, J.J.; Markou, T.; Cooper, S.T.E.; Glennon, P.E.; Fuller, S.J.; Sugden, P.H.; Clerk, A. Cardiomyocyte BRAF is a key signalling intermediate in cardiac hypertrophy in mice. *Clin. Sci.* **2022**, *136*, 1661–1681. [CrossRef] [PubMed]
36. Holschneider, D.P.; Scremin, O.U.; Roos, K.P.; Chialvo, D.R.; Chen, K.; Shih, J.C. Increased baroreceptor response in mice deficient in monoamine oxidase A and B. *Am. J. Physiol. Heart Circ. Physiol.* **2002**, *282*, H964–H972. [CrossRef] [PubMed]

37. Lairez, O.; Calise, D.; Bianchi, P.; Ordener, C.; Spreux-Varoquaux, O.; Guilbeau-Frugier, C.; Escourrou, G.; Seif, I.; Roncalli, J.; Pizzinat, N.; et al. Genetic deletion of MAO-A promotes serotonin-dependent ventricular hypertrophy by pressure overload. *J. Mol. Cell. Cardiol.* **2009**, *46*, 587–595. [CrossRef] [PubMed]
38. Neumann, J.; Grobe, J.M.; Weisgut, J.; Schwelberger, H.G.; Fogel, W.A.; Marušáková, M.; Wache, H.; Bähre, H.; Buchwalow, I.B.; Dhein, S.; et al. Histamine can be Formed and Degraded in the Human and Mouse Heart. *Front. Pharmacol.* **2021**, *12*, 582916. [CrossRef] [PubMed]
39. Saheera, S.; Potnuri, A.G.; Guha, A.; Palaniyandi, S.S.; Thandavarayan, R.A. Histamine 2 receptors in cardiovascular biology: A friend for the heart. *Drug Discov. Today* **2022**, *27*, 234–245. [CrossRef]
40. Potnuri, A.G.; Allakonda, L.; Appavoo, A.; Saheera, S.; Nair, R.R. Association of histamine with hypertension-induced cardiac remodeling and reduction of hypertrophy with the histamine-2-receptor antagonist famotidine compared with the beta-blocker metoprolol. *Hypertens. Res. Off. J. Jpn. Soc. Hypertens.* **2018**, *41*, 1023–1035. [CrossRef]
41. Leary, P.J.; Kronmal, R.A.; Bluemke, D.A.; Buttrick, P.M.; Jones, K.L.; Kao, D.P.; Kawut, S.M.; Krieger, E.V.; Lima, J.A.; Minobe, W.; et al. Histamine H2 Receptor Polymorphisms, Myocardial Transcripts, and Heart Failure (from the Multi-Ethnic Study of Atherosclerosis and Beta-Blocker Effect on Remodeling and Gene Expression Trial). *Am. J. Cardiol.* **2018**, *121*, 256–261. [CrossRef] [PubMed]
42. Tanijiri, H. Cardiac hypertrophy in spontaneously hypertensive rats. *Jpn. Heart J.* **1975**, *16*, 174–188. [CrossRef]
43. Pino, R.; Failli, P.; Mazzetti, L.; Buffoni, F. Monoamine oxidase and semicarbazide-sensitive amine oxidase activities in isolated cardiomyocytes of spontaneously hypertensive rats. *Biochem. Mol. Med.* **1997**, *62*, 188–196. [CrossRef]
44. Fischer, Y.; Thomas, J.; Kamp, J.; Jüngling, E.; Rose, H.; Carpéné, C.; Kammermeier, H. 5-hydroxytryptamine stimulates glucose transport in cardiomyocytes via a monoamine oxidase-dependent reaction. *Biochem. J.* **1995**, *311*, 575–583. [CrossRef] [PubMed]
45. Universal child immunization by 1990: A realistic goal. *World Ir. Nurs.* **1986**, *15*, 5–6.
46. Shao, D.; Tian, R. Glucose Transporters in Cardiac Metabolism and Hypertrophy. *Compr. Physiol.* **2015**, *6*, 331–351. [CrossRef]
47. Gibb, A.A.; Lorkiewicz, P.K.; Zheng, Y.-T.; Zhang, X.; Bhatnagar, A.; Jones, S.P.; Hill, B.G. Integration of flux measurements to resolve changes in anabolic and catabolic metabolism in cardiac myocytes. *Biochem. J.* **2017**, *474*, 2785–2801. [CrossRef]
48. Spruijt, O.A.; de Man, F.S.; Groepenhoff, H.; Oosterveer, F.; Westerhof, N.; Vonk-Noordegraaf, A.; Bogaard, H.-J. The effects of exercise on right ventricular contractility and right ventricular-arterial coupling in pulmonary hypertension. *Am. J. Respir. Crit. Care Med.* **2015**, *191*, 1050–1057. [CrossRef]
49. Wang, Z.; Schreier, D.A.; Hacker, T.A.; Chesler, N.C. Progressive right ventricular functional and structural changes in a mouse model of pulmonary arterial hypertension. *Physiol. Rep.* **2013**, *1*, e00184. [CrossRef]
50. Esfandiary, A.; Kutsche, H.S.; Schreckenberg, R.; Weber, M.; Pak, O.; Kojonazarov, B.; Sydykov, A.; Hirschhäuser, C.; Wolf, A.; Haag, D.; et al. Protection against pressure overload-induced right heart failure by uncoupling protein 2 silencing. *Cardiovasc. Res.* **2019**, *115*, 1217–1227. [CrossRef]
51. Canton, M.; Skyschally, A.; Menabò, R.; Boengler, K.; Gres, P.; Schulz, R.; Haude, M.; Erbel, R.; Di Lisa, F.; Heusch, G. Oxidative modification of tropomyosin and myocardial dysfunction following coronary microembolization. *Eur. Heart J.* **2006**, *27*, 875–881. [CrossRef] [PubMed]
52. Sharma, K.; Kass, D.A. Heart Failure with Preserved Ejection Fraction: Mechanisms, Clinical Features, and Therapies. *Circ. Res.* **2014**, *115*, 79–96. [CrossRef] [PubMed]
53. Mamazhakypov, A.; Veith, C.; Schermuly, R.T.; Sydykov, A. Surgical protocol for pulmonary artery banding in mice to generate a model of pressure-overload-induced right ventricular failure. *STAR Protoc.* **2023**, *4*, 102660. [CrossRef] [PubMed]
54. Bøtker, H.E.; Hausenloy, D.; Andreadou, I.; Antonucci, S.; Boengler, K.; Davidson, S.M.; Deshwal, S.; Devaux, Y.; Di Lisa, F.; Di Sante, M.; et al. Practical guidelines for rigor and reproducibility in preclinical and clinical studies on cardioprotection. *Basic Res. Cardiol.* **2018**, *113*, 39. [CrossRef] [PubMed]
55. Alack, K.; Weiss, A.; Krüger, K.; Höret, M.; Schermuly, R.; Frech, T.; Eggert, M.; Mooren, F.-C. Profiling of human lymphocytes reveals a specific network of protein kinases modulated by endurance training status. *Sci. Rep.* **2020**, *10*, 888. [CrossRef] [PubMed]
56. Weiss, A.; Neubauer, M.C.; Yerabolu, D.; Kojonazarov, B.; Schlueter, B.C.; Neubert, L.; Jonigk, D.; Baal, N.; Ruppert, C.; Dorfmuller, P.; et al. Targeting cyclin-dependent kinases for the treatment of pulmonary arterial hypertension. *Nat. Commun.* **2019**, *10*, 2204. [CrossRef] [PubMed]
57. Yerabolu, D.; Weiss, A.; Kojonazarov, B.; Boehm, M.; Schlueter, B.C.; Ruppert, C.; Günther, A.; Jonigk, D.; Grimminger, F.; Ghofrani, H.-A.; et al. Targeting Jak-Stat Signaling in Experimental Pulmonary Hypertension. *Am. J. Respir. Cell Mol. Biol.* **2021**, *64*, 100–114. [CrossRef]

International Journal of
*Molecular Sciences*

MDPI

*Article*

# Investigation of Strategies to Block Downstream Effectors of AT1R-Mediated Signalling to Prevent Aneurysm Formation in Marfan Syndrome

**Irene Valdivia Callejon [1,†], Lucia Buccioli [1,†], Jarl Bastianen [1], Jolien Schippers [1], Aline Verstraeten [1], Ilse Luyckx [1,2], Silke Peeters [1], A. H. Jan Danser [3], Roland R. J. Van Kimmenade [4], Josephina Meester [1] and Bart Loeys [1,3,*]**

1 Centre of Medical Genetics, Antwerp University Hospital, University of Antwerp, 2650 Antwerp, Belgium; irene.valdiviacallejon@uantwerpen.be (I.V.C.); lucia.buccioli@uantwerpen.be (L.B.); jarl.bastianen@uantwerpen.be (J.B.); jolien.schippers@uantwerpen.be (J.S.); aline.verstraeten@uantwerpen.be (A.V.); ilse.luyckx@uantwerpen.be (I.L.); silke.peeters@uantwerpen.be (S.P.)
2 Department of Human Genetics, Radboud University Medical Center, 6525 GA Nijmegen, The Netherlands
3 Department of Internal Medicine, Erasmus Medical Center, 3015 GD Rotterdam, The Netherlands; a.danser@erasmusmc.nl
4 Department of Cardiology, Radboud University Medical Center, 6525 GA Nijmegen, The Netherlands; roland.vankimmenade@radboudumc.nl
* Correspondence: bart.loeys@uantwerpen.be; Tel.: +32-3-275-97-74
† These authors contributed equally to this work.

**Abstract:** Cardiovascular outcome in Marfan syndrome (MFS) patients most prominently depends on aortic aneurysm progression with subsequent aortic dissection. Angiotensin II receptor blockers (ARBs) prevent aneurysm formation in MFS mouse models. In patients, ARBs only slow down aortic dilation. Downstream signalling from the angiotensin II type 1 receptor (AT1R) is mediated by G proteins and β-arrestin recruitment. AT1R also interacts with the monocyte chemoattractant protein-1 (MCP-1) receptor, resulting in inflammation. In this study, we explore the targeting of β-arrestin signalling in MFS mice by administering TRV027. Furthermore, because high doses of the ARB losartan, which has been proven beneficial in MFS, cannot be achieved in humans, we investigate a potential additive effect by combining lower concentrations of losartan (25 mg/kg/day and 5 mg/kg/day) with barbadin, a β-arrestin blocker, and DMX20, a C-C chemokine receptor type 2 (CCR2) blocker. A high dose of losartan (50 mg/kg/day) slowed down aneurysm progression compared to untreated MFS mice (1.73 ± 0.12 vs. 1.96 ± 0.08 mm, $p = 0.0033$). TRV027, the combination of barbadin with losartan (25 mg/kg/day), and DMX-200 (90 mg/kg/day) with a low dose of losartan (5 mg/kg/day) did not show a significant beneficial effect. Our results confirm that while losartan effectively halts aneurysm formation in *Fbn1*$^{C1041G/+}$ MFS mice, neither TRV027 alone nor any of the other compounds combined with lower doses of losartan demonstrate a notable impact on aneurysm advancement. It appears that complete blockade of AT1R function, achieved by administrating a high dosage of losartan, may be necessary for inhibiting aneurysm progression in MFS.

**Keywords:** Marfan syndrome; AT1R antagonism; β-arrestin; aortic aneurysm

## 1. Introduction

Marfan syndrome (MFS) is a heritable disorder characterized by abnormalities in the connective tissues, primarily affecting the ocular, skeletal, and cardiovascular systems [1]. It is caused by pathogenic variants in the *FBN1* gene and follows an autosomal dominant inheritance pattern, although a very rare number of MFS families have been reported to harbour autosomal recessive variants [2]. While MFS typically segregates within families, around 30% of cases are caused by de novo variants [1,3].

Amongst all clinical features, the cardiovascular manifestations of MFS contribute most significantly to the observed morbidity and mortality. The primary cardiovascular characteristic is the enlargement of the aortic root, specifically at the sinus of Valsalva, a feature present in the majority of patients. There can be considerable variability in the rate of this dilation among affected individuals. As a result of the aortic root enlargement, aortic valve insufficiency with subsequent regurgitation can develop. Dilation may progress to an aneurysm, potentially leading to a type A dissection or rupture. A small percentage of patients (10–20%) show dilation of the descending and abdominal aorta, which can result in a type B dissection [3].

At the mechanistic level, an increase in transforming growth factor beta (TGF-β) pathway activity has been observed in aortic wall samples from individuals with syndromic thoracic aortic aneurysm (TAA), as well as in mouse models of MFS [3–5]. The *FBN1* gene encodes for the fibrillin-1 protein, a structural macromolecule that forms microfibrils through polymerisation. These microfibrils are found in the connective tissue, contributing to the creation of tissue-specific frameworks and serving as reservoirs for growth factors, such as TGF-β. Due to pathogenic variants in the *FBN1* gene, these fibrillin microfibrils fail to adequately sequester TGF-β within latent TGF-β binding complexes connected to fibrillin-1 [6]. This disruption is believed to lead to the observed dysregulation of the TGF-β pathway. Since this pathway regulates several processes in cells, its dysregulation may eventually contribute to other molecular features described in the aortic walls of TAA samples, including the phenotypic switching of vascular smooth muscle cells (VSMCs), endothelial dysfunction, changes in the extracellular matrix (ECM), and inflammation [5,7,8].

Despite recent advancements in our understanding of TAA pathomechanisms, the available pharmacological options remain quite limited. β-blockers are widely acknowledged as the standard of care required to decrease the aortic dilation rate by reducing blood pressure. Nevertheless, the trials conducted so far do not demonstrate a full arrest of aortic growth, highlighting the necessity of additional studies to determine the effectiveness of β-blockers in this context [9]. The use of more cardiac-specific β-blockers might improve tolerability, but certain individuals may still experience intolerance to medications within this class of drugs. Hence, for this group of patients, the use of angiotensin II type 1 receptor blockers (ARB) can be considered an alternative or adjunctive therapy [10]. This class of medication has been used for treating renal disorders characterised by elevated levels of TGFβ, providing additional evidence for their use in other conditions associated with excessive TGFβ signalling [11]. While the use of the ARB losartan selectively antagonises the angiotensin II type 1 signalling only, debate about the contribution of angiotensin II type 1 receptor (AT1R) and angiotensin II type 2 receptor (AT2R) to aortic aneurysm is still ongoing. It is suggested that the two receptors exert opposite effects: the former appears to contribute to pathogenesis, while the latter is suggested to have a protective effect [4].

Despite the very encouraging results from the preclinical losartan study in a murine MFS model [12], the ARB losartan did not show superior effects compared to standard therapy (i.e., β-blockade) in a human comparative trial [13]. The largest trial showed that a high dose of β-blockers (up to four-fold of the normal dose) performed equally well as a low dose of losartan [13,14]. A meta-analysis also suggested that treatment with β-blockers and losartan could have an additive effect, although the study did not have enough power to prove it [10]. Furthermore, subsequent trials have shown that higher doses of sartans perform better (personal communication: Dr. Harry C. Dietz). The latter was also shown in the MFS mouse model: a high dose of losartan performed much better than low or medium doses of losartan. The human equivalent dose of this high oral dose of losartan would be 4–8 mg/kg/day, which is much higher than the doses commonly used in a human setting. Currently used doses in MFS patients are up to 2 mg/kg/day in children and 100–150 mg per day (1.4–2.1 mg/kg/day) in adults. These higher doses in humans may result in more side effects.

Although current therapies targeting AT1R do not completely prevent aortic dilation and dissection, the involvement of AT1R in TAA seems to be evident. Exploring new approaches for targeting AT1R remains a promising strategy to improve treatment for TAA. In particular, AT1R blocking strategies that would reinforce the use of lower doses of losartan would be very attractive. AT1R signalling involves two different components: G protein activation, which promotes vasoconstriction, and β-arrestin recruitment, which leads to enhanced cardiomyocyte contractility and cell survival. Classical ARBs, such as losartan, inhibit both pathways [15]. Nevertheless, emerging evidence indicates that biased ligands can selectively target these pathways independently, engaging specific subsets of the normal signalling repertoire of the receptor, particularly the β-arrestin pathway. However, the precise effects of such selectivity are yet to be fully understood [16].

TRV027 is a β-arrestin-biased AT1R ligand that competitively inhibits G-protein-dependent signalling while promoting β-arrestin recruitment. In rats, it has been shown to reduce arterial pressure similarly to an ARB, but unlike an ARB, it also enhances cardiac contractility and maintains stroke volume [17]. In both in vitro and in vivo settings, TRV027 has been demonstrated to have antiapoptotic and inotropic effects in wild-type (WT) mice but not in β-arrestin knock-out mice [17]. In a dog model of acute heart failure, TRV027 has acted as a potent and balanced vasodilator, improving cardiac output and offering renal benefits [18]. In a $ApoE^{-/-}$ mouse model in which aortic aneurysms were induced by angiotensin II type 1 (Ang II), the co-infusion of Ang II with TRV027 prevented the formation of these aneurysms. TRV027 halted some detrimental effects such as aortic dilation, asymmetric wall thickening, inflammation, vascular fibrosis, elastolysis, and the seepage of blood [19]. These findings suggest that TRV027 could offer therapeutic benefits by blocking G-protein-mediated vasoconstriction and other damaging processes while simultaneously promoting cardio- and reno-protective β-arrestin-mediated signalling [18–20].

Although the above-mentioned studies have shown that β-arrestin-biased stimulation may result in positive cardiovascular effects, the literature also suggests positive effects following its deletion. A murine MFS model with genetic β-arrestin (βarr2) deletion, i.e., $Fbn1^{C1041G/+}$/βarr2$^{-/-}$ mice, exhibited delayed aortic root dilation compared to $Fbn1^{C1041G/+}$ mice. Moreover, the aortas of $Fbn1^{C1041G/+}$/*βarr2*$^{-/-}$ mice showed reduced mRNA and protein expression of key mediators involved in TAA formation, including matrix metalloproteinase (MMP)-2 and -9, along with decreased activation of ERK1/2, compared to $Fbn1^{C1041G/+}$ mice [21]. Using primary aortic root smooth muscle cells where β-arrestin was targeted through small interfering RNAs, it was demonstrated that the induction of MMP-2 and -9 expression by Ang II relies on βarr2. This pathway involves the activation of ERK1/2 and transactivation of epidermal growth factor receptor (EGFR). These results suggest that β-arrestin may play a non-canonical role in TAA formation in MFS by regulating the ERK1/2-dependent expression of pro-aneurysmal genes and proteins downstream of the AT1R [21]. β-arrestin deficiency has also been shown to attenuate abdominal aortic aneurysm (AAA) in an $ApoE^{-/-}$ mouse model in which AAA was induced by Ang II infusion [22].

Another emerging important player in TAA development is vascular inflammation [23–25]. Tieu et al. showed that Ang II infusion in C57BL/6J mice led to the production of monocyte chemoattractant protein 1 (MCP-1). The activation of MCP-1-mediated pathways resulted in the progression of aneurysm growth and dissection following Ang II infusion, while MCP-1 knock-out mice showed delayed aneurysmal growth [26]. Previous studies have also suggested an interplay between the AT1R and the receptor for MCP-1, known as a C-C chemokine receptor type 2 (CCR2). Functional heteromers of the AT1R and CCR2 resulted in the CCR2–G protein coupling, sensitive to AT1R activation, as well as to apparent enhanced β-arrestin recruitment with agonist co-stimulation. Moreover, in a rat model of chronic renal disease, it was observed that combined treatment with AT1R and CCR2 selective inhibitors was synergistically beneficial [27].

Thus, the aim of this study is to explore alternative strategies for targeting AT1R-mediated signalling that could be beneficial in MFS. To achieve this, we first investigate the effect of targeting β-arrestin downstream signalling by administering TRV027, which promotes β-arrestin recruitment. Secondly, we evaluate the combined effects of the biased

AT1R ligands barbadin and DMX200 with low doses of the ARB losartan as a potential strategy to enhance effectiveness while potentially minimizing side effects (Figure 1).

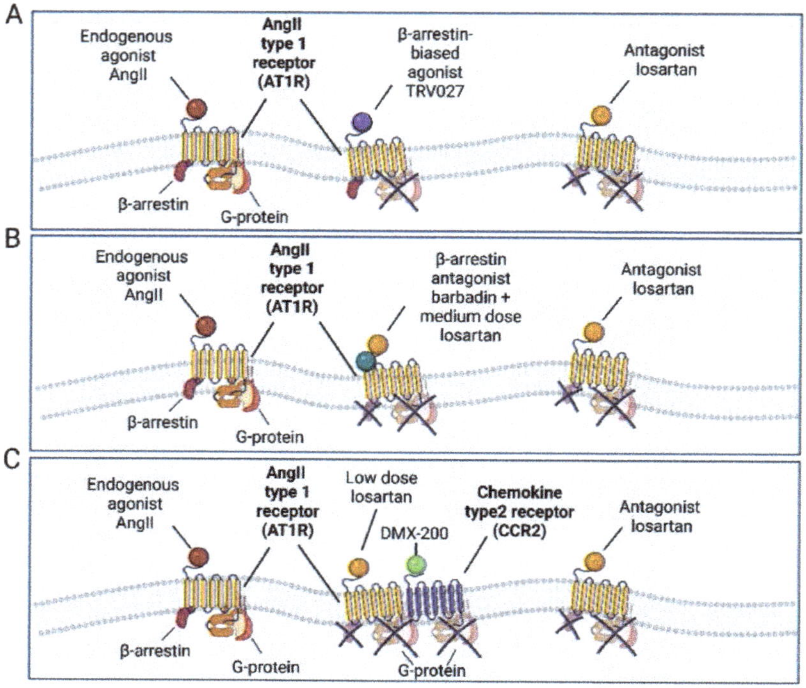


**Figure 1.** Mechanism of action of the tested compounds. (**A**) TRV027 functions as a β-arrestin-biased ligand, competitively inhibiting G-protein-dependent signalling while promoting β-arrestin recruitment. (**B**) Barbadin selectively blocks β-arrestin. The combination of barbadin with losartan, a AngII antagonist, may result in the potentiated inhibition of AT1R. (**C**) DMX-200 acts as a CCR2 inhibitor. CCR2 couples with AT1R, triggering inflammatory cascades. By combining losartan with DMX-200, AT1R blocking may be further potentiated by also inhibiting subsequent inflammatory pathways. Image created with BioRender.com.

## 2. Results

### 2.1. TRV027 Infusion Does Not Show Any Beneficial Effect on Ascending Aorta and Aortic Root Dilation in a MFS Mouse Model

Based on its previously reported positive effect on blood pressure and aortic distensibility [17–19], we hypothesised that TRV027 infusion in MFS mice may have a positive effect in preventing aortic dilation. Small osmotic pumps were subcutaneously implanted in male littermate MFS (*Fbn1*$^{C1041G/+}$) and WT mice, achieving a TRV027 infusion rate of 10 µg/kg/min. A control group was also included, in which mice were infused with saline solution. At 4 weeks of age, just before the start of the treatment, MFS mice already showed dilation of the aortic root compared to WT (Figure 2), while no dilation of the ascending aorta was detected (Figure S1). At 8 and 12 weeks of age, after 4 and 8 weeks of treatment, respectively, no significant differences were observed on aortic root and ascending aorta diameters between MFS and WT mice infused with TRV027 and saline solution. MFS mice from both groups showed significant dilation of the aortic root at 8 and 12 weeks compared to their WT counterparts (at 8 weeks, $p < 0.0001$ for TRV027-treated group and $p = 0.0016$ for the saline control group; at 12 weeks, $p < 0.0001$ for both TRV027 and the control group) (Figure 2). These results suggest that TRV027 treatment has no impact on the aortic root or the ascending aorta enlargement of MFS or WT mice.

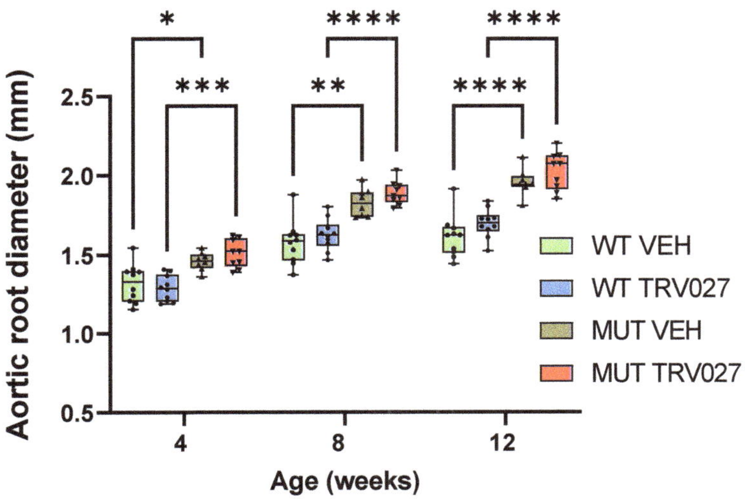


**Figure 2.** TRV027 does not have any effect on the aortic root diameter of MFS mice. Echocardiography data at 12 weeks show that the aortic root diameter of MFS mice treated with TRV027 during a period of 8 weeks are not significantly different to those treated with vehicle solution. MFS mice from both groups (vehicle and TRV027-treated mice) show a significant dilation before the start of the treatment (4 weeks), during the treatment (8 weeks), and after the end of the treatment (12 weeks). Data are represented as boxplots. Statistical test analysis: One-way ANOVA and Tukey's post-test per timepoint. **** $p \leq 0.0001$, *** $p \leq 0.001$, ** $p \leq 0.01$ and * $p \leq 0.05$. VEH: vehicle; WT: wild type; MUT: mutant.

### 2.2. *Combined Treatment of Barbadin and Losartan Does Not Show a Significant Effect on Ascending Aorta and Aortic Root Diameter Compared to Losartan Alone*

Based on the previous observation that the knock-out of β-arrestin-2 improved aortic dilation in the MFS *Fbn1*$^{C1041G/+}$ mouse model, we hypothesised that combined treatment with the biased β-arrestin blocker barbadin, alongside a lower dose of losartan, may have similar effects on the prevention of aortic dilation to a high dose of losartan by itself. This idea raises the possibility of achieving superior efficacy without facing side effects associated with high doses of losartan in humans. To test the hypothesis, we gave a combined treatment of both losartan (25 mg/kg/day) and barbadin (0.9 mg/kg/day) to WT and MFS mice and compared them with MFS and WT treated with losartan alone (25 mg/kg/day). The treatment started at 4 weeks of age and continued until 12 weeks of age. Our results showed that aortic root diameters were already increased in MFS mice from 4 weeks of age, just before the start of the treatment (Figures 3 and S7). When comparing both treated groups, combined treatment with both losartan (25 mg/kg/day) and barbadin (0.9 mg/kg/day) did not show any significant improvement in the aortic root dilation over time compared to using losartan (25 mg/kg/day) alone (Figures 3 and S7). Both treated MFS mice groups showed significant dilation compared to their WT counterparts at 8 and 12 weeks (at 8 weeks, $p < 0.0001$ for barbadin- and losartan-treated mice and $p = 0.061$ for losartan-treated mice; at 12 weeks, $p = 0.0087$ for barbadin- and losartan-treated mice and $p < 0.0001$ for losartan-treated mice) (Figures 3 and S7). These findings suggest that an additive treatment with barbadin does not exert any effect on the aortic phenotype of MFS mice. It is noteworthy that this experiment does not include a group treated with barbadin alone. Although we observed no synergistic effect from the combined treatment of barbadin and losartan, it cannot be concluded what the impact of barbadin alone is on aortic dilatation in MFS mice. Additional interactions with losartan could potentially influence the effect of barbadin when administered together. A subtle decrease in aortic diameter in both treated WT and MFS mice was observed at

8 and 12 weeks compared with their respective genotypes in the untreated group (Figure 3). However, these differences did not reach statistical significance in either case. Since the same decrease was observed in both treated groups, this suggests that losartan alone may be responsible for this effect. This finding suggests that while losartan at a dose of 25 mg/kg/day has some beneficial effect on aortic growth in both MFS and WT mice, it is not enough to prevent dilation in MFS mice. As for the ascending aorta measurements, there was no evident dilation in MFS mice at the ages when diameters were measured (4, 8, and 12 weeks of age), and no distinction was observed between the two treated groups (Figure S3). Additionally, there were no differences in ascending aortic diameter between treated and untreated mice (Figure S3).

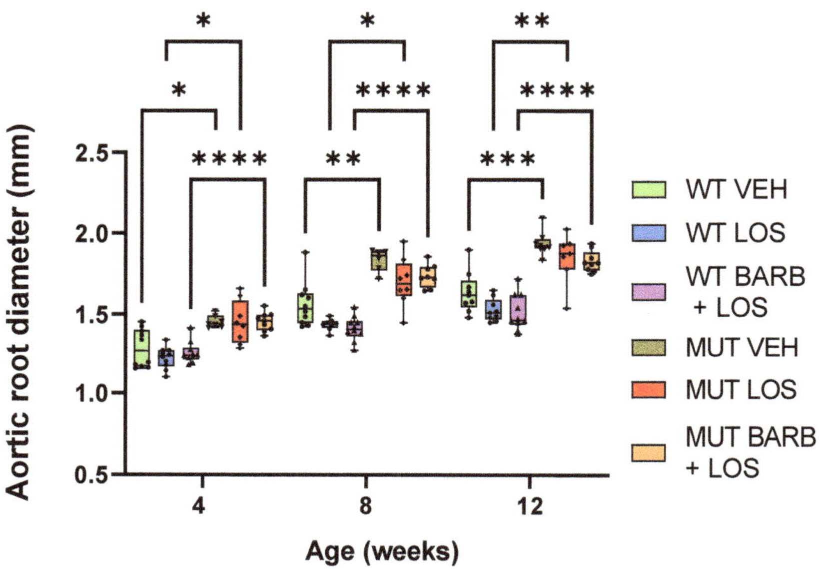


**Figure 3.** Combined treatment with barbadin and losartan does not show any significant effect on the aortic root diameter of MFS mice compared to losartan alone. Echocardiography data at 12 weeks show that aortic root diameters of MFS mice treated during a period of 8 weeks with barbadin and losartan are not significantly different to those treated with losartan alone. MFS mice from both groups (barbadin and losartan and losartan alone) show a significant dilation compared to the WT controls before the start of the treatment (4 weeks), during the treatment (8 weeks), and after the end of the treatment (12 weeks). A slight decrease in diameter is observed at 8 and 12 weeks for MFS and WT mice of both treated groups compared with the vehicle, but this difference lacks statistical significance. Data are represented as boxplots. Statistical test analysis: One-way ANOVA and Tukey's post-test per timepoint. **** $p \leq 0.0001$, *** $p \leq 0.001$, ** $p \leq 0.01$ and * $p \leq 0.05$. VEH: vehicle; LOS: losartan; BARB: barbadin; WT: wild type; MUT: mutant.

*2.3. Combined Treatment of DMX-200 and Low Dose of Losartan Does Not Show Any Significant Effect on the Ascending Aorta and Aortic Root Diameter of MFS Mice Compared to Both Treated and Untreated WT Groups*

Based on prior studies in the context of chronic kidney disease, we hypothesised that co-inhibition of AT1R and CCR2 may result in a decrease in β-arrestin recruitment but also highly affect CCR2–G protein signalling, possibly through allosteric modulation [27]. Another study has assessed the combinatory effect of DMX200 with an ARB in the context of COVID-19 to potentially achieve a synergistic anti-inflammatory effect [28]. We propose that combinatory blockage of AT1R and CCR2 may result in selective β-arrestin inhibition while also preventing inflammatory cascades, having to a positive effect on aneurysm formation. From these observations, we aimed to find if the combination of CCR2blocker DMX-200 with a low dose of losartan could produce comparable outcomes in averting aortic dilation as a high dose of losartan alone. Hence, WT and MFS mice received a combined treatment of a low dose of losartan (5 mg/kg/day) and DMX-200 (90 mg/kg/day) and were compared with a group treated with high dose of losartan alone (50 mg/kg/day). As shown in Figure 4, we observed that, at both 8 and 12 weeks of age, the MFS group treated with high dose of losartan showed no statistically significant difference with the treated WT group. This suggests that a high dose of losartan can prevent aneurysm development in our murine MFS model. Moreover, a reduction in aortic diameter was noted in the WT group upon treatment with a high dose of losartan at 8 and 12 weeks, as opposed to its untreated counterpart. This can also be observed when considering the growth rate of the aortic root diameter, with treatment with a high dose of losartan resulting in a lower growth rate in both WT and MFS mice compared to both untreated groups (Figure 5). On the other hand, the MFS group treated with low dose of losartan and DMX-200 showed a statistically significant difference ($p < 0.0209$) in aortic diameter when compared to the WT counterpart at 12 weeks. However, when compared to the MFS group treated with high dose of losartan, it was observed that the difference was not statistically significant. This might also suggest a trend in decreased aortic dilation in the group treated with losartan and DMX-200. It is important to note that DMX200 was not tested alone, and the effect of the low dose of losartan (5 mg/kg/day) is also unknown. We can conclude that combination of these two compounds at the tested doses does not exhibit any significant positive effect on aortic dilatation in MFS mice. However, we cannot determine the individual effects of DMX200 and the low dose of losartan, as a potential interaction between the two could be influencing the outcome when administered together.

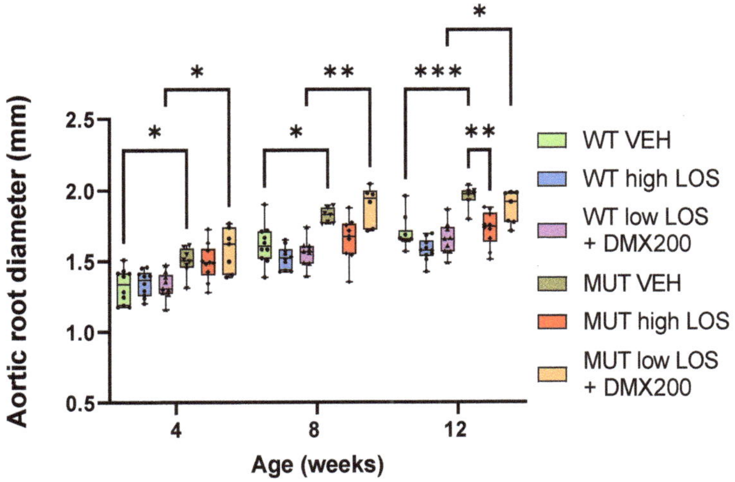


**Figure 4.** Combined treatment of DMX-200 and low dose of losartan does not show any significant effect on the aortic root diameter of MFS mice compared to both treated and untreated WT groups. Echocardiography data at 12 weeks show that aortic root diameters of MFS mice treated during a period of 8 weeks with DMX-200 and low dose of losartan are significantly different to those in the WT group receiving the same treatment. On the other hand, MFS mice treated with high dose of losartan alone show no difference to the treated WT group during the whole treatment period (8 weeks in total). Similarly, no significant difference is observed between the MFS group treated with high dose of losartan and vehicle WT group at 12 weeks. A slight decrease in aortic diameter is also observable in WT group treated with high dose of losartan when compared to WT vehicle group. Data are represented as boxplots. Statistical test analysis: One-way ANOVA and Tukey post-test per timepoint *** $p \leq 0.001$, ** $p \leq 0.01$ and * $p \leq 0.05$. VEH: vehicle; LOS: losartan; WT: wild type; MUT: mutant.

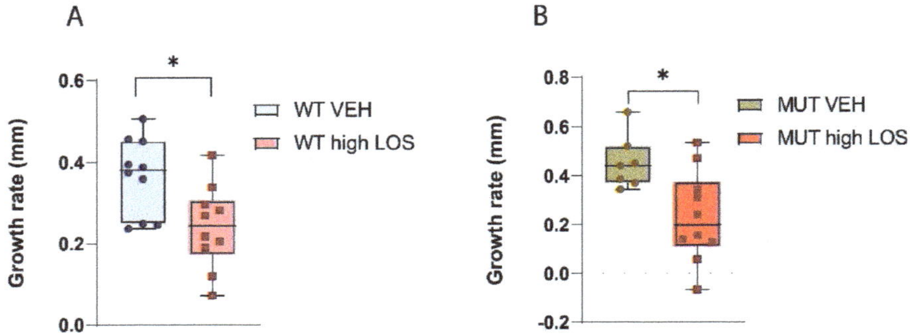


**Figure 5.** Growth rate of mice upon no treatment and high dose of losartan. The growth rate is represented by the difference in the aortic root diameter between 4 and 12 weeks of age. A decrease in the aortic root diameter of WT (**A**) and MFS (**B**) mice is observed upon treatment with a high dose of losartan when compared to the untreated group. Statistical test analysis: Mann–Whitney U-Test. * $p \leq 0.05$. LOS: losartan; VEH: vehicle; WT: wild type; MUT: mutant..

## 3. Discussion

Despite very promising results in an MFS mouse model, human clinical studies pursuing AT1R blockade achieved less convincing results [13]. A meta-analysis of all ARB/β-blocker human studies showed that both ARBs and β-blockers slow down aortic root dilation to some extent but do not provide full rescue of the phenotype as in mice [10]. One of the possible explanations is that the doses of ARBs in the mouse trials were very high and, thus, might not have been tolerable in humans. AT1R activation triggers two different signalling components: G protein cascade, which promotes vasoconstriction, and β-arrestin recruitment, which leads to enhanced cardiomyocyte contractility and cell survival [29]. Interestingly, recent findings suggest that biased ligands can selectively target these pathways independently, engaging specific subsets of the normal signalling repertoire of the receptor, particularly the β-arrestin pathway. However, the precise effects of such selectivity remain not fully understood [16]. As such, we tested different pharmaceutical strategies aiming at the modulation of the AT1R/β-arrestin signalling pathway in the *Fbn1*$^{C1041G/+}$ MFS mouse model. Existing data on the role of β-arrestin have been ambiguous. In this study, we explored if biased inhibition or stimulation plays a positive, negative, or neutral role in the AT1R effects. Furthermore, it has been shown that the MCP-1 -CCR2 pathway plays a role in aneurysm development [26]. Hence, this study also aimed to investigate if the combined inhibition of the AT1R and CCR2 may represent a therapeutic avenue for aortic aneurysm development.

TRV027 has shown cardio- and reno-protective and antiapoptotic effects in different animal models [17–19]. At a dose of 10 μg/kg/day, TRV027 exhibits potent G protein inhibition with β-arrestin recruitment, resulting in lowered arterial pressure and increased cardiac contractility in rats [17]. Despite these promising outcomes, our findings indicate that TRV027 infusion at a rate of 10 μg/kg/day does not influence the aortic diameter in the MFS model *Fbn1*$^{C1041G/+}$. This could suggest that solely inhibiting G-protein-dependent signalling is not sufficient to prevent TAA in MFS. The potential counteraction of any G protein blocking effect by β-arrestin recruitment, promoted by TRV027, could also contribute to this lack of effect on aortic root diameter in the MFS model. Additionally, the dose used in this study, while effective in producing cardiac-related benefits in rats, might not be sufficient to exert any observable effect on aortic dilation in MFS mice. Increasing the concentration of TRV027, however, presents challenges due to the limited solubility of TRV027 in water (300 mg/mL). Furthermore, other limitations include insufficient pharmacological data on the efficacy and stability of TRV027 administration via infusion with minipumps.

Our findings also indicate that barbadin, a known selective inhibitor of β-arrestin, did not confer any benefits in mitigating aortic aneurysm progression in *Fbn1*$^{C1041G/+}$ mice. The anticipated positive outcome was based on previous studies demonstrating delayed aortic root dilation in *Fbn1*$^{C1041G/+}$/*βarr2*$^{-/-}$ mice, suggesting that inhibiting β-arrestin with a compound might result in a similar effect. Research about the effects of different concentrations of barbadin is limited but has shown that a dosage of 0.3 mg/kg/day is able to potentiate the effects of the weight-loss drug lorcaserin in male mice [30]. Nevertheless, our study found that, even with a higher dose, there is no impact on the aortic diameter of MFS mice. The observed lack of efficacy could potentially be attributed to the need for a more potent β-arrestin blockade, considering that gene knock-out represents a more robust intervention compared to receptor blocking using a compound. Higher doses of barbadin might be required to exert a therapeutic effect. However, such doses may deviate from clinically relevant levels (no data are currently available), raising concerns about the translatability of these findings to human patients.

Losartan has already been reported to prevent aortic dilation in the MFS model *Fbn1*$^{C1041G/+}$ [12], and with our study, we confirmed the efficacy of a high dose of losartan (50 mg/kg/day). AT1R and CCR2 were reported to functionally interact, and combined treatment with the CCR2 inhibitor DMX-200 (30 mg/kg/day) and irbesartan, another AT1R antagonist, synergistically decreased β-arrestin recruitment and inflammatory cascades in chronic kidney disease [27]. However, in our study, the combination of a lower dose of losartan and DMX-200 (90 mg/kg/day) did not show any effect in preventing or arresting aortic dilation, and a further increase in the DMX-200 doses would not be translatable to humans [31,32].

This study has some limitations. While we investigated the combined treatment of barbadin and DMX200 with losartan, we did not evaluate their individual effects. Consequently, definitive conclusions regarding the impact of barbadin and DMX200 alone on aortic aneurysm in MFS mice cannot be drawn. However, the current purpose of our study was to show an additive positive effect of these compounds on top of lower losartan doses, as this would have clinical relevance. We addressed this question and observed no significant synergistic effect, at least with the doses tested in this study.

In conclusion, we observed that none of the tested compounds targeting downstream effectors of AT1R signalling exerted a significant effect on the aneurysms in the MFS mouse model *Fbn1*$^{C1041G/+}$. In contrast, a high dose of losartan (50 mg/kg/day), which blocks the AT1R, was able to prevent the development of aortic dilation. This finding could indicate that full AT1R blockade might be required for the inhibition of aneurysm progression in MFS. Such inhibition could be obtained via an additive effect of another compound on top of a lower losartan dose. [10]. Alternatively, the observed effects of the high dose of losartan on aortic diameter may be independent of the AT1R, indicating alternative mechanisms. Indeed, a potential additive effect of treatment with losartan and β-blockers, which primarily target blood pressure and are not AT1R-related, has been suggested [10]. However, due to insufficient statistical power in the studies, conclusive evidence has not been established. Alternatively, another strategy could involve more site-directed losartan-mediated AT1R blockade based on nanoparticles designed to target the aneurysmal aortic area. The latter could potentially provide a more localised and stronger effect of locally delivered losartan. This approach could lead to better outcomes without the accompanying side effects of using high doses.

## 4. Materials and Methods

### 4.1. Mice

*Fbn1*$^{C1041G/+}$ and WT mice on a C57Bl6J background were used in this study. The total number of mice used per treatment group and genotype was N = 10. However, for plot representation, the number of mice included decreased to N = 7 due to animal loss from non-cardiovascular-related causes or the insufficient quality of the echocardiography imaging data. *Fbn1*$^{C1041G/+}$ and WT mice were housed together up to a maximum of

8 mice per cage. All mice had ad libitum access to food and water. Mice were treated for 8 weeks, starting at 4 weeks of age. After the end of the treatment, mice were euthanised via $CO_2$ inhalation.

### 4.2. TRV027 Treatment

TRV027 (Trevena Inc., Chesterbrook, PA, USA) was infused on *Fbn1*$^{C1041G/+}$ and WT mice using micro-osmotic pumps (Alzet model 1004, Durect corporation, Cupertino, CA USA) at an infusion rate of 10 µg/kg/min. The pumps were implanted at 4 weeks of age and only changed once, at 8 weeks of age. In the control group, osmotic pumps were implanted with saline solution following the same procedures.

### 4.3. Barbadin and Losartan (25 mg/kg/day) Treatment

*Fbn1*$^{C1041G/+}$ and WT received daily intraperitoneal injections with either barbadin (0.9 mg/kg/day) or DMSO solution (control). Barbadin (2774, Axon Medchem B.V., Groningen, The Netherlands) was first dissolved in DMSO (15408099, Fisher Scientific B.V., Brussel Belgium) at a concentration of 3.33 mg/mL and subsequently diluted in PBS-Tween80 (11590476, Fisher Scientific B.V., Brussel, Belgium and P1754, Merck, Hoeilaart, Belgium) at a dose of 0.9 mg/kg/day, accounting for the weight of each animal. The same procedure was applied for the control group, but DMSO solution was used instead of barbadin. The injections started at 4 weeks and continued until 12 weeks of age. All mice were additionally treated with losartan at a dose of 25 mg/kg/day. Losartan was dissolved in drinking water at a concentration of 0.3g/L, and drinking bottles were replaced once per week.

### 4.4. DMX-200 and Losartan (5 mg/kg/day and 50 mg/kg/day) Treatment

*Fbn1*$^{C1041G/+}$ and WT were treated with a high dose of losartan (50 mg/kg/day) or low dose of losartan (5 mg/kg/day) and DMX-200 (396265, Merck, Hoeilaart, Belgium) (90 mg/kg/day) in drinking water. Losartan was dissolved in drinking water at a concentration of 0.6 g/L to obtain the high dose of losartan and 0.06 g/L for the low dose of losartan. DMX-200 was dissolved at a concentration of 0.36 g/L. The solutions were changed once per week.

### 4.5. Transthoracic Echocardiography

To visualise the aorta, the hairs of unsedated mice were removed with "Veet sensitive skin" cream. Next, mice were weighted (Figures S4–S6) and subjected to echocardiography using a VisualSonics Vevo 2100 imaging system (FUJIFILM VisualSonics, Inc., Toronto, Canada) and a 30 MHz transducer. Both aortic root and ascending aorta (Figures S1–S3) were imaged in B-mode and PLAX view. Measurements of the aortic root were taken in the sinus of Valsalva, while measurements of the ascending aorta were taken in the mid-ascending aorta. Three independent measurements from the maximal internal aortic dimensions were averaged. All data acquisition steps and measurements were performed while blind to genotype and treatment.

### 4.6. Statistics

Data are represented in interquartile range (IQR) boxplots, in which the error bars represent minimum and maximum values, the horizontal bars indicate median values, and the extremities of the boxes indicate interquartile ranges. Comparison of the aortic diameters between multiple groups within each timepoint was performed with a one-way ANOVA analysis, followed by a post hoc analysis with Tukey's multiple comparisons test (Tables S1–S4). Differences in aortic growth between the two groups were tested with a Mann–Whitney U-Test. A value of $p \leq 0.05$ was considered statistically significant. Data analysis and plotting were carried out using GraphPad Prism software version 9.0.0.

**Supplementary Materials:** The following supporting information can be downloaded via this link: https://www.mdpi.com/article/10.3390/ijms25095025/s1.

**Author Contributions:** Conceptualisation, B.L., R.R.J.V.K. and A.V.; methodology, J.M., I.L. and S.P.; formal analysis, I.V.C., L.B. and J.M.; investigation, I.V.C., L.B., J.B. and J.S.; writing—original draft preparation, I.V.C. and L.B.; writing—review and editing, B.L., A.V., J.M., I.L., S.P., A.H.J.D. and R.R.J.V.K.; visualisation, I.V.C. and L.B.; supervision, B.L., R.R.J.V.K., A.V. and J.M.; project administration, B.L., J.M. and I.L.; funding acquisition, B.L., A.H.J.D. and R.R.J.V.K. All authors have read and agreed to the published version of the manuscript.

**Funding:** The research was funded by Health Holland–Stichting LSH-TKI, grant number LSHM22018. Irene Valdivia Callejon and Lucia Buccioli are PhD fellows supported by FWO (Research Foundation Flanders, 11I2423N and 11J3123N). Josephina Meester and Silke Peeters are postdoctoral fellows supported by FWO; 12AO124N and 12X5422N), and Dr. Loeys holds a consolidator grant from the European Research Council (Genomia–ERC-COG-2017-771945). Dr. Loeys and Dr. Verstraeten are members of the European Reference Network on rare multisystemic vascular disorders (VASCERN–project ID: 769036 partly co-funded by the European Union Third Health Programme).

**Institutional Review Board Statement:** All mouse breedings and experiments were conducted in accordance with the ethical guidelines set by the Ethical Committee of Animal Testing of the University of Antwerp and received formal ethical clearance in August 2022 (approval code 2022-50).

**Data Availability Statement:** The original contributions presented in this study are included in the article/Supplementary Material; further inquiries can be directed to the corresponding author. The raw data supporting the conclusions of this article will be made available by the authors on request.

**Acknowledgments:** TRV027 was kindly donated by Trevena.

**Conflicts of Interest:** The authors declare no conflicts of interest. The funders had no role in the design of the study; the collection, analyses, or interpretation of data; the writing of the manuscript; or the decision to publish the results.

## References

1. Dietz, H.C.; Cutting, C.R.; Pyeritz, R.E.; Maslen, C.L.; Sakai, L.Y.; Corson, G.M.; Puffenberger, E.G.; Hamosh, A.; Nanthakumar, E.J.; Curristin, S.M.; et al. Marfan syndrome caused by a recurrent de novo missense mutation in the fibrillin gene. *Nature* **1991**, *352*, 337–339. [CrossRef]
2. Arnaud, P.; Hanna, N.; Aubart, M.; Leheup, B.; Dupuis-Girod, S.; Naudion, S.; Lacombe, D.; Milleron, O.; Odent, S.; Faivre, L.; et al. Homozygous and compound heterozygous mutations in the FBN1 gene: Unexpected findings in molecular diagnosis of Marfan syndrome. *J. Med. Genet.* **2017**, *54*, 100–103. [CrossRef] [PubMed]
3. Meester, J.A.N.; Verstraeten, A.; Schepers, D.; Alaerts, M.; Van Laer, L.; Loeys, B.L. Differences in manifestations of Marfan syndrome, Ehlers-Danlos syndrome, and Loeys-Dietz syndrome. *Ann. Cardiothorac. Surg.* **2017**, *6*, 582–594. [CrossRef]
4. Habashi, J.P.; Doyle, J.J.; Holm, T.M.; Aziz, H.; Schoenhoff, F.; Bedja, D.; Chen, Y.; Modiri, A.N.; Judge, D.P.; Dietz, H.C. Angiotensin II type 2 receptor signaling attenuates aortic aneurysm in mice through ERK antagonism. *Science* **2011**, *332*, 361–365. [CrossRef] [PubMed]
5. Crosas-Molist, E.; Meirelles, T.; López-Luque, J.; Serra-Peinado, C.; Selva, J.; Caja, L.; Del Blanco, D.G.; Uriarte, J.J.; Bertran, E.; Mendizábal, Y.; et al. Vascular smooth muscle cell phenotypic changes in patients with Marfan syndrome. *Arter. Thromb. Vasc. Biol.* **2015**, *35*, 960–972. [CrossRef] [PubMed]
6. Sakai, L.Y.; Keene, D.R.; Renard, M.; De Backer, J. FBN1: The disease-causing gene for Marfan syndrome and other genetic disorders. *Gene* **2016**, *591*, 279–291. [CrossRef]
7. Yang, X.; Xu, C.; Yao, F.; Ding, Q.; Liu, H.; Luo, C.; Wang, D.; Huang, J.; Li, Z.; Shen, Y.; et al. Targeting endothelial tight junctions to predict and protect thoracic aortic aneurysm and dissection. *Eur. Heart J.* **2023**, *44*, 1248–1261. [CrossRef]
8. Harky, A.; Fan, K.S.; Fan, K.H. The genetics and biomechanics of thoracic aortic diseases. *Vasc. Biol.* **2019**, *1*, R13–R25. [CrossRef]
9. Gersony, D.R.; McClaughlin, M.A.; Jin, Z.; Gersony, W.M. The effect of beta-blocker therapy on clinical outcome in patients with Marfan's syndrome: A meta-analysis. *Int. J. Cardiol.* **2007**, *114*, 303–308. [CrossRef]
10. Pitcher, A.; Spata, E.; Emberson, J.; Davies, K.; Halls, H.; Holland, L.; Wilson, K.; Reith, C.; Child, A.H.; Clayton, T.; et al. Angiotensin receptor blockers and β blockers in Marfan syndrome: An individual patient data meta-analysis of randomised trials. *Lancet* **2022**, *400*, 822–831. [CrossRef]
11. Houlihan, C.A.; Akdeniz, A.; Tsalamandris, C.; Cooper, M.E.; Jerums, G.; Gilbert, R.E. Urinary transforming growth factor-beta excretion in patients with hypertension, type 2 diabetes, and elevated albumin excretion rate: Effects of angiotensin receptor blockade and sodium restriction. *Diabetes Care* **2002**, *25*, 1072–1077. [CrossRef] [PubMed]
12. Habashi, J.P.; Judge, D.P.; Holm, T.M.; Cohn, R.D.; Loeys, B.L.; Cooper, T.K.; Myers, L.; Klein, E.C.; Liu, G.; Calvi, C.; et al. Losartan, an AT1 antagonist, prevents aortic aneurysm in a mouse model of Marfan syndrome. *Science* **2006**, *312*, 117–121. [CrossRef]

13. Groenink, M.; Hartog, A.W.D.; Franken, R.; Radonic, T.; de Waard, V.; Timmermans, J.; Scholte, A.J.; Berg, M.P.v.D.; Spijkerboer, A.M.; Marquering, H.A.; et al. Losartan reduces aortic dilatation rate in adults with Marfan syndrome: A randomized controlled trial. *Eur. Heart J.* **2013**, *34*, 3491–3500. [CrossRef] [PubMed]
14. Brooke, B.S.; Habashi, J.P.; Judge, D.P.; Patel, N.; Loeys, B.; Dietz, H.C., III. Angiotensin II blockade and aortic-root dilation in Marfan's syndrome. *N. Engl. J. Med.* **2008**, *358*, 2787–2795. [CrossRef] [PubMed]
15. Ma, Z.; Viswanathan, G.; Sellig, M.; Jassal, C.; Choi, I.; Garikipati, A.; Xiong, X.; Nazo, N.; Rajagopal, S. β-Arrestin-Mediated Angiotensin II Type 1 Receptor Activation Promotes Pulmonary Vascular Remodeling in Pulmonary Hypertension. *JACC Basic. Transl. Sci.* **2021**, *6*, 854–869. [CrossRef]
16. Van Gastel, J.; Hendrickx, J.O.; Leysen, H.; Santos-Otte, P.; Luttrell, L.M.; Martin, B.; Maudsley, S. β-Arrestin Based Receptor Signaling Paradigms: Potential Therapeutic Targets for Complex Age-Related Disorders. *Front. Pharmacol.* **2018**, *9*, 1369. [CrossRef]
17. Violin, J.D.; DeWire, S.M.; Yamashita, D.; Rominger, D.H.; Nguyen, L.; Schiller, K.; Whalen, E.J.; Gowen, M.; Lark, M.W. Selectively engaging β-arrestins at the angiotensin II type 1 receptor reduces blood pressure and increases cardiac performance. *J. Pharmacol. Exp. Ther.* **2010**, *335*, 572–579. [CrossRef]
18. Boerrigter, G.; Whalen, E.J.; Lark, M.; Burnett, J.C. Cardiorenal actions of TRV120027, a novel ß-arrestin-biased ligand at the angiotensin II type I receptor, in healthy and heart failure canines: A novel therapeutic strategy for acute heart failure. *Circ. Heart Fail.* **2011**, *4*, 770–778. [CrossRef]
19. Jara, Z.P.; Harford, T.J.; Singh, K.D.; Desnoyer, R.; Kumar, A.; Srinivasan, D.; Karnik, S.S. Distinct Mechanisms of β-Arrestin-Biased Agonist and Blocker of AT1R in Preventing Aortic Aneurysm and Associated Mortality. *Hypertension* **2023**, *80*, 385–402. [CrossRef]
20. Soergel, D.G.; Subach, R.A.; Cowan, C.L.; Violin, J.D.; Lark, M.W. First clinical experience with TRV027: Pharmacokinetics and pharmacodynamics in healthy volunteers. *J. Clin. Pharmacol.* **2013**, *53*, 892–899. [CrossRef]
21. Wisler, J.W.; Harris, E.M.; Raisch, M.; Mao, L.; Kim, J.; Rockman, H.A.; Lefkowitz, R.J. The role of β-arrestin2-dependent signaling in thoracic aortic aneurysm formation in a murine model of Marfan syndrome. *Am. J. Physiol. Heart Circ. Physiol.* **2015**, *309*, H1516–H1527. [CrossRef]
22. Trivedi, D.B.; Loftin, C.D.; Clark, J.; Myers, P.; DeGraff, L.M.; Cheng, J.; Zeldin, D.C.; Langenbach, R. β-Arrestin-2 deficiency attenuates abdominal aortic aneurysm formation in mice. *Circ. Res.* **2013**, *112*, 1219–1229. [CrossRef]
23. Liu, X.; Chen, W.; Zhu, G.; Yang, H.; Li, W.; Luo, M.; Shu, C.; Zhou, Z. Single-cell RNA sequencing identifies an Il1rn(+)/Trem1(+) macrophage subpopulation as a cellular target for mitigating the progression of thoracic aortic aneurysm and dissection. *Cell Discov.* **2022**, *8*, 11. [CrossRef]
24. Li, Y.; Ren, P.; Dawson, A.; Vasquez, H.G.; Ageedi, W.; Zhang, C.; Luo, W.; Chen, R.; Li, Y.; Kim, S.; et al. Single-Cell Transcriptome Analysis Reveals Dynamic Cell Populations and Differential Gene Expression Patterns in Control and Aneurysmal Human Aortic Tissue. *Circulation* **2020**, *142*, 1374–1388. [CrossRef]
25. Hirakata, S.; Aoki, H.; Ohno-Urabe, S.; Nishihara, M.; Furusho, A.; Nishida, N.; Ito, S.; Hayashi, M.; Yasukawa, H.; Imaizumi, T.; et al. Genetic Deletion of Socs3 in Smooth Muscle Cells Ameliorates Aortic Dissection in Mice. *JACC Basic. Transl. Sci.* **2020**, *5*, 126–144. [CrossRef]
26. Tieu, B.C.; Lee, C.; Sun, H.; LeJeune, W.; Recinos, A.; Ju, X.; Spratt, H.; Guo, D.C.; Milewicz, D.; Tilton, R.G.; et al. An adventitial IL-6/MCP1 amplification loop accelerates macrophage-mediated vascular inflammation leading to aortic dissection in mice. *J. Clin. Investig.* **2009**, *119*, 3637–3651. [CrossRef]
27. Ayoub, M.A.; Zhang, Y.; Kelly, R.S.; See, H.B.; Johnstone, E.K.M.; McCall, E.A.; Williams, J.H.; Kelly, D.J.; Pfleger, K.D.G. Functional interaction between angiotensin II receptor type 1 and chemokine (C-C motif) receptor 2 with implications for chronic kidney disease. *PLoS ONE* **2015**, *10*, e0119803. [CrossRef]
28. Lawler, P.R.; Derde, L.P.G.; van de Veerdonk, F.L.; McVerry, B.J.; Huang, D.T.; Berry, L.R.; Lorenzi, E.; van Kimmenade, R.; Gommans, F.; Vaduganathan, M.; et al. Effect of Angiotensin-Converting Enzyme Inhibitor and Angiotensin Receptor Blocker Initiation on Organ Support-Free Days in Patients Hospitalized With COVID-19: A Randomized Clinical Trial. *Jama* **2023**, *329*, 1183–1196.
29. Van Gucht, I.; Meester, J.A.; Bento, J.R.; Bastiaansen, M.; Bastianen, J.; Luyckx, I.; Van Den Heuvel, L.; Neutel, C.H.; Guns, P.J.; Vermont, M.; et al. A human importin-β-related disorder: Syndromic thoracic aortic aneurysm caused by bi-allelic loss-of-function variants in IPO8. *Am. J. Hum. Genet.* **2021**, *108*, 1115–1125. [CrossRef]
30. He, Y.; Liu, H.; Yin, N.; Yang, Y.; Wang, C.; Yu, M.; Liu, H.; Liang, C.; Wang, J.; Tu, L.; et al. Barbadin Potentiates Long-Term Effects of Lorcaserin on POMC Neurons and Weight Loss. *J. Neurosci.* **2021**, *41*, 5734–5746. [CrossRef]
31. Dimerix. DMX-200 for Focal Segmental Glomerulosclerosis (FSGS). Available online: https://dimerix.com/products/dmx-200-for-focal-segmental-glomerulosclerosis/ (accessed on 15 March 2024).
32. ISRCTN Registry. Efficacy and Safety of DMX-200 in Patients with Focal Segmental Glomerulosclerosis. Available online: https://www.isrctn.com/ISRCTN72772236 (accessed on 15 March 2024).

International Journal of
*Molecular Sciences*

MDPI

*Review*

# Catestatin in Cardiovascular Diseases

**Joanna Kulpa [1], Jarosław Paduch [1], Marcin Szczepanik [1], Anna Gorący-Rosik [2], Jakub Rosik [1,*], Magdalena Tchórz [1], Andrzej Pawlik [1] and Jarosław Gorący [2]**

1 Department of Physiology, Pomeranian Medical University, 70-111 Szczecin, Poland; joanna.h.kulpa@gmail.com (J.K.); jaroslawpaduch@hotmail.com (J.P.); marcin.t.szczepanik@gmail.com (M.S.); magdalena.tchorz98@gmail.com (M.T.); pawand@poczta.onet.pl (A.P.)

2 Independent Laboratory of Invasive Cardiology, Pomeranian Medical University, 70-111 Szczecin, Poland; ania.goracy@gmail.com (A.G.-R.); jaroslaw.goracy@pum.edu.pl (J.G.)

* Correspondence: jakubrosikjr@gmail.com

**Abstract:** Cardiovascular diseases are one of the leading causes of mortality and morbidity worldwide. The pathogenesis of this group of disorders is highly complex and involves interactions between various cell types and substances, among others, catestatin (CTS). In recent years, numerous researchers have expanded our knowledge about CTS's role in development and its potential for the treatment of a variety of diseases. In this review, the authors discuss CTS's importance in the pathogenesis of arterial hypertension, coronary artery disease, and heart failure. Moreover, we present CTS's influence on heart and vessel function.

**Keywords:** arterial hypertension; cardiovascular disease; catestatin; coronary artery diseases; heart failure; molecular mechanisms

## 1. Introduction

Cardiovascular diseases (CVDs) account for one of the leading causes of morbidity and mortality worldwide. As CVDs remain the dominant cause of death in Europe, ischemic heart disease (IHD) has been the leading reason [1]. IHD is a group of related syndromes exhibiting myocardial ischemia and dysfunction due to an imbalance between the supply and the heart's demand for blood. In most cases, atherosclerosis is the main factor contributing to stenosis in the coronary arteries and, therefore, a reduction in blood flow [2,3]. Atherosclerosis is a chronic inflammatory disease in which cholesterol-rich plaques deposit inside the arterial walls. A complex multicellular process characterizes it. Both genetic and environmental factors facilitate the expansion of the atherosclerotic plaques in various arterial territories [4]. Several risk factors for IHD have been established, including hypercholesteremia, diabetes, arterial hypertension (HA), obesity, and metabolic syndrome [5]. Developing and implementing novel preventive and treatment methods should be crucial to mitigate CVD risk factors and reduce the public health burden of CVDs in all countries. Chromogranin A (CgA) is a pro-protein found in neuroendocrine organs, more precisely, in the secretory granules of chromaffin cells. Proteolytic cleavage of CgA generates several biologically active peptides, such as pancreastatin, an inhibitor of glucose-induced insulin secretion; WE14, a peptide that is an antigen for diabetogenic CD4+ T-cell clones; serpinin, an adrenergic peptide; vasostatin a vasodilating, antiadrenergic, and antiangiogenic peptide, and catestatin (CTS) [6,7]. Based on the previous research, CgA and some of its bioactive fragments, CTS, among others, appear to act as a potential biomarker for various neoplastic and inflammatory diseases as well as CVDs [8–11]. CTS has emerged as a pleiotropic peptide, providing several cardioprotective effects. CTS may exhibit different functions by binding to various receptors and then causing subsequent activation of many signaling pathways [8,10,11].

Furthermore, previous data demonstrated that single nucleotide polymorphisms (SNPs) appearing in the CTS-expressing region of the *CHGA* gene might result in different variants of this molecule, which may exert different activity [8]. CTS's cardiovascular effects include suppression of beta-adrenergic activation and consequently acting in a negative inotropic and chronotropic way, stimulating angiogenesis and proliferation of vascular smooth muscle cells, lowering endothelial thrombogenicity, and suppressing atherosclerosis and inflammation [8]. To this date, many studies have demonstrated the role of CTS in the pathogenesis and development of various CVDs, such as HA, diabetes mellitus (DM), atherosclerosis, and coronary heart disease [2,6]. Its levels appeared to also be different in heart failure (HF) [12]. Although there are a large amount of data from existing basic and clinical research, many mechanisms underlying CVD pathogenesis have yet to be explained. However, the role of CTS as a biomarker in this group of diseases is still conceptive. The most promising CTS measurement application seems to be in patients with HF [8]. It may deliver additive prognostic input to natriuretic peptides and be useful in cardiovascular events risk stratification regarding HF. Assessing many HF markers simultaneously may enhance better management and prognosis in HF patients. Higher CTS levels may help clinicians better manage and more aggressively treat these patients. Yet, such findings are still based on limited data. Aside from CTS serving as a diagnostic or prognostic tool, it also shows encouraging potency as an innovative therapeutic agent in many pathological conditions associated with chronic inflammation, including autoimmune, metabolic disorders, and CVDs [13]. The fact should be further thoroughly investigated as some preclinical documentation has already demonstrated CTS's positive cardioprotective and hemodynamic effects [8,14]. Nonetheless, more large-scale studies are needed to confirm these findings. In this review, the authors aim to discuss the role of CTS in the development of different CVDs and to determine whether this molecule could serve as a prognostic biomarker as well as a potential therapeutic target for these diseases in the future.

The authors searched for relevant manuscripts in scientific databases (Google Scholar and PubMed) using the following search queries: "catestatin" AND "cardiovascular disease"; "catestatin" AND "arterial hypertension"; "catestatin" AND "coronary artery disease"; "catestatin" AND "heart failure".

## 2. Catestatin in the Regulation of Blood Pressure

HA is one of the most prevalent cardiovascular diseases worldwide. While it is the most common modifiable risk factor associated with cardiovascular morbidity and mortality, many patients still do not achieve recommended blood pressure (BP) targets. This occurs even though proven approaches to lower BP, like diet modification and pharmacotherapy, are widely available [15]. It is, then, crucial to implement novel, adequate strategies for HA management.

CTS has been established as a pleiotropic peptide that regulates the cardiovascular system, inflammatory processes, autoimmune reactions, and metabolic homeostasis [14]. Previous research on this molecule has suggested that CTS may contribute to the pathogenesis of HA. The mechanisms that have already emerged for cardiovascular effects of CTS are autocrine inhibition of catecholamine (CA) secretion from adrenal medullary chromaffin cells and adrenergic neurons, paracrine stimulation of histamine release from mast cells, and modulation of sympathetic and parasympathetic activities by acting at the baroreceptor center of the nucleus tractus solitarius [16]. CTS has been reported to suppress the release of CAs by acting as a noncompetitive mediator of the nicotinic cholinergic stimulation of chromaffin cells and by adenylate cyclase-activating polypeptide stimulation. CTS not only limits CA secretion but also inhibits the release of other chromaffin cell neurotransmitters, such as neuropeptide Y and adenosine triphosphate [7]. CTS may reduce BP, regulating baroreflex sensitivity and heart rate variability [8]. CTS involvement in various molecular pathways is presented in Figure 1.

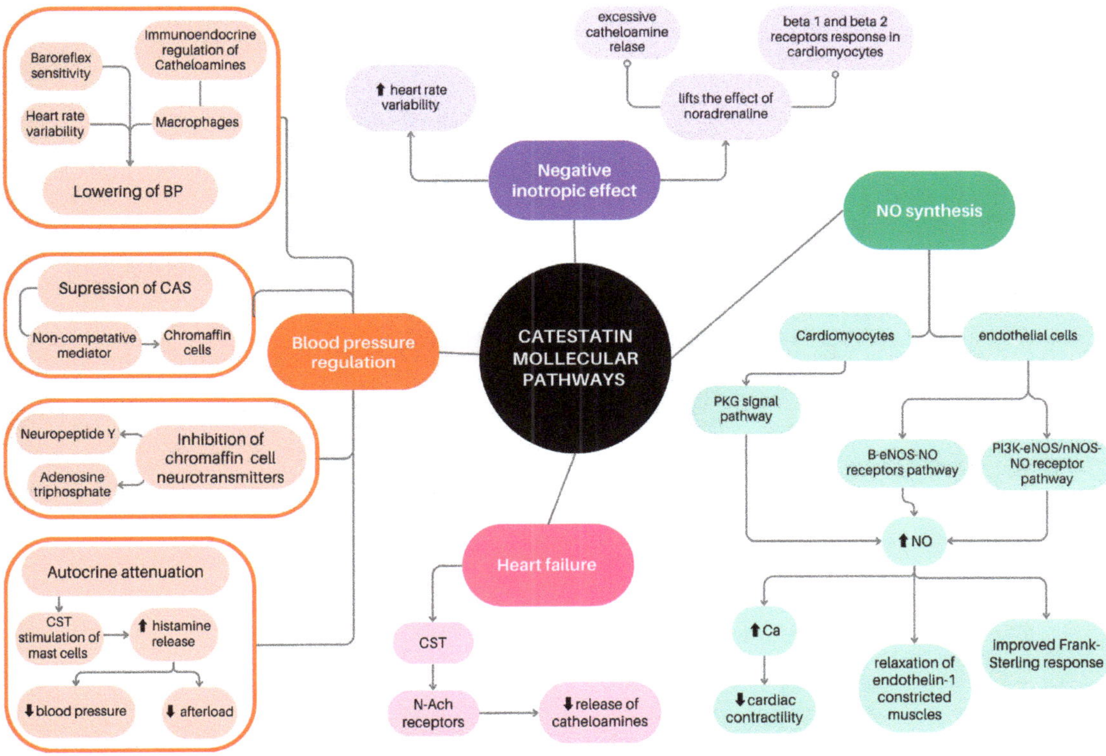


**Figure 1.** Catestatin molecular pathways [7,17] and involvement in cardiovascular diseases: blood pressure regulation [7,8,18–21], inotropic effect [22,23], and heart failure [24].

Ying et al., in their study, generated mice with knockout (KO) of the region of the *CgA* gene that only coded CTS (CTS-KO). The CTS-KO mice were hypertensive. The administration of exogenous CTS rescued their high BP, whereas CTS did not alter mice with normal BP. Furthermore, they demonstrated that a raised abundance of macrophage infiltrates in the adrenal medulla coexisting with increased levels of CAs leads to the subsequent development of HA in CTS-KO mice. These results may suggest that CTS regulates the BP by influencing the immunoendocrine regulation of CA secretion via macrophages, which may play a key role as effector cells for the antihypertensive activity of CTS and, alongside the chromaffin cells, be the primary source of circulating CTS. Elevated BP in CTS-KO mice is implied to result from a lack of CTS, known as the endogenous inhibitor of nicotine-evoked catecholamine ejection [18]. Previous research has shown the association between nicotine-induced elevated BP and increased CA secretion [19]. Ying et al. proved it using chlorisondamine, which lowered BP in CTS-KO mice [18]. Hence, CTS is assumed to be the autocrine attenuator of cardiac inflammation in HA. The most important recent animal and in vitro studies on CTS are summarized in Table 1.

**Table 1.** Most important recent animals and in vitro studies on catestatin.

| Study and Its Reference | Methodology | Results Summary |
| --- | --- | --- |
| Zhang et al. 2024 [25] | Fourty-two male adult Sprague–Dawley rats randomly divided into CTS and NON-CTS equinumerous groups. Ventricular arrhythmias were induced by ligation of the LAD and electrical stimulation. | **CTS notably reduced induced ventricular arrhythmia caused by ischaemia and electric stimulation in rats.** In the CTS group: ↑Ito, ↑IK, ↑IK1, and ↓ICa-L activity. |
| Lener et al. 2023 [26] | Matrigel assays; human coronary artery endothelial cells (HCAECs) and human coronary artery smooth muscle cells (HCASMCs) | **CTS induces chemotaxis of HCAECs** (relative CI CTS 1 nM 1.79 ± 0.1, n = 3, $p < 0.01$ vs. control) similar to VEGF (relative CI VEGF 50 ng/mL 2.13 ± 0.09, n = 3, $p < 0.01$ vs. control). **CTS stimulates HCAEC proliferation** (relative proliferation CTS 1 nM 1.62 ± 0.05, n = 5, $p < 0.01$). **CTS had similar effect on capillary tube formation as VEGF** (relative tube formation CTS 1 nM 2.42 ± 0.1, n = 3, $p < 0.001$ vs. control; relative tube formation VEGF 50 ng/mL 2.06 ± 0.1, n = 3, $p < 0.001$ vs. control). **CTS activated ERK 1/2 signaling pathway in HCAEC.**<br>**CTS stimulates HCASMC proliferation** (relative proliferation CTS 10 nM 1.6 ± 0.09, n = 3, $p < 0.001$ vs. control, relative proliferation CTS 1 nM 1.27± 0.04, n = 3, $p$ = n.s. control)<br>**CTS stimulated ERK 1/2 signaling and activation of the PI3-kinase-Akt pathway in HCASMC.**<br>**CTS reduced $H_2O_2$ induced apoptosis** (relative apoptosis CTS 1 nM 0.76 ± 0.05, $p < 0.05$ vs. control) |
| Qiu et al. 2023 [24] | SG: C57BL/six male mice with TAC/DOCA induced HFpEF<br>G1: 28 days of CTS treatment, n = 8<br>G2: 28 days of placebo, n = 8<br><br>C57BL/six male mice after thoracotomy without TAC/DOCA<br>G3: 28 days of CTS treatment, n = 8<br>G4: 28 days of placebo, n = 8<br><br>ECHO evaluation, conductance catheter pressure-volume analysis, microscopy and genetic analysis of cardiomyocytes 7 weeks after surgery. | **CTS Protects Diastolic Dysfunction in mice with HFpEF.**<br>In the TAC/DOCA (G2) group, ↑E/A velocity ratio; ↓E wave deceleration time; ↑E/e′ suggesting diastolic dysfunction. ↑LV end-systolic pressure, ↑LV chamber stiffness, ↑LV concentric hypertrophy, ↑Volume of cardiomyocytes.<br><br>**Those effects were reduced in TAC/DOCA + CTS (G1) group.**<br><br>**Mitochondrial ROS generation reduction in TAC/DOCA + CTS (G1) compared to G2).**<br><br>**Restoration of mitochondrial respiratory chain by CTS treatment.** |
| Bralewska et al. 2023 [27] | HTR-8/SVneo (CRL-3271) and BeWO (CCL-98) trophoblast cell lines incubated in pre-eclamptic environment (hypoxia, pro-inflammatory, oxidative stress) | Throphoblast cells produce CgA and CTS. Pre-eclamptic environment promotes ↓CHGA gene expression ($p < 0.001$); ↓CTS level in trophoblast; ↑apoptosis.<br>**There is negative correlation between CTS level and apoptotic index** for both HTR-8/SVneo (R = 0.4) and BeWo cells (R = 0.5).<br>**CTS acts as antiapoptotic factor in vitro.** |

**Table 1.** *Cont.*

| Study and Its Reference | Methodology | Results Summary |
|---|---|---|
| Muntjewerff et al. 2022 [28] | Transwell migration assays on human blood monocytes and neutrophiles;<br>aortic ring model from Cx3cr1+/gfp transgenic mice. | **CTS itself has a weak chemotactic effect on monocytes and neutrophiles, however it counteracts the chemoattraction** of leukocytes by inflammatory chemokines CCL2, CXCL2, and IL-8<br>**CTS promotes angiogenesis.** |
| Ying et al. 2021 [18] | TG: CTS-KO C57BL/6 male mice, n = 8<br>CG: CTS-WT C57BL/6 male mice, n = 8 | In CTS-KO mice: **↑SBP, ↑DSB, ↑MAP**.<br>**Pro-inflammatory: serum cytokines ↑TNF-α, ↑IFN-γ, ↑CCL2, ↑CCL3, ↑CXCL**, genes upregulation: Tnfa, Ifng, Emr1, Itgam, Itgax, Nos2a, IL12b CcL2, and CxcL1. **Anti-inflammatory: serum ↓IL-10**, genes downregulation: IL10, IL4, Mrc1, Arg1, Clec7a and Clec10a. **Adrenal and plasma ↑catecholamines. ↑Sympathetic nerve activity.** Those effects were reversed by exogenous CTS administration. IPC-induced cardioprotection impairment. ↑phosphorylation (Ser177/181) of IKK-β (inflammatory NF-κB signaling pathway). Macrophages production: ↓TNF-α, ↓CCL-2, ↓CCL-3, ↓CXCL-1, ↓IL-1β, ↑IL-10 after CTS administration. **Macrophages themselves produce CgA and CTS.** |
| Alam et al. 2020 [22] | H9c2 myoblasts stimulated for sarcomere reorganization by Troponin T antibodies and norepinephrine. | **CTS attenuates myoblasts hypertrophy and suppresses the generation of ROS induced by norepinephrine; however, CTS does not protect cells from apoptotic signalling induced by norepinephrine.** |
| Chu et al. 2020 [29] | 8 weeks old male Sprague–Dawley rats<br>TG: Langendorff global ischemia/reperfusion model;<br>CG: healthy rats;<br>primary culture of cardiomyocytes from neonatal rats. | Average LV infarct size was33.66 ± 3.61% in I/R group. **Posttreatment with CTS reduced infarct size** to 20.25 ± 3.23% ($p$ = 0.011 vs. I/R group). ↓LDH in myocardium in I/R group than in CG (6843.5 ± 1136.0) U/g vs. (102,470.0 ± 1066.1)U/g, $p < 0.001$. **CTS intervention reduced the ↓LDH** (8994.4 ± 963.8)U/g vs. (6843.5 ± 1136.0)U/g, $p < 0.001$). **CTS post-treatment decreased oxidative-stress and reduced apoptosis of cardiomyocytes after I/R. CTS reduced apoptosis in cardiomyocytes culture induced by $H_2O_2$ through activating the β2 adrenergic receptor and PI3K/Akt pathway.** |
| Chen et al. 2019 [30] | Atherosclerosis model: 8 weeks old male ApoE-KO mice divided into groups:<br>G1 (CG) PBS i.p.<br>G2: CTS i.p.<br>G3 CTS + DX600 (ACE2 inhibitor) i.p.<br>Human aortic endothelial cells (HAECs) and human umbilical vein endothelial cells (HUVECs) | **CTS reduced TNF-α-induced expression of IL-6, MMP-2 and adhesion molecules** (ICAM-1, VCAM-1, and E-selectin) in HAECs. **CTS promoted expression and activity of ACE2 in HUVECs. CTS reduced adhesion events and increased the rolling velocity of leukocytes,** those effect were blocked by ACE2 inhibitor though.<br>**Plaque area of the aorta was reduced in the CTS-treated mice** vs. controls. |

**Table 1.** *Cont.*

| Study and Its Reference | Methodology | Results Summary |
|---|---|---|
| Chen et al. 2019 [31] | Eight-week-old C57/BL6 mice randomly divided into control (n = 20), control CTS (n = 20), APE (n = 20), and APE CTS (n = 20) Human pulmonary artery endothelial cells (HPAECs). APE in mice was induced by injection of collagen and epinephrine through the inferior vena cava | **Plasma CTS lower in APE** than in CG ($p < 0.01$) Negative correlation between CTS and Platelets level (Pearson correlation test r = 0.6732). **Survival rate 30 min after APE onset was higher in APE CTS** than in APE group (80% vs. 30%, $p < 0.01$). **CTS had anti-thrombotic activity in APE mice.** CTS inhibited APE-induced release of inflammatory neutrophils and macrophages. **CTS blocked TLR-4 p38 phosphorylation in HPAECs.** |

APE—acute pulmonary embolism; CCL—C-C motif chemokine ligand; CG—control group; CgA—chromogranin A; CTS—catestatin; CXCL—C-X-C motif chemokine ligand; DSB—diastolic blood pressure; DOCA—deoxycorticosterone acetate; G—group; HR—hear rate; ICa-L—L-type calcium current; IFN-γ—Interferon gamma; IKK-β—inhibitor of nuclear factor kappa-B kinase subunit beta; IK—delayed rectifier potassium current; IK1—inward rectifier current; IL—Interleukin; IPC—Ischemic pre-conditioning; I/R—ischaemia/reperfuson; Ito—transient outward potassium current; KO—knockout; LAD—left anterior descending coronary artery; MAP—mean arterial pressure; ns.—not statistically significant; SBP—systolic blood pleasure; SHR—Spontaneously Hypertensive Rats; TAC—transverse aortic constriction; TG—test group; TNF-α—tumor necrosis factor alpha; VEGF—vascular endothelial growth factor; WKY—Wistar-Kyoto rats; WT—wild type.

Jianqiang et al. explored wavering levels of plasma CTS in Spontaneously Hypertensive Rats (SHRs) and their littermates, Wistar–Kyoto rats (WKYs), and found that they were significantly higher in SHRs than in WKYs. Along with HA development and progression in SHR, plasma CTS levels gradually rose. The reduction in the heart rate of SHRs after exogenous administration of CTS may prove that it inhibits sympathetic activity in hypertensive individuals [20]. CTS acted as a potent vasodilator, and this effect was mediated by an elevated release of histamine. Kruger et al. showed that the most likely mechanism by which CTS might lead to histamine release would be stimulating mast cells [21]. Considering the central nervous system (CNS), CTS is reported to exert both sympathoexcitatory and cholinergic effects [8]. Administering CTS into the rat rostral ventrolateral medulla (RVLM), which is responsible for BP control in the brain stem, led to sympathoexcitatory effects, increased barosensitivity, and depletion of chemosensitivity and the somatosympathetic reflex, resulting in high BP [32]. On the contrary, the injection of CTS into the rat caudal ventrolateral medulla (CVLM) contributed to decreased barosensitivity and the depletion of the peripheral chemoreflex, resulting in low BP [33]. Furthermore, the injection of CTS into the rat's central amygdala also decreased BP providing protection against vascular dementia and neurodegeneration [34]. These reports indicate that CTS plays a significant role in cardiorespiratory control in the CNS. Reduced plasma CTS levels and increased CgA levels have been noted in humans with HA [35]. O'Connor et al., in their study, demonstrated decreased plasma levels of CTS in normotensive offspring with a family history of HA. Dysregulation in the processing of CgA to CTS and CTS lowering may appear in the early stages of HA development, even though categorization into normotensive and hypertensive groups did not reveal significant differences in plasma CTS levels [35].

The most important recent human studies on CTS are summarized in Table 2. Durakoglugil et al. found that the difference in CTS levels between untreated patients with HA and healthy individuals was irrelevant after adjusting for age, gender, height, and weight [36]. Meng et al. explored levels of plasma CTS in HA patients and the relationship between CTS and left ventricular hypertrophy (LVH). Their results showed elevated CTS plasma levels in the HA group and lower CTS to norepinephrine ratio in patients with LVH compared to normal controls, suggesting that this molecule might participate in developing HA and LVH [37]. In contrast, in another study by O'Connor et al., CTS was significantly reduced in patients suffering from HA [38]. Choi et al. identified CTS genotypes and performed a genotype–phenotype association analysis on 343 participants from the Japanese population. The results showed that BP was higher in Gly/Ser subjects than in the wild-type Gly/Gly individuals. Accordingly, it was demonstrated that the CTS variant allele Ser-364 may be associated with higher BP in the Japanese population [39]. On the contrary, the Ser-364 allele was linked with lowered diastolic BP levels only in males in a Southern California study population. It seemed to reduce the risk of developing HA in men as well. The effect was not observed for SBP or in women [40].

**Table 2.** Most important recent clinical trials on catestatin.

| Study and Its Reference | Methodology | Results Summary |
|---|---|---|
| **Coronary artery disease** | | |
| Xu et al. 2022 [41] | Cohort study among 165 patients with AMI;<br>4 years follow-up for MACEs<br>MACEs group n = 24.<br><br>Young = age <60 years old<br>Elderly = age ≥60 years old | **Lower CTS level in MACEs group** (0.74 ± 0.49 ng/mL vs. 1.10 ± 0.79 ng/mL, $p = 0.033$); MACEs rate was higher in the elderly group than in the young group (23.8% [15/63] vs. 8.8% [9/102], $p = 0.008$). **CTS level was lower in the MACEs group than in the non-MACEs group** (0.76 ± 0.50 ng/mL vs. 1.31 ± 0.77 ng/mL, $p = 0.012$) and **CTS was associated with MACEs (Kaplan Meier, $p = 0.007$) among the elderly group, but not in the young group** (Kaplan Meier, $p = 0.893$). **In the Cox proportional hazards regression CTS was independent factor for MACEs in elderly patients** (hazard ratio 0.19, 95% confidence interval 0.06–0.62, $p = 0.006$). |
| Chen et al. 2019 [30] | Cross-sectional study<br>Stage 1<br>TG: 224 patients with CAD and<br>CG: 204 healthy controls<br>Stage 2<br>association between CTS and atherosclerosis severity in 921 CAD patients | **CTS lower in CAD patients than in CG** 1.14 (1.05–1.24) ng/mL vs. 2.15 (1.92–2.39) ng/mL, $p < 0.001$.<br>**Negative correlation between CTS and atherosclerosis** severity ($r = -0.208$, $p < 0.001$) |
| Xu et al. 2017 [42] | Cohort study among 170 patients with suspected ACS who underwent coronarography<br>TG: STEMI n = 46; UAP n = 89<br>CG: No CAD n = 35<br>2 years follow-up for MACEs | **Plasma CTS in STEMI group (0.80 ± 0.62 ng/mL) and UAP group (0.99 ± 0.63 ng/mL) were lower than in CG** (1.38 ± 0.98 ng/mL; $p = 0.001$). In multivariable linear regression, body mass index, presence of hypertension, and type of **CAD** were independently related to the plasma CTS level. However, there were no significant differences in MACEs between patients with high and low levels of CTS |
| Liu et al. 2013 [43] | Cohort study on 120 CAD and<br>TG: SAP n = 15; UAP n = 47; NSTEMI n = 22; STEMI n = 36<br>CG: 30 healthy individuals<br>CTS measurement at admission<br>Median follow-up time: 1045 days | **CTS higher in TG than in CG** (1.02 ± 0.70 vs. 0.41 ± 0.14, $p < 0.05$)<br>**CTS higher in SAP than in CG** (0.72 ± 0.50 vs. 0.41 ± 0.14, $p < 0.05$)<br>**CTS higher in UAP than in CG** (0.88 ± 0.58 vs. 0.41 ± 0.14, $p < 0.05$)<br>**CTS higher in NSTEMI than in CG** (1.05 ± 0.48 vs. 0.41 ± 0.14, $p < 0.05$)<br>**CTS higher in STEMI than in CG** (1.31 ± 0.91 vs. 0.41 ± 0.14, $p < 0.05$)<br>**CTS correlated positively with NE** (Spearman correlation coefficient $r = 0.51$, $p = 0.00$) **and NTproBNP** ($r = 0.24$, $p = 0.01$)<br>**Plasma CTS on admission was not associated with adverse cardiovascular events.** |

**Table 2.** *Cont.*

| Study and Its Reference | Methodology | Results Summary |
|---|---|---|
| **Heart failure** | | |
| Chu et al. 2024 [44] | A cohort study on 199 HF patients according to modified Framingham criteria. LVEF $\leq$ 40% n = 100; LVEF > 40% n = 102; Determination of CTS predictive value in HFrEF and HFmrEF/HFpEF respestively. | **Plasma CTS level had a moderate predictive ability for CV death** with a C statistic of 0.59 (95% CI 0.45–0.74), sensitivity 51.8%, specificity 71.8% in the HFrEF population and **better prognostic value** with C statistic of 0.72 (95% CI 0.59–0.85), sensitivity 70,8%, specificity 71,8% **in the HFmrEF/HFpEF population.** |
| Qiu et al. 2023 [24] | A cross-sectional study on 81 patients with HFpEF and 76 non–heart failure controls. | **Serum CTS level was higher in HFpEF group than in CG** (11.21 [interquartile range, 6.81–19.12] ng/mL vs. 23.62 [interquartile range, 11.53–34.81] ng/mL; $p < 0.001$). **Serum CTS level was positively correlated with NT-proBNP** level (r = 0.41; $p < 0.001$) and E/e′ ratio (r = 0.25; $p = 0.002$). |
| Borovac et al. 2020 [45] | Cohort study on 96 acute decompensated HF followed up until discharge Survivors n = 90 Non-survivors n = 6 | **Serum CTS higher in non-survivors than in survivors** 19.8 (IQR 9.9–28.0) vs. 5.6 (IQR 3.4–9.8) ng/mL, $p < 0.001$. **CTS was an independent predictor of in-hospital death** (FC 6.58, 95% CI 1.66–21.78, $p = 0.003$). In ROC analysis CTS AUC (0.905 95% CI 0.792–1.000, $p < 0.001$). |
| Zhu et al. 2017 [46] | A cohort study on 72 patients with STEMI followed-up for 65 months and 30 healthy controls. Serum CTS measurement. ECHO. | **CTS levels correlated with the changes of LVEDD ($p < 0.0001$), EF ($p = 0.0002$), E ($p = 0.0003$), A ($p < 0.0001$), E′ ($p < 0.0001$), E/A ($p < 0.0001$), as well as E/E′ ($p < 0.0001$).** |
| **Pre-eclampsia** | | |
| Palmrich et al. 2023 [47] | Cross-sectional study among 100 pregnant women. TG: 50 pre-eclamptic singleton pregnancy patients CG: 50 healthy pregnant women Serum CTS level comparison. | **CTS serum level in pre-eclamptic group lower than in CG** (median CTS: 3.03 ng/mL, IQR [1.24–7.21 ng/mL] vs. 4.82 ng/mL, IQR [1.82–10.02 ng/mL]; $p = 0.010$). |
| Tüten et al. 2022 [48] | Cross-sectional study among 2oo pregnant women. TG: 50 women with mild preeclampsia, 50 women with severe preeclampsia, CG: 100 healthy pregnant women | **Mean serum CTS increased in the preeclampsia group** than in CG (290.7 ± 95.5 pg/mL vs. 182.8 ± 72.0 pg/mL). No significant differences in CTS level between mild and severe preeclampsia groups (282.7 ± 97.9 pg/mL vs. 298.7 ± 93.4 pg/mL, $p = 0.431$). Serum CTS had positive correlations with systolic and diastolic blood pressure, urea, uric acid, and creatinine. |
| Bralewska et al. 2021 [49] | A cohort study of 205 pregnant women. TG: 102 pre-eclamtic patients CG: 103 healthy pregnant women Placental expression of the CgA gene and placental CTS level comparison. | **Placental expression of chromogranin A higher in pre-eclamptic patients** than in CG (−0.25 ± 1.7 vs. −0.82 ± 1.5, $p = 0.011$). **Mean CTS level lower in pre-eclamptic group** than in CG (6.4 ± 1.0 vs. 6.7 ± 1.4, $p = 0.04$). |
| Özalp et al. 2021 [50] | Cross-sectional study TG: 27 women with early-onset pre-eclampsia, 28 women with late-onset pre-eclampsia CG: 28 healthy pregnant women. Maternal serum CTS measurement and fetal ECHO. | **The fetal E/A ratio positively correlated with the maternal serum CTS levels in both the pre-eclampsia group and CG** ($p < 0.001$, $p < 0.001$). **Fetal isovolumetric relaxation time and MPI values negatively correlated with maternal CTS** in pre-eclampsia and CG ($p < 0.001$, $p = 0.001$, $p < 0.001$, and $p = 0.002$, respectively). **No significant CTS level difference between TG and CG.** |

**Table 2.** *Cont.*

| Study and Its Reference | Methodology | Results Summary |
|---|---|---|
| **Chronic kidney disease** | | |
| Luketin et al. 2021 [51] | Cross-sectional study<br>TG: 91 adult patients with end-term chronic kidney disease haemodialyzed > 1 year<br>CG: 70 healthy adult individuals | **Plasma CTS higher in HD than CG** (32.85 ± 20.18 vs. 5.39 ± 1.24 ng/mL, $p < 0.001$). **Positive correlations between CTS and AGEs** (r = 0.492, $p < 0.001$) and **between CTS and both the Dialysis Malnutrition Score** (r = 0.295, $p = 0.004$) **and Malnutrition-Inflammation Score** (r = 0.290, $p = 0.005$). |
| **Pulmonary embolism** | | |
| Izci et al. 2020 [52] | Prospectively study<br>TG: 160 patients with contrasted CT-confirmed pulmonary embolism<br>female n = 76; male n = 84<br>CTS measurement within 24 h after admission.<br>CG: 97 healthy individuals<br>female n = 55; male n = 42<br>Serum CTS measurement 0, 3 and 7 days after admission; ECHO | **Plasma CTS higher in APE group than in CG** (17.5 ± 6.1 ng/mL vs. 27.3 ± 5.7 ng/mL, $p < 0.001$). **Plasma CTS higher in the sPESI ≥ 1** (n = 72) than in the patients with sPESI < 1 (37.3 ± 6.1 vs. 24.2 ± 5.3 ng/mL, $p < 0.001$). **Positive correlation between CTS level and sPESI score** (±0.581, $p < 0.001$). ROC curve analysis with cut-off level of 31.2 ng/mL, and the **CTS level predicted mortality with a sensitivity of 100% and specificity of 52.6%** (AUC = 0.883, 95% CI: 0.689–0.921). **CTS level correlated with right ventricular dysfunction. Negative endpoints were associated with higher CTS levels after admission.** |
| **Rheumatoid arthritis** | | |
| Pàmies et al. 2024 [53] | A cohort study of 199 rheumatoid arthritis patients.<br>female n = 132; male n = 67 | **RF-positive patients had higher CTS levels than RF-negative patients** ($p < 0.001$). Positive correlations between: CTS and LDL-C (ρ = 0.32, $p = 0.009$); CTS and IL-32 (ρ = 0.20, $p = 0.003$); CTS and fet-a (ρ = 0.20, $p = 0.004$). **↑CTS in women with DM2** ($p = 0.04$). |

A—peak late diastolic mitral flow velocity; ACS—acute coronary syndrome; AGEs—advanced glycation end products; AMI—acute myocardial infarction; APE—acute pulmonary embolism; CG—control group; CTS—catestatin; DM2—type 2 diabetes mellitus; E—peak early diastolic mitral flow velocity; E′—doppler-derived peak early diastolic mitral flow velocity; ECHO—echocardiography; FC—firth coefficient; fet-a—fetuin-A; HD—haemodialysis; HF—heart failure; HFmrEF—heart failure with mildly reduced ejection fraction; HFpEF—heart failure with preserved ejection fraction; HFrEF—heart failure with reduced ejection fraction; IVS—Interventricular septal end diastolic dimension; MACEs—major adverse cardiovascular events; MPI—myocardial performance index; NE- norepinephrine; NT-proBNP—N-terminal pro-B-type natriuretic peptide; LVPW—left ventricular posterior wall thickness; RF—rheumatoid factor; SAP—stable angina pectoris; sPESI—Simplified Pulmonary Embolism Severity Index; STEMI—myocardial infarction wit ST-elevation; TG—test group; UAP—unstable angina pectoris.

Furthermore, due to its neuroprotective potential, CTS might be a novel target for the treatment and prevention of HA [34]. However, the results of the aforementioned studies are inconsistent, and it is too soon to draw conclusions.

## 3. Inotropic Effect of Catestatin and Association Between Catestatin and Heart Rate Variability

CTS has a negative inotropic effect, improves cardiomyocyte condition after episodes of ischemic reperfusion, and lengthens cardiomyocyte survival rate. CTS protects the heart and bloodstream from overdrive caused by the excessive release of catecholamines, such as norepinephrine and epinephrine. Furthermore, CTS lifts the effects of noradrenalin on Beta1 and Beta2 receptors in cardiomyocytes. Moreover, it prevents excessive myocardial remodeling and reduces the formation of reactive oxygen species [22].

Higher heart rate variability (HRV) values are commonly found during a relaxed state. However, people with a high HRV may have better stress resilience or better cardiovascular function [54]. Researchers performed several studies to explore the influence of CTS on the HRV. In patients with acutely decompensated HF, a higher level of CTS meant a lower heart rate [55]. A different study demonstrated that CTS regulates and prevents tachycardia in mice. In that study, researchers supplemented CgAknockout mice with CTS, which protected the subjects from high systolic blood pressure and decreased HRV [23]. Confirmation of these reports is provided by Dev et al. They performed an experiment in which supplementation of CTS increased HRV [56].

## 4. Catestatin İnfluence on NO Synthesis and Metabolism

CTS increases NO synthesis both from cardiomyocytes and endothelial cells. NO is produced in the NO(NOS)-NO-cGMP-cGMP-protein kinase (PKG) signal pathway and released as a result of the endothelial receptor B-eNOS-NO or PI3K-eNOS/nNOS-NO pathway [7,17]. Released NO reduces cellular Ca2+, which results in decreased cardiac contractility and negative lusitropic and inotropic effects. Moreover, NO causes the relaxation of endothelin-1 preconstricted coronaries and smooth muscles in the endothelium of other vessels. CTS increases the rate of positive Frank–Starling responses by increasing NO production. Because of that, it is suggested that CTS is a NO-dependent modulator of the regulation of the heart [7,17]. This correlation between NO synthesis and its impact on the cardiovascular system is crucial based on the results of several recent studies on CTS and its antiadrenergic effect. The antiadrenergic effects of CTS on the heart depend not only on a direct impact on papillary muscle cells but also on an increased rate of NO synthesis by endothelial cells [17]. Using inhibitors for specific points of the NO synthesis pathway in cardiomyocytes eliminates the negative inotropic effect exhibited by CTS [17]. Furthermore, using a selective eNOS inhibitor abolished the effect of CTS, and a neuronal NOS inhibitor only reduced this effect, suggesting that the effect of CTS on cardiac contractility is mediated primarily through the eNOS pathway.

Kiranmayi et al. determined a significant connection between the Ser-364 allele and HA in Chennai and the Chandigarh populations. They demonstrated that the Ser-364 individuals have higher BP levels. The CTS Ser-364 peptide, via altered interactions with ADRB2 and differential activation of extracellular regulated kinase (ERK) and endothelial nitric oxide synthase (eNOS) phosphorylation, led to greater modulation of the endothelial NO pathway compared with the CTS-WT peptide. This may contribute to the increased risk of HA in Ser-364 allele carriers, possibly due to lower NO production, which leads to diminished endothelial vasodilation. In conclusion, the Ser-364 allele appears to play an important role in HA development in Indian populations [57]. Another study on an Indian population found no association of this SNP with BP [58]. The Ser-364 allele seems to exert comparable effects on BP in different Asian populations but not in Caucasians. These contradictory associations determine heterogeneity in different populations. Thus, this proves the need to conduct more studies on diverse ethnic populations. Further studies concerning the underlying mechanisms of action of this physiological antihypertensive peptide may bring significant insights into the pathogenesis of cardiovascular diseases.

A study using cardiomyocytes isolated from rat papillary muscles and endothelial cells obtained from buffalo aortas showed, in even more detail, that CTS affects myocardial contractility mainly through NO synthesis. They used inhibitors of individual points of the NO synthesis pathway (PI3K, NOS) in the presence of CTS, which abolished its antiadrenergic effect. The endocardial endothelium layer was also removed and treated with CTS, which did not abolish the inotropic effect induced by the β adrenoceptor agonist-isoproterenol under basal conditions but completely abolished the antiadrenergic effect induced by wild-type CTS [59].

## 5. Catestatin in Coronary Artery Disease and Atherosclerosis Development

Despite significant advances in the diagnosis and treatment of coronary artery disease (CAD), it remains a leading cause of mortality and Disability Adjusted Life Years (DALYs) worldwide, resulting in 7 million deaths and loss of 129 million DALYs each year, which places CAD among the most serious threats to global health [60].

Atherosclerosis is the primary and most widely studied risk factor for CVDs, including CAD. It is an intravascular chronic inflammation process initiated by malfunction and increased endothelium permeability in the luminal layer of arteries or the intima. An elevated concentration of low-density lipoproteins (LDLs) in the serum promotes their accumulation in the intimal subendothelial space, followed by biochemical modifications, including oxidation, acetylation, glycosylation, glycoxidation, carbamylation, and others, primarily related to exposure to reactive oxygen species [61]. Modified LDLs in the presence of other atherogenic factors, such as nicotine addiction, DM, turbulent blood flow, and HA, lead to endothelial cell activation and monocyte recruitment in the intima [62,63]. LDLs are phagocyted by macrophages and accumulate in vascular smooth muscle cells, which migrate from the tunica media, and macrophages, forming foam cells and fatty streaks [64]. Over time, apoptosis is impaired, and cell necrosis predominates. A necrotic core, cell-impoverished, rich-in-lipids region characterized by an inflammatory microenvironment is formed. A fibrous cap protects the arterial lumen from the necrotic core's prothrombotic properties. At this point, atherosclerotic plaque is well-developed and unlikely to disappear. Further expansion of the necrotic core results in arterial lumen reduction and blood flow disorders, rising atherosclerotic plaque instability, and the possibility of plaque rupture, eventually causing potentially lethal complications such as myocardial infarction (MI), ischemic stroke, and peripheral arterial disease [65].

The impact of CTS on atherosclerosis and CAD development and course is multi-faceted and, despite the great interest of scientists in the last years, not yet completely clarified. It has been proven that CTS modulates the immune system, suppresses tissue inflammation, reduces reactive oxygen species production, attenuates stimulatory effects of catecholamines, reduces oxidative stress-induced cardiomyocyte apoptosis, and plays an essential role in monocyte migration and differentiation [22,29,66]. In addition, CTS has a beneficial influence on endothelial NO synthesis, glycemia, serum lipids levels, and arterial blood pressure, which has been extensively described in other sections of this article.

Although CTS by itself exhibits weak chemotactic properties in vitro and in vivo in rodent models, it has been shown to block monocyte and neutrophil migration that is dependent on chemotactic factors such as CCL2, CXCL2, and IL-8 [28].

Moreover, CTS-KO mice were characterized by increased serum levels of proinflammatory cytokines, including TNF-$\alpha$, IFN-$\gamma$, CCL2, CCL3, and CXCL-1, whereas the anti-inflammatory IL-10 concentration was decreased. Furthermore, CTS-KO mouse cardiomyocytes had a lower expression of anti-inflammatory genes for IL10, IL4, Mrc1, Arg1, Clec7a, and Clec10a and upregulated proinflammatory genes, such as Tnfa, Ifng, Emr1, Itgam, Itgax, Nos2a, IL12b CcL2, and CxcL1. Those effects were fully reversible by exogenous CTS supply. Additionally, intensified phosphorylation (Ser177/181) of part of IKK-$\beta$, the inflammatory NF-$\kappa$B signaling pathway, was observed in CTS-KO mice [18].

### 5.1. Catestatin and Angiogenesis After Myocardial Infarction

Lener et al. showed that CTS stimulates the proliferation of human coronary artery smooth muscle cells and endothelial cells in Matrigel assays. The capillary-like tube formation stimulated by CTS was comparable to the effect acquired after administering the vascular endothelial growth factor (VEGF). Furthermore, CTS demonstrated antiapoptotic potential for human cardiomyocytes in vitro [26].

Moreover, the association between higher serum CTS levels and better collateral development in patients with chronic total occlusion of the coronary artery (CTO) has been reported. A total of 38 CTO patients and 38 healthy individuals were included in the study. The plasma CTS level was lower in the control group than in the CTO group ($1.36 \pm 0.97$ ng/mL vs. $1.97 \pm 1.01$ $p = 0.009$). Within the CTO group, patients with good collateral development had higher CTS and VEGF levels than patients with unsatisfactory collateral development ($2.36 \pm 0.73$ vs. $1.61 \pm 1.12$ ng/mL, $p = 0.018$; $425.23 \pm 140.10$ vs. $238.48 \pm 101.00$ pg/mL, $p < 0.001$) [67].

### 5.2. Catestatin Antiarrhythmic Potential

The occurrence of ventricular tachycardia (VT) or ventricular fibrillation (VF) is a common and serious threat to patients after MI, especially in an early phase of ischemia. Ventricular arrhythmias can be a direct cause of sudden cardiac death (SCD) and worsen short- and long-term prognosis for patients in general [68].

Zhang et al. described an experiment in a rodent model in which the left anterior descending coronary artery was ligated and electrically stimulated, resulting in ventricular tachyarrhythmia. The rodents were randomly divided into the control and the CTS administration groups. The patch-clamp technique was used to monitor the action potential, transient outward potassium current, delayed rectifier potassium current, inward rectifying potassium current, and L-type calcium current in rodent cardiomyocytes. Intensified Ito, IK, and IK1 activity was observed in the CTS group. Simultaneously, CTS inhibited ICa-L, shortening action potential time and reducing ventricular arrhythmia [25]. However, Pei et al. showed that a higher CTS level was associated with increased malignant arrhythmia incidence during hospitalization in patients with acute MI [69].

### 5.3. Catestatin as a Potential Coronary Artery Disease Course Marker

Xu et al. measured plasma CTS levels on admission in 170 patients with a suspected acute coronary syndrome (ACS) who had an emergency coronary angiography, of which 46 had MI with ST-segment elevation (STEMI), 89 had unstable angina pectoris (UAP), and 35 did not have CAD. The patients were followed up for two years to check if there would be any major adverse cardiovascular events (MACEs), such as recurrent acute MI, rehospitalization for HF, revascularization, and death due to cardiovascular causes. The plasma CTS level was higher in individuals without CAD ($1.38 \pm 0.98$ ng/mL; $p = 0.001$) than in the STEMI ($0.80 \pm 0.62$ ng/mL) and UAP ($0.99 \pm 0.63$ ng/mL) groups. In multivariable linear regression analysis, body mass index, the presence of HA, and the type of CAD were independently associated with plasma CTS levels. There was no significant difference in the occurrence of MACEs between high and low CTS levels groups [42].

To explore this issue further, Xu et al. followed up on 165 patients with acute MI for four years. At the beginning of the study, the plasma CTS level and relevant medical data were collected. During the observation time, 24 patients had MACEs. The group experiencing MACEs exhibited notably lower plasma CTS levels ($0.74 \pm 0.49$ ng/mL compared to $1.10 \pm 0.79$ ng/mL, $p = 0.033$) and had a higher average age ($59.0 \pm 11.4$ years vs. $53.2 \pm 12.8$ years, $p = 0.036$). The incidence of MACEs was significantly greater in the older population (aged 60 and above) compared to the younger group (under 60 years old) (23.8% [15 out of 63] vs. 8.8% [9 out of 102], $p = 0.008$). Additionally, CTS levels were significantly lower in the MACEs group than in those without MACEs ($0.76 \pm 0.50$ ng/mL vs. $1.31 \pm 0.77$ ng/mL, $p = 0.012$). Among older people, CTS levels were significantly linked to MACEs (Kaplan Meier, $p = 0.007$), whereas this association was not observed in the younger group (Kaplan Meier, $p = 0.893$). In the Cox proportional hazards regression analysis, elevated CTS levels emerged as an independent predictor of MACEs after controlling for other risk factors (hazard ratio [HR] = 0.19, 95% confidence interval [95% CI] = 0.06–0.62, $p = 0.006$) in older patients [41].

Chen et al. compared CTS serum concentrations between 224 patients with CAD and 204 healthy control group members in multistage research. The CTS serum level was lower in CAD patients than in healthy individuals [1.14 (1.05–1.24) ng/mL vs. 2.15 (1.92–2.39) ng/mL, $p < 0.001$]. A correlation between CTS level and CAD severity among 921 CAD patients was studied in the subsequent step. It turned out to be inversely correlated with CAD severity ($r = -0.208$, $p < 0.001$) [30]. On the other hand, higher serum CTS levels in CAD patients were reported.

Furthermore, some studies showed an association between higher serum CTS levels and worse prognosis. Zhu et al. followed up on 72 patients with STEMI for 65 months. CTS serum levels were measured at admission and on day 3 and day 7 after STEMI. Echocardiographic parameters were evaluated on day 3 and at the follow-up's end. Patients with higher serum CTS levels at day 3 had worse left ventricular function and a higher risk of left ventricular remodeling [46]. Zhu et al. evaluated 100 patients after MI and successful percutaneous coronary intervention. The patients were followed up for 65 months for endpoints such as death from cardiovascular causes, readmission with the ACS, or admission with congestive HF. Negative endpoints were associated with higher CTS levels after admission. Liu et al. followed up on 120 patients with coronary heart disease for 1045 days. No association between CTS levels on admission and MACEs was observed in this study [43].

Scientific reports on CTS levels in patients with CAD are inconclusive. Both increased and decreased levels of CTS have been reported as a positive prognostic factor in CAD patients. The studies' results might be puzzling due to differences in the study group selection and different endpoints reported. Multicenter prospective clinical trials in this area could address this issue.

## 6. Catestatin in Heart Failure

HF is an ineffective ability of the heart to pump blood efficiently through the body. According to a recent report from 2024, the number of Americans over the age of 20 who suffer from diseases of the heart, including HF, is estimated to be around 127.9 million, which is nearly 50% of the country's total population. However, in 2019, a much lower number was suggested globally. Across 204 countries, the number of people living with HF was around 56.2 million [70]. Neurohormonal changes aim to optimize the cardiac function of patients with HF. The activation of the renin–angiotensin–aldosterone axis and the sympathetic nervous system at first allows the body to manage impaired cardiac output by optimizing preload and afterload levels. However, over time, the process is counterproductive due to excessive peripheral vascular resistance.

Many researchers have discovered that CTS is an important member of a group of substances regulating homeostasis in the human organism [47,53,71]. CTS is stored and released, along with catecholamines, in the neuronal endings of the sympathetic nervous system. CTS affects N-Ach receptors, which decrease the release of catecholamines. Additionally, CTS causes the release of histamine, which directly causes vasodilatation. These processes lower blood pressure and, in turn, reduce the afterload. A lower afterload decreases the severity of HF [24].

A study showed that higher levels of CTS were noted in patients who died than in patients who survived, and the authors concluded that CTS plays an important protective role in managing cardiac output in HF patients. Additionally, patients suffering from

HF with ischemic origin present higher concentrations of CTS compared to non-ischemic HF [45]. A higher concentration of CTS was found in patients in the acute phase of decompensated HF of ischemic etiology. In light of this information, CTS can be viewed as a marker of sympathetic activation [72]. Moreover, researchers concluded that CTS could be a better, more sensitive marker for patients with HF whose ejection fraction was mildly reduced and preserved than patients with a reduced fraction [44]. It was confirmed that the higher the level of CTS, the more likely CV death may occur [24,44]. Researchers performed tests on animal models and proved that CTS has a protective influence on cardiomyocytes in mice with HFpEF. In this study, injections with CTS protected against the decline of ejection fraction and remodeling of cardiomyocytes [24].

In one study, the perfusion rate and left ventricular pressure (LVP) were monitored in rat hearts to further understand CTS's role in the process. In isolated cardiomyocytes, CTS improved the survival rates of cardiomyocytes by 65% after a simulated HF [73]. Furthermore, a similar treatment in rats with HF after MI gave promising results. CTS caused better capture of Ca2+ by atriums and lowered the probability of atrial fibrillation [74].

However, an elevated CTS level may correspond to acute worsening of HF with ischemic etiology and is most likely a way of compensating for increased sympathetic activity caused by the disease [45]. Another piece of research confirmed that the concentration of CTS is higher in HFrEF with a higher NYHA class, and correlates with NT-proBNP. The authors estimate that CTS could be a prognosis marker in HF because a higher level of CTS was linked with the worse long-term state of patients and all-cause death [11].

## 7. Catestatin in Other Diseases

### 7.1. Catestatin in Pre-Eclampsia

Pre-eclampsia is a serious and potentially lethal condition that may occur during pregnancy. With its significant impact on mortality and morbidity of mothers and newborns, it is a serious threat to global health. In women who have experienced pre-eclampsia, increased risks of metabolic and CVDs, such as DM, stroke, and MI, are observed, which results in reduced life expectancy. Moreover, children from pregnancies with pre-eclampsia are at increased risk of preterm birth and perinatal death. They are also in danger of neurodevelopment disability, metabolic issues, and CVDs in the future. Pre-eclampsia is a multiple-systems disease characterized by suddenly emerging HA after the 20th week of pregnancy and at least one of the following conditions: proteinuria, uteroplacental, or another organ dysfunction. A dysfunctional placenta releases a variety of proinflammatory cytokines into the circulatory system, which results in systemic inflammation, endothelial dysfunction, and, finally, the occurrence of seizures—eclampsia. Pre-eclampsia can be treated symptomatically with hypotensive drugs and anticonvulsants, which are not indifferent to the child. However, it can be cured only by termination of pregnancy, in many cases prematurely, and removal of the defective placenta [75]. The production and secretion of CgA and, indirectly, CTS occur in placental cells. CTS regulates vascular blood pressure and angiogenesis in the placenta as well as the proliferation, migration, and apoptosis of placental cells [76].

Bralewska et al. tested the expression of the CgA gene and the CTS levels in placentas among 102 patients with pre-eclampsia and 103 healthy pregnant women as a control group. Although the expression of placental CgA was significantly higher in the pre-eclampsia group ($-0.25 \pm 1.7$ vs. $-0.82 \pm 1.5$, $p = 0.011$), the mean CTS level in placentas in the pre-eclampsia group was lower than in the control group ($6.4 \pm 1.0$ vs. $6.7 \pm 1.4$, $p = 0.04$) [49]. A study on HTR-8/SVneo and BeWo Trophoblastic Cell Lines showed that CTS acts as an antiapoptotic factor. As a blood pressure-lowering agent, it also acts as a compensatory mechanism. It appears that CTS deficiency in trophoblasts may favor the occurrence of pre-eclampsia [27].

Tüten et al. compared serum CTS levels among 50 pregnant women with mild pre-eclampsia, 50 pregnant women with severe pre-eclampsia, and 100 healthy women with uncomplicated pregnancy and matched gestational age as a control group. Mean serum CTS was significantly higher in the pre-eclampsia group compared to the control group (290.7 ± 95.5 pg/mL vs. 182.8 ± 72.0 pg/mL; $p < 0.001$). Mean serum CTS was not significantly different in mild and severe pre-eclampsia groups (282.7 ± 97.9 pg/mL vs. 298.7 ± 93.4 pg/mL, $p = 0.431$) [48].

On the other hand, Palmrich et al. reported that CTS serum levels among 50 women with pre-eclamptic pregnancies were lower than in 50 healthy controls with matched gestational age (median CTS: 3.03 ng/mL, IQR [1.24–7.21 ng/mL] vs. 4.82 ng/mL, IQR [1.82–10.02 ng/mL]; $p = 0.01$). The difference in CTS levels between early- and late-onset pre-eclampsia groups was not statistically significant.

Özalp et al. also compared maternal serum CTS levels in 27 early-onset pre-eclampsia individuals and 28 late-onset pre-eclampsia with a control group of 28 healthy pregnant women. However, the differences between the groups were not statistically significant. Interestingly, researchers found correlations between maternal serum CTS levels and fetal echocardiographic parameters: The fetal E/A ratio was positively correlated with maternal CTS levels ($p < 0.001$) in both the pre-eclampsia and control groups. Fetal isovolumetric relaxation time and fetal myocardial performance index values were negatively correlated with CTS levels in both the pre-eclampsia ($p < 0.001$, $p = 0.001$, respectively) and control group ($p < 0.001$, $p = 0.002$, respectively) [50].

CTS plays an important, but not fully understood, role in the pathogenesis of pre-eclampsia. Early diagnosis and initiation of treatment are crucial to ensuring the safety of mother and child, which is often possible only in highly referential obstetric clinics. CTS appears to be a promising marker in this disease as well as other maternal and fetal CVDs. However, scientific reports in this area are not conclusive, and further research is necessary.

### *7.2. Catestatin in Acute Pulmonary Embolism*

Acute pulmonary embolism is a dangerous condition characterized by the narrowing or occlusion of a pulmonary artery or its branches by embolic material. It is most often formed by thromboses in the deep veins of the lower extremities. However, in rare cases, embolism from fat, amniotic fluid, neoplasm cells, or air may be the cause. To estimate the patient's risk of death within 30 days of the onset of acute pulmonary embolism, the validated Pulmonary Embolism Severity Index (PESI) score can be used. Depending on the severity of the disease and the capacity of the medical center, it can be treated pharmacologically by administering anticoagulants or invasive mechanical embolectomy [77,78].

It was demonstrated, in a rodent model, that CTS has an important protective role in acute pulmonary embolism. Eight-week-old C57/BL6 mice were divided into the following groups: the control group with no CTS treatment, the control group with CTS treatment, the induced acute pulmonary embolism with CTS pre-treatment group, and the induced acute pulmonary embolism without CTS pre-treatment group. Pulmonary embolism was induced by administration of collagen and epinephrine. CTS levels were decreased, while platelet numbers, von Willebrand factor, E-selectin, P-selectin, myeloperoxidase, and monocyte chemoattractant protein 1 serum levels were increased in the pulmonary embolism group compared to the control group. Interestingly, this growth was less intense in the group after CTS pre-treatment. Platelet numbers were negatively correlated with CTS levels ($r = 0.6732$, $p = 0.002$). Survival rates 30 min after acute pulmonary embolism onset were 30% in the group without CTS pre-treatment and 80% in the group with CTS pre-treatment, respectively. Lung samples were collected and assessed in histopathological examination. In acute pulmonary embolism groups, pulmonary embolization, damaged walls of alveoli, and immune cell infiltration were observed. However, pathological changes in the group after CTS pre-treatment were less severe. In mouse models, CTS mitigates endothelial inflammation and promotes thrombus resolution in acute pulmonary embolism. According to the authors, the beneficial effects of CTS are multi-causal and include inhibition of endothelial inflammation by the TLR-4-p38 signaling pathway [31].

On the other hand, Izci et al. analyzed 84 men and 76 women with acute pulmonary embolism. The sPESI scores were calculated for all patients. CTS levels in plasma among patients with acute pulmonary embolism were higher than in healthy individuals (17.5 ± 6.1 ng/mL vs. 27.3 ± 5.7 ng/mL, $p < 0.001$). In a group with sPESI ≥ 1, plasma CTS levels were higher than in a group with sPESI < 1 (37.3 ± 6.1 vs. 24.2 ± 5.3 ng/mL, $p < 0.001$). Furthermore, a positive correlation was found between CTS plasma level and sPESI score ($p < 0.001$). During the hospitalization, 20 deaths in the group with sPESI ≥ 1 (mortality 27.7%) and 9 deaths in the group with sPESI < 1 (mortality 10.2%) were reported ($p = 0.01$). Receiver operating characteristic (ROC) curve analysis with a cut-off CTS serum level of 31.2 ng/mL was performed. It showed that CTS level predicted mortality with 100% sensitivity and 52.6% specificity (AUC = 0.883, 95% CI: 0.689–0.921). CTS level was correlated with dysfunction of the right ventricle [52].

Based on available publications, CTS positively affects the course of acute pulmonary embolism in an animal model and in vitro and may be a promising disease course marker. However, the available data on CTS levels during acute pulmonary embolism are inconclusive. The differences that occurred might be related to the methods of measurement, the time between disease onset and sample collection, or differences in CTS metabolism between humans and rodents. There are limited data available in this area and further research is necessary.

### 7.3. Catestatin in Chronic Kidney Disease

Chronic kidney disease (CKD) is characterized by progressive loss of renal function. Sodium retention, volume expansion, inflammation, oxidative stress, overactivity of the sympathetic nervous system, mineral bone disorders, hormonal disorders, and uremic toxins play important roles in CKD progression and cardiovascular complications occurrence. Generalized atherosclerosis with a significant tendency to intense calcification is typical for CKD. Patients with CKD have an increased risk of HF, arrhythmias, CAD, or SCD. CVDs remain the first cause of death in G5 CKD patients on hemodialysis, with 20 times higher CVD-related mortality than in the general population. Furthermore, many patients with CKD die due to CVD complications before reaching end-stage renal dysfunction [79,80].

Luketin et al. compared plasma CTS levels and advanced glycation end products (AGEs) among 91 hemodialysis patients and 70 healthy individuals. AGEs promote atherosclerosis in an inflammatory environment, leading to cardiovascular events among patients. The study showed that plasma CTS levels were increased in the hemodialysis group compared to the control group (32.85 ± 20.18 vs. 5.39 ± 1.24 ng/mL, $p < 0.001$). The CTS serum level was positively correlated with AGE levels ($r = 0.492$, $p < 0.001$). Moreover, CTS serum levels were also positively correlated with both the Dialysis Malnutrition Score ($r = 0.295$, $p = 0.004$) and the Malnutrition-Inflammation Score ($r = 0.290$, $p = 0.005$) [51].

Sun et al. followed up on 330 hemodialysis patients for 36 months. In that period, 29 deaths due to CVD and 28 deaths due to other diseases were reported, and one patient was lost. The multivariate regression analysis among 272 alive patients showed an association between plasma CTS level ≥ 1.9 ng/mL and increased risk of cardiac death (RR 6.13, 95% CI 2.54, 18.45). Furthermore, survival analysis showed an increased rate of cardiac death in the group with plasma CTS levels $\geq$ 1.9 ng/mL than in the group with plasma CTS levels < 1.9 ng/mL ($p < 0.001$). Also, the overhydration to total body weight ratio and daily diuresis were linearly correlated with plasma CTS level (r = 0.502, $p < 0.001$ and r = −0.338, $p < 0.001$) [79].

According to the available scientific data, increased CTS serum levels are typical for hemodialyzed patients and can be linked to a worse prognosis. It should be assumed that CTS plays an important role in the pathophysiology of cardiovascular complications in CKD hemodialysis patients, and it could be a promising novel marker of disease severity. However, the availability of scientific evidence is limited, and confirmation in further research is necessary.

## 8. Conclusions and Future Perspectives

To conclude, cardiovascular disease pathogenesis is an important and still not fully understood field of study. Studies on CTS are promising and show the possibility of using CTS as a treatment for many CVDs. Additionally, CTS increases the production of NO in cardiomyocytes and endothelial cells. CTS's role in the development of CVDs and its potential as a therapeutic target are interesting topics in cardiology. The aforementioned studies suggest the importance of this peptide for the development of common diseases, which remain a serious and expensive problem for healthcare systems worldwide. Moreover, CTS can be a biochemical marker for patients with preserved and mildly reduced HF. On the other hand, there are diseases in which the role of CTS is still not fully understood, e.g., atherosclerosis—studies about the influence of CTS on the remodeling of the vessel walls are inconsistent. Some of the reported studies are just preliminary to the subsequent phases of research on CTS involvement in CVDs. However, it is too soon to generalize CTS's importance for CAD, HF, and AH.

**Author Contributions:** Conceptualization, J.R. and A.G.-R.; methodology, J.R.; formal analysis, J.G.; writing—original draft preparation, J.K., M.S., J.P. and M.T.; writing—review and editing, J.R. and A.G.-R.; supervision, A.P. and J.G. All authors have read and agreed to the published version of the manuscript.

**Funding:** This research received no external funding.

**Institutional Review Board Statement:** Not applicable.

**Informed Consent Statement:** Not applicable.

**Data Availability Statement:** Not applicable.

**Conflicts of Interest:** The authors declare no conflicts of interest.

## Abbreviations

The following abbreviations are used in this manuscript:

| | |
|---|---|
| 95% CI | 95% confidence interval |
| ACS | Acute coronary syndrome |
| ANS | Autonomic nervous system |
| CAD | Coronary artery disease |
| CgA | Chromogranin A |
| CKD | Chronic kidney disease |
| CTS | Catestatin |
| CTO | Chronic total occlusion of coronary artery |
| CVD | Cardiovascular disease |
| DM | Diabetes mellitus |

| | |
|---|---|
| HA | Arterial hypertension |
| HF | Heart failure |
| HR | Hazard ratio |
| HRV | Heart rate variability |
| LDL | Low-density lipoproteins |
| LVP | Left ventricular pressure |
| MACE | Major adverse cardiovascular events |
| MI | Myocardial infarction |
| SCD | Sudden cardiac death |
| STEMI | Myocardial infarction with ST-segment elevation |
| UAP | Unstable angina pectoris |
| VEGF | vascular endothelial growth factor |
| VF | Ventricular fibrillation |
| VT | Ventricular tachycardia |

## References


1. Townsend, N.; Kazakiewicz, D.; Lucy Wright, F.; Timmis, A.; Huculeci, R.; Torbica, A.; Gale, C.P.; Achenbach, S.; Weidinger, F.; Vardas, P. Epidemiology of cardiovascular disease in Europe. *Nat. Rev. Cardiol.* **2022**, *19*, 133–143. [CrossRef] [PubMed]
2. Watanabe, T. The Emerging Roles of Chromogranins and Derived Polypeptides in Atherosclerosis, Diabetes, and Coronary Heart Disease. *Int. J. Mol. Sci.* **2021**, *22*, 6118. [CrossRef] [PubMed]
3. Pandolfi, S.; Chirumbolo, S.; Franzini, M.; Tirelli, U.; Valdenassi, L. Oxygen-ozone therapy for myocardial ischemic stroke and cardiovascular disorders. *Med. Gas Res.* **2025**, *15*, 36–43. [CrossRef]
4. Gorący, J.; Gorący, A.; Wójcik-Grzeszczuk, A.; Gorący, I.; Rosik, J. Analysis of Genetic Variants in the Glucocorticoid Receptor Gene. *Biomedicines* **2022**, *10*, 1912. [CrossRef] [PubMed]
5. Tan, S.C.W.; Zheng, B.B.; Tang, M.L.; Chu, H.; Zhao, Y.T.; Weng, C. Global Burden of Cardiovascular Diseases and its Risk Factors, 1990-2021: A Systematic Analysis for the Global Burden of Disease Study 2021. *QJM* **2025**, hcaf022. [CrossRef]
6. Garg, R.; Agarwal, A.; Katekar, R.; Dadge, S.; Yadav, S.; Gayen, J.R. Chromogranin A-derived peptides pancreastatin and catestatin: Emerging therapeutic target for diabetes. *Amino Acids* **2023**, *55*, 549–561. [CrossRef]
7. Zalewska, E.; Kmieć, P.; Sworczak, K. Role of Catestatin in the Cardiovascular System and Metabolic Disorders. *Front. Cardiovasc. Med.* **2022**, *9*, 909480. [CrossRef]
8. Bozic, J.; Kumric, M.; Ticinovic Kurir, T.; Urlic, H.; Martinovic, D.; Vilovic, M.; Tomasovic Mrcela, N.; Borovac, J.A. Catestatin as a Biomarker of Cardiovascular Diseases: A Clinical Perspective. *Biomedicines* **2021**, *9*, 1757. [CrossRef]
9. Meng, Q.H.; Halfdanarson, T.R.; Bornhorst, J.A.; Jann, H.; Shaheen, S.; Shi, R.Z.; Schwabe, A.; Stade, K.; Halperin, D.M. Circulating Chromogranin A as a Surveillance Biomarker in Patients with Carcinoids-The CASPAR Study. *Clin. Cancer Res.* **2024**, *30*, 5559–5567. [CrossRef]
10. Vanli Tonyali, N.; Karabay, G.; Arslan, B.; Aktemur, G.; Tokgoz Cakir, B.; Seyhanli, Z.; Demir Çendek, B.; Yilmaz Ergani, S.; Eroglu, H.; Mermi, S.; et al. Maternal Serum Catestatin Levels in Gestational Diabetes Mellitus: A Potential Biomarker for Risk Assessment and Diagnosis. *J. Clin. Med.* **2025**, *14*, 435. [CrossRef]
11. Wołowiec, Ł.; Banach, J.; Budzyński, J.; Wołowiec, A.; Kozakiewicz, M.; Bieliński, M.; Jaśniak, A.; Olejarczyk, A.; Grześk, G. Prognostic Value of Plasma Catestatin Concentration in Patients with Heart Failure with Reduced Ejection Fraction in Two-Year Follow-Up. *J. Clin. Med.* **2023**, *12*, 4208. [CrossRef]
12. Loh, Y.P.; Cheng, Y.; Mahata, S.K.; Corti, A.; Tota, B. Chromogranin A and derived peptides in health and disease. *J. Mol. Neurosci.* **2012**, *48*, 347–356. [CrossRef]
13. Muntjewerff, E.M.; Christoffersson, G.; Mahata, S.K.; van den Bogaart, G. Putative regulation of macrophage-mediated inflammation by catestatin. *Trends Immunol.* **2022**, *43*, 41–50. [CrossRef] [PubMed]
14. Bourebaba, Y.; Mularczyk, M.; Marycz, K.; Bourebaba, L. Catestatin peptide of chromogranin A as a potential new target for several risk factors management in the course of metabolic syndrome. *Biomed. Pharmacother.* **2021**, *134*, 111113. [CrossRef]
15. Al Ghorani, H.; Kulenthiran, S.; Lauder, L.; Böhm, M.; Mahfoud, F. Hypertension trials update. *J. Hum. Hypertens.* **2021**, *35*, 398–409. [CrossRef] [PubMed]
16. Mahapatra, N.R. Catestatin is a novel endogenous peptide that regulates cardiac function and blood pressure. *Cardiovasc. Res.* **2008**, *80*, 330–338. [CrossRef] [PubMed]
17. Angelone, T.; Quintieri, A.M.; Brar, B.K.; Limchaiyawat, P.T.; Tota, B.; Mahata, S.K.; Cerra, M.C. The antihypertensive chromogranin a peptide catestatin acts as a novel endocrine/paracrine modulator of cardiac inotropism and lusitropism. *Endocrinology* **2008**, *149*, 4780–4793. [CrossRef]

18. Ying, W.; Tang, K.; Avolio, E.; Schilling, J.M.; Pasqua, T.; Liu, M.A.; Cheng, H.; Gao, H.; Zhang, J.; Mahata, S.; et al. Immunosuppression of Macrophages Underlies the Cardioprotective Effects of CST (Catestatin). *Hypertension* **2021**, *77*, 1670–1682. [CrossRef]
19. Kübler, W.; Haass, M. Cardioprotection: Definition, classification, and fundamental principles. *Heart* **1996**, *75*, 330–333. [CrossRef]
20. Jianqiang, G.; Li, S.; Guo, J. GW27-e0952 Effect Of Catecholamine Release-Inhibitory Peptide Catestatin on Sympathetic Activity Of Hypertension. *J. Am. Coll. Cardiol.* **2016**, *68*, C33–C34. [CrossRef]
21. Krüger, P.-G.; Mahata, S.K.; Helle, K.B. Catestatin (CgA344–364) stimulates rat mast cell release of histamine in a manner comparable to mastoparan and other cationic charged neuropeptides. *Regul. Pept.* **2003**, *114*, 29–35. [CrossRef] [PubMed]
22. Alam, M.J.; Gupta, R.; Mahapatra, N.R.; Goswami, S.K. Catestatin reverses the hypertrophic effects of norepinephrine in H9c2 cardiac myoblasts by modulating the adrenergic signaling. *Mol. Cell Biochem.* **2020**, *464*, 205–219. [CrossRef]
23. Dev, N.B.; Gayen, J.R.; O'Connor, D.T.; Mahata, S.K. Chromogranin a and the autonomic system: Decomposition of heart rate variability and rescue by its catestatin fragment. *Endocrinology* **2010**, *151*, 2760–2768. [CrossRef] [PubMed]
24. Qiu, Z.; Fan, Y.; Wang, Z.; Huang, F.; Li, Z.; Sun, Z.; Hua, S.; Jin, W.; Chen, Y. Catestatin Protects Against Diastolic Dysfunction by Attenuating Mitochondrial Reactive Oxygen Species Generation. *J. Am. Heart Assoc.* **2023**, *12*, e029470. [CrossRef]
25. Zhang, Y.; Chen, H.; Ma, Q.; Jia, H.; Ma, H.; Du, Z.; Liu, Y.; Zhang, X.; Guan, Y. Electrophysiological Mechanism of Catestatin Antiarrhythmia: Enhancement of. *J. Am. Heart Assoc.* **2024**, *13*, e035415. [CrossRef]
26. Lener, D.; Noflatscher, M.; Kirchmair, E.; Bauer, A.; Holfeld, J.; Gollmann-Tepeköylü, C.; Kirchmair, R.; Theurl, M. The angiogenic neuropeptide catestatin exerts beneficial effects on human coronary vascular cells and cardiomyocytes. *Peptides* **2023**, *168*, 171077. [CrossRef] [PubMed]
27. Bralewska, M.; Pietrucha, T.; Sakowicz, A. Reduction in CgA-Derived CST Protein Level in HTR-8/SVneo and BeWo Trophoblastic Cell Lines Caused by the Preeclamptic Environment. *Int. J. Mol. Sci.* **2023**, *24*, 7124. [CrossRef]
28. Muntjewerff, E.M.; Parv, K.; Mahata, S.K.; van Riessen, N.K.; Phillipson, M.; Christoffersson, G.; van den Bogaart, G. The anti-inflammatory peptide Catestatin blocks chemotaxis. *J. Leukoc. Biol.* **2022**, *112*, 273–278. [CrossRef]
29. Chu, S.Y.; Peng, F.; Wang, J.; Liu, L.; Meng, L.; Zhao, J.; Han, X.N.; Ding, W.H. Catestatin in defense of oxidative-stress-induced apoptosis: A novel mechanism by activating the beta2 adrenergic receptor and PKB/Akt pathway in ischemic-reperfused myocardium. *Peptides* **2020**, *123*, 170200. [CrossRef]
30. Chen, Y.; Wang, X.; Yang, C.; Su, X.; Yang, W.; Dai, Y.; Han, H.; Jiang, J.; Lu, L.; Wang, H.; et al. Decreased circulating catestatin levels are associated with coronary artery disease: The emerging anti-inflammatory role. *Atherosclerosis* **2019**, *281*, 78–88. [CrossRef]
31. Chen, H.; Liu, D.; Ge, L.; Wang, T.; Ma, Z.; Han, Y.; Duan, Y.; Xu, X.; Liu, W.; Yuan, J.; et al. Catestatin prevents endothelial inflammation and promotes thrombus resolution in acute pulmonary embolism in mice. *Biosci. Rep.* **2019**, *39*, BSR20192236. [CrossRef] [PubMed]
32. Gaede, A.H.; Pilowsky, P.M. Catestatin in rat RVLM is sympathoexcitatory, increases barosensitivity, and attenuates chemosensitivity and the somatosympathetic reflex. *Am. J. Physiol. Regul. Integr. Comp. Physiol.* **2010**, *299*, R1538–R1545. [CrossRef]
33. Gaede, A.H.; Pilowsky, P.M. Catestatin, a chromogranin A-derived peptide, is sympathoinhibitory and attenuates sympathetic barosensitivity and the chemoreflex in rat CVLM. *Am. J. Physiol. Regul. Integr. Comp. Physiol.* **2012**, *302*, R365–R372. [CrossRef]
34. Avolio, E.; Mahata, S.K.; Mantuano, E.; Mele, M.; Alò, R.; Facciolo, R.M.; Talani, G.; Canonaco, M. Antihypertensive and neuroprotective effects of catestatin in spontaneously hypertensive rats: Interaction with GABAergic transmission in amygdala and brainstem. *Neuroscience* **2014**, *270*, 48–57. [CrossRef]
35. O'Connor, D.T.; Kailasam, M.T.; Kennedy, B.P.; Ziegler, M.G.; Yanaihara, N.; Parmer, R.J. Early decline in the catecholamine release-inhibitory peptide catestatin in humans at genetic risk of hypertension. *J. Hypertens.* **2002**, *20*, 1335–1345. [CrossRef] [PubMed]
36. Durakoğlugil, M.E.; Ayaz, T.; Kocaman, S.A.; Kırbaş, A.; Durakoğlugil, T.; Erdoğan, T.; Çetin, M.; Şahin, O.Z.; Çiçek, Y. The relationship of plasma catestatin concentrations with metabolic and vascular parameters in untreated hypertensive patients: Influence on high-density lipoprotein cholesterol. *Anatol. J. Cardiol.* **2015**, *15*, 577–585. [CrossRef]
37. Meng, L.; Ye, X.J.; Ding, W.H.; Yang, Y.; Di, B.B.; Liu, L.; Huo, Y. Plasma catecholamine release-inhibitory peptide catestatin in patients with essential hypertension. *J. Cardiovasc. Med.* **2011**, *12*, 643–647. [CrossRef]
38. O'Connor, D.T.; Zhu, G.; Rao, F.; Taupenot, L.; Fung, M.M.; Das, M.; Mahata, S.K.; Mahata, M.; Wang, L.; Zhang, K.; et al. Heritability and genome-wide linkage in US and australian twins identify novel genomic regions controlling chromogranin a: Implications for secretion and blood pressure. *Circulation* **2008**, *118*, 247–257. [CrossRef]
39. Choi, Y.; Miura, M.; Nakata, Y.; Sugasawa, T.; Nissato, S.; Otsuki, T.; Sugawara, J.; Iemitsu, M.; Kawakami, Y.; Shimano, H.; et al. A common genetic variant of the chromogranin A-derived peptide catestatin is associated with atherogenesis and hypertension in a Japanese population. *Endocr. J.* **2015**, *62*, 797–804. [CrossRef]
40. Rao, F.; Wen, G.; Gayen, J.R.; Das, M.; Vaingankar, S.M.; Rana, B.K.; Mahata, M.; Kennedy, B.P.; Salem, R.M.; Stridsberg, M.; et al. Catecholamine release-inhibitory peptide catestatin (chromogranin A(352-372)): Naturally occurring amino acid variant Gly364Ser

causes profound changes in human autonomic activity and alters risk for hypertension. *Circulation* **2007**, *115*, 2271–2281. [CrossRef]
41. Xu, W.X.; Fan, Y.Y.; Song, Y.; Liu, X.; Liu, H.; Guo, L.J. Prognostic differences of catestatin among young and elderly patients with acute myocardial infarction. *World J. Emerg. Med.* **2022**, *13*, 169–174. [CrossRef]
42. Xu, W.; Yu, H.; Wu, H.; Li, S.; Chen, B.; Gao, W. Plasma Catestatin in Patients with Acute Coronary Syndrome. *Cardiology* **2017**, *136*, 164–169. [CrossRef]
43. Liu, L.; Ding, W.; Zhao, F.; Shi, L.; Pang, Y.; Tang, C. Plasma levels and potential roles of catestatin in patients with coronary heart disease. *Scand. Cardiovasc. J.* **2013**, *47*, 217–224. [CrossRef]
44. Chu, S.Y.; Peng, F.; Wang, J.; Liu, L.; Zhao, J.; Han, X.N.; Ding, W.H. Catestatin as a predictor for cardiac death in heart failure with mildly reduced and preserved ejection fraction. *ESC Heart Fail.* **2025**, *12*, 517–524. [CrossRef]
45. Borovac, J.A.; Glavas, D.; Susilovic Grabovac, Z.; Supe Domic, D.; Stanisic, L.; D'Amario, D.; Kwok, C.S.; Bozic, J. Circulating sST2 and catestatin levels in patients with acute worsening of heart failure: A report from the CATSTAT-HF study. *ESC Heart Fail.* **2020**, *7*, 2818–2828. [CrossRef]
46. Zhu, D.; Xie, H.; Wang, X.; Liang, Y.; Yu, H.; Gao, W. Catestatin-A Novel Predictor of Left Ventricular Remodeling After Acute Myocardial Infarction. *Sci. Rep.* **2017**, *7*, 44168. [CrossRef]
47. Palmrich, P.; Schirwani-Hartl, N.; Haberl, C.; Haslinger, P.; Heinzl, F.; Zeisler, H.; Binder, J. Catestatin-A Potential New Therapeutic Target for Women with Preeclampsia? An Analysis of Maternal Serum Catestatin Levels in Preeclamptic Pregnancies. *J. Clin. Med.* **2023**, *12*, 5931. [CrossRef]
48. Tüten, N.; Güralp, O.; Gök, K.; Hamzaoglu, K.; Oner, Y.O.; Makul, M.; Bulut, H.; Irmak, K.; Tüten, A.; Malik, E. Serum catestatin level is increased in women with preeclampsia. *J. Obstet. Gynaecol.* **2022**, *42*, 55–60. [CrossRef]
49. Bralewska, M.; Biesiada, L.; Grzesiak, M.; Rybak-Krzyszkowska, M.; Huras, H.; Gach, A.; Pietrucha, T.; Sakowicz, A. Chromogranin A demonstrates higher expression in preeclamptic placentas than in normal pregnancy. *BMC Pregnancy Childbirth* **2021**, *21*, 680. [CrossRef]
50. Özalp, M.; Yaman, H.; Demir, Ö.; Aytekin Garip, S.; Aran, T.; Osmanağaoğlu, M.A. The role of maternal serum catestatin in the evaluation of preeclampsia and fetal cardiac functions. *Turk. J. Obstet. Gynecol.* **2021**, *18*, 272–278. [CrossRef] [PubMed]
51. Luketin, M.; Mizdrak, M.; Boric-Skaro, D.; Martinovic, D.; Tokic, D.; Vilovic, M.; Supe-Domic, D.; Ticinovic Kurir, T.; Bozic, J. Plasma Catestatin Levels and Advanced Glycation End Products in Patients on Hemodialysis. *Biomolecules* **2021**, *11*, 456. [CrossRef]
52. Izci, S.; Acar, E.; Inanir, M. Plasma catestatin level predicts sPESI score and mortality in acute pulmonary embolism. *Arch. Med. Sci. Atheroscler. Dis.* **2020**, *5*, e49–e56. [CrossRef] [PubMed]
53. Pàmies, A.; Llop, D.; Ibarretxe, D.; Rosales, R.; Girona, J.; Masana, L.; Vallvé, J.C.; Paredes, S. Enhanced Association of Novel Cardiovascular Biomarkers Fetuin-A and Catestatin with Serological and Inflammatory Markers in Rheumatoid Arthritis Patients. *Int. J. Mol. Sci.* **2024**, *25*, 9910. [CrossRef]
54. Tiwari, R.; Kumar, R.; Malik, S.; Raj, T.; Kumar, P. Analysis of Heart Rate Variability and Implication of Different Factors on Heart Rate Variability. *Curr. Cardiol. Rev.* **2021**, *17*, e160721189770. [CrossRef]
55. Borovac, J.A.; Glavas, D.; Susilovic Grabovac, Z.; Supe Domic, D.; D'Amario, D.; Bozic, J. Catestatin in Acutely Decompensated Heart Failure Patients: Insights from the CATSTAT-HF Study. *J. Clin. Med.* **2019**, *8*, 1132. [CrossRef]
56. Dev, N.B.; Mir, S.A.; Gayen, J.R.; Siddiqui, J.A.; Mustapic, M.; Vaingankar, S.M. Cardiac Electrical Activity in a Genomically "Humanized" Chromogranin A Monogenic Mouse Model with Hyperadrenergic Hypertension. *J. Cardiovasc. Transl. Res.* **2014**, *7*, 483–493. [CrossRef]
57. Kiranmayi, M.; Chirasani, V.R.; Allu, P.K.; Subramanian, L.; Martelli, E.E.; Sahu, B.S.; Vishnuprabu, D.; Kumaragurubaran, R.; Sharma, S.; Bodhini, D.; et al. Catestatin Gly364Ser Variant Alters Systemic Blood Pressure and the Risk for Hypertension in Human Populations via Endothelial Nitric Oxide Pathway. *Hypertension* **2016**, *68*, 334–347. [CrossRef]
58. Sahu, B.S.; Obbineni, J.M.; Sahu, G.; Allu, P.K.; Subramanian, L.; Sonawane, P.J.; Singh, P.K.; Sasi, B.K.; Senapati, S.; Maji, S.K.; et al. Functional genetic variants of the catecholamine-release-inhibitory peptide catestatin in an Indian population: Allele-specific effects on metabolic traits. *J. Biol. Chem.* **2012**, *287*, 43840–43852. [CrossRef]
59. Bassino, E.; Fornero, S.; Gallo, M.P.; Ramella, R.; Mahata, S.K.; Tota, B.; Levi, R.; Alloatti, G. A novel catestatin-induced antiadrenergic mechanism triggered by the endothelial PI3K-eNOS pathway in the myocardium. *Cardiovasc. Res.* **2011**, *91*, 617–624. [CrossRef] [PubMed]
60. Ralapanawa, U.; Sivakanesan, R. Epidemiology and the Magnitude of Coronary Artery Disease and Acute Coronary Syndrome: A Narrative Review. *J. Epidemiol. Glob. Health* **2021**, *11*, 169–177. [CrossRef] [PubMed]
61. Alique, M.; Luna, C.; Carracedo, J.; Ramírez, R. LDL biochemical modifications: A link between atherosclerosis and aging. *Food Nutr. Res.* **2015**, *59*, 29240. [CrossRef]
62. Jebari-Benslaiman, S.; Galicia-García, U.; Larrea-Sebal, A.; Olaetxea, J.R.; Alloza, I.; Vandenbroeck, K.; Benito-Vicente, A.; Martín, C. Pathophysiology of Atherosclerosis. *Int. J. Mol. Sci.* **2022**, *23*, 3346. [CrossRef] [PubMed]

63. Theofilis, P.; Sagris, M.; Oikonomou, E.; Antonopoulos, A.S.; Siasos, G.; Tsioufis, C.; Tousoulis, D. Inflammatory Mechanisms Contributing to Endothelial Dysfunction. *Biomedicines* **2021**, *9*, 781. [CrossRef] [PubMed]
64. Guijarro, C.; Cosín-Sales, J. LDL cholesterol and atherosclerosis: The evidence. *Clin. Investig. Arterioscler.* **2021**, *33* (Suppl. S1), 25–32. [PubMed]
65. Libby, P. The changing landscape of atherosclerosis. *Nature* **2021**, *592*, 524–533. [CrossRef]
66. Muntjewerff, E.M.; Dunkel, G.; Nicolasen, M.J.T.; Mahata, S.K.; van den Bogaart, G. Catestatin as a Target for Treatment of Inflammatory Diseases. *Front. Immunol.* **2018**, *9*, 2199. [CrossRef]
67. Xu, W.; Yu, H.; Li, W.; Gao, W.; Guo, L.; Wang, G. Plasma Catestatin: A Useful Biomarker for Coronary Collateral Development with Chronic Myocardial Ischemia. *PLoS ONE* **2016**, *11*, e0149062. [CrossRef]
68. Bhar-Amato, J.; Davies, W.; Agarwal, S. Ventricular Arrhythmia after Acute Myocardial Infarction: 'The Perfect Storm'. *Arrhythm. Electrophysiol. Rev.* **2017**, *6*, 134–139. [CrossRef]
69. Pei, Z.; Ma, D.; Ji, L.; Zhang, J.; Su, J.; Xue, W.; Chen, X.; Wang, W. Usefulness of catestatin to predict malignant arrhythmia in patients with acute myocardial infarction. *Peptides* **2014**, *55*, 131–135.
70. Martin, S.S.; Aday, A.W.; Almarzooq, Z.I.; Anderson, C.A.M.; Arora, P.; Avery, C.L.; Baker-Smith, C.M.; Barone Gibbs, B.; Beaton, A.Z.; Boehme, A.K.; et al. 2024 Heart Disease and Stroke Statistics: A Report of US and Global Data from the American Heart Association. *Circulation* **2024**, *149*, e347–e913.
71. Mahata, S.K.; Kiranmayi, M.; Mahapatra, N.R. Catestatin: A Master Regulator of Cardiovascular Functions. *Curr. Med. Chem.* **2018**, *25*, 1352–1374. [CrossRef]
72. Polyakova, E.A.; Mikhaylov, E.N.; Sonin, D.L.; Cheburkin, Y.V.; Galagudza, M.M. Neurohumoral, cardiac and inflammatory markers in the evaluation of heart failure severity and progression. *J. Geriatr. Cardiol.* **2021**, *18*, 47–66. [PubMed]
73. Penna, C.; Alloatti, G.; Gallo, M.P.; Cerra, M.C.; Levi, R.; Tullio, F.; Bassino, E.; Dolgetta, S.; Mahata, S.K.; Tota, B.; et al. Catestatin improves post-ischemic left ventricular function and decreases ischemia/reperfusion injury in heart. *Cell Mol. Neurobiol.* **2010**, *30*, 1171–1179. [CrossRef] [PubMed]
74. Yan, M.; Liu, T.; Zhong, P.; Xiong, F.; Cui, B.; Wu, J.; Wu, G. Chronic catestatin treatment reduces atrial fibrillation susceptibility via improving calcium handling in post-infarction heart failure rats. *Peptides* **2023**, *159*, 170904. [CrossRef] [PubMed]
75. Dimitriadis, E.; Rolnik, D.L.; Zhou, W.; Estrada-Gutierrez, G.; Koga, K.; Francisco, R.P.V.; Whitehead, C.; Hyett, J.; da Silva Costa, F.; Nicolaides, K.; et al. Pre-eclampsia. *Nat. Rev. Dis. Primers* **2023**, *9*, 8. [CrossRef]
76. Bralewska, M.; Pietrucha, T.; Sakowicz, A. The Role of Catestatin in Preeclampsia. *Int. J. Mol. Sci.* **2024**, *25*, 2461. [CrossRef]
77. Hassine, M.; Kallala, M.Y.; Mahjoub, M.; Boussaada, M.; Bouchahda, N.; Gamra, H. Pulmonary embolism: The Pulmonary Embolism Severity Index (PESI) score and mortality predictors. *Pan Afr. Med. J.* **2023**, *45*, 48.
78. Shah, I.K.; Merfeld, J.M.; Chun, J.; Tak, T. Pathophysiology and Management of Pulmonary Embolism. *Int. J. Angiol.* **2022**, *31*, 143–149. [CrossRef]
79. Cozzolino, M.; Mangano, M.; Stucchi, A.; Ciceri, P.; Conte, F.; Galassi, A. Cardiovascular disease in dialysis patients. *Nephrol. Dial. Transplant.* **2018**, *33* (Suppl. S3), iii28–iii34. [CrossRef]
80. Zoccali, C.; Mallamaci, F.; Adamczak, M.; de Oliveira, R.B.; Massy, Z.A.; Sarafidis, P.; Agarwal, R.; Mark, P.B.; Kotanko, P.; Ferro, C.J.; et al. Cardiovascular complications in chronic kidney disease: A review from the European Renal and Cardiovascular Medicine Working Group of the European Renal Association. *Cardiovasc. Res.* **2023**, *119*, 2017–2032. [CrossRef]

International Journal of
*Molecular Sciences*

MDPI

*Review*

# The Interaction of Vasopressin with Hormones of the Hypothalamo–Pituitary–Adrenal Axis: The Significance for Therapeutic Strategies in Cardiovascular and Metabolic Diseases

**Ewa Szczepanska-Sadowska *, Katarzyna Czarzasta, Wiktor Bogacki-Rychlik and Michał Kowara**

Department of Experimental and Clinical Physiology, Laboratory of Centre for Preclinical Research, Medical University of Warsaw, 02-097 Warsaw, Poland
* Correspondence: eszczepanska@wum.edu.pl

**Abstract:** A large body of evidence indicates that vasopressin (AVP) and steroid hormones are frequently secreted together and closely cooperate in the regulation of blood pressure, metabolism, water–electrolyte balance, and behavior, thereby securing survival and the comfort of life. Vasopressin cooperates with hormones of the hypothalamo–pituitary–adrenal axis (HPA) at several levels through regulation of the release of corticotropin-releasing hormone (CRH), adrenocorticotropic hormone (ACTH), and multiple steroid hormones, as well as through interactions with steroids in the target organs. These interactions are facilitated by positive and negative feedback between specific components of the HPA. Altogether, AVP and the HPA cooperate closely as a coordinated functional AVP-HPA system. It has been shown that cooperation between AVP and steroid hormones may be affected by cellular stress combined with hypoxia, and by metabolic, cardiovascular, and respiratory disorders; neurogenic stress; and inflammation. Growing evidence indicates that central and peripheral interactions between AVP and steroid hormones are reprogrammed in cardiovascular and metabolic diseases and that these rearrangements exert either beneficial or harmful effects. The present review highlights specific mechanisms of the interactions between AVP and steroids at cellular and systemic levels and analyses the consequences of the inappropriate cooperation of various components of the AVP-HPA system for the pathogenesis of cardiovascular and metabolic diseases.

**Keywords:** AVP; CRH; ACTH; cardiac failure; glucocorticoids; mineralocorticoids; androgens; estrogens; hypertension; stress

## 1. Introduction

The first experimental studies showing the close functional relationship between vasopressin and steroid hormones were published over 60 years ago. In 1960 Hilton et al. reported that arterial perfusion of the adrenal glands with arginine vasopressin (AVP) or lysine vasopressin (LVP) potently stimulates the release of cortisol [1]. Subsequently, it has been found that in many instances vasopressin and steroid hormones are secreted together and closely cooperate in the regulation of blood pressure, metabolism, water–electrolyte balance, and behavior in a manner securing survival and the comfort of life. It has also been shown that cellular stress combined with hypoxia, disturbances of metabolism, cardiovascular and respiratory disorders, neurogenic stress, and inflammation may disorganize the cooperation between AVP and steroid hormones. Under homeostatic, unstressed conditions both steroid hormones and vasopressin are released in characteristic diurnal rhythms. The expression of AVP mRNA in the suprachiasmatic nucleus shows a distinct endogenous circadian rhythm. This diurnal rhythm of secretion is also typical for adrenocorticotropic hormone (ACTH) and glucocorticoids (GCs). Interestingly, it appears that the regulation of AVP synthesis by steroids also manifests circadian rhythmicity [2].

Vasopressin interacts with the hypothalamo–pituitary–adrenal axis (HPA) at various levels, i.e., in the hypothalamic nuclei, affecting release of CRH; in the pituitary gland, enhancing ACTH release; and in the adrenal glands, modulating the release or action of multiple steroid hormones (Figure 1). These interactions are facilitated by the positive and negative feedback occurring between specific components of the HPA system with the engagement of other neurotransmitting and neuropeptidergic pathways.

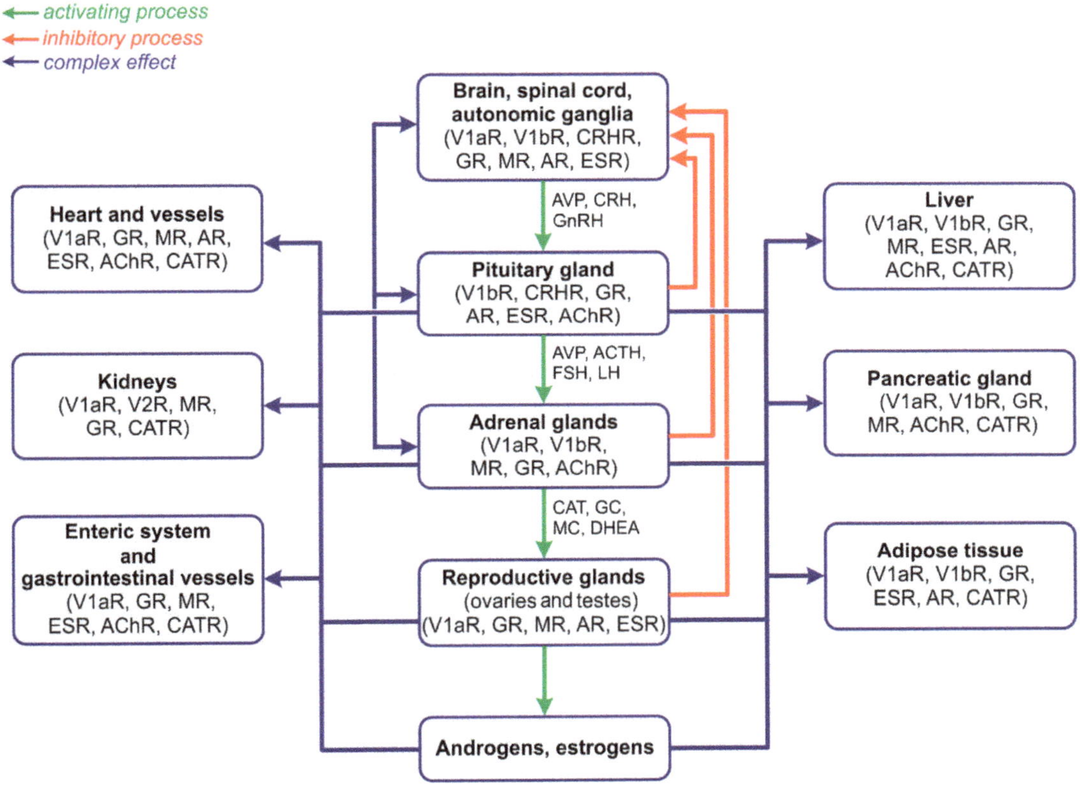


**Figure 1.** Flow diagram showing the sites of the synthesis, action, and interaction of vasopressin and hormones of the hypothalamo–pituitary–adrenal axis in the central nervous system and peripheral organs. Green arrows—stimulatory effects; red arrows—inhibitory effects. Blue arrows—complex effects (stimulatory or inhibitory, see text for explanations). Abbreviations: AChR—aceytolcholine receptor; ACTH—adrenocorticotropic hormone; AR—androgen receptor; CAT—catecholamine; CATR—catecholamine receptor; CRHR—corticotropin releasing hormone receptor; ESR—estrogen receptor; FSH—follicle-stimulating hormone; GC—glucocorticoid; GnRH—gonadotropin-releasing hormone; GR—glucocorticoid receptor; LH—luteinizing hormone; MC—mineralocorticoid; MR—mineralocorticoid receptor; V1aR—vasopressin V1a receptor; V1bR—vasopressin V1b receptor; V2R—vasopressin V2 receptor.

Vasopressin and CRH are synthesized mainly in the hypothalamic nuclei; however, both these peptides can also be produced in other regions of the brain. In the hypothalamus, vasopressin and CRH are released partly by the same cells of the paraventricular nucleus (PVN). Immunostaining studies have shown that virtually all parvocellular CRH neurons in the PVN are stained positively for vasopressin [3] and are co-packaged in neurosecretory vesicles of the hypothalamic–pituitary axons in the median eminence [4]. and in the hypophyseal portal circulation [5–7]. There is also evidence that AVP may be necessary for the appropriate action of glucocorticoids, as it was shown that the maximal binding capacities for corticosterone and dexamethasone in the hippocampus and the anterior pituitary are significantly lower in homozygous diabetes insipidus (HDI) rats that do not synthesize vasopressin than in their nondiabetic counterparts. Importantly, the difference could be eliminated by AVP treatment [8]. On the other hand, glucocorticoids were found to exert a negative effect on the release and action of vasopressin. For instance, in rats, corticosterone was found to inhibit the release of AVP from the explants containing a supraoptic nucleus (SON) and sending projections to the neural lobe of the pituitary [9]. Experiments on Sprague Dawley rats have shown that the release of AVP from the hypothalamic slices encompassing PVN and SON neurons could be inhibited by corticosterone, cortisol, testosterone, and 17-beta estradiol in a dose-dependent manner, whereas dexamethasone, aldosterone, and progesterone were not effective [10]. The chronic application of dexamethasone, which is an agonist of the corticosterone receptor, significantly decreased plasma corticosterone and ACTH concentrations and elicited differential changes in the expressions of CRH, ACTH, and AVP in several brain regions, including the cortex, the hippocampus, the hypothalamus, and the cerebellum [11]. There is also evidence that, in the rat, AVP exerts a weak stimulatory effect on the secretion of aldosterone from the glomerulosa cells of the adrenal medulla [12].

It is highly probable that some steroid hormones can modulate vasopressin's action through actions exerted at the level of the AVP receptors. For instance, experiments on rats have shown that adrenalectomy, which eliminates the main source of circulating steroid hormones, reduces vasopressin V1a receptors (V1aR) in the hippocampus, the dorsolateral septum, and the bed nucleus of the stria terminalis (BNST). Moreover, the effects of adrenalectomy on V1aR in the hippocampus and the BNST can be reduced by treatment with corticosterone and aldosterone [13–15]. At the same time, it should be noted that glucocorticoids may act in the opposite way on the expression of V1b vasopressin receptors (V1bR) because the administration of dexamethasone was found to increase V1bR mRNA in the pituitary, whereas adrenalectomy reduced V1bR mRNA. The latter effect could be reversed by the administration of dexamethasone [16].

It appears that negative feedback between glucocorticoids and vasopressin arises in early life. In the rat, applications of dexamethasone and aldosterone to fetal hypothalamic cell cultures have been found to inhibit the release of CRH, AVP, and oxytocin through mechanisms involving the activation of protein kinase A (PKA) and protein kinase C (PKC) [17]. In addition, the administration of cortisol to the medium bathing of the hypothalamic neurons of fetal sheep significantly inhibited the potassium-induced secretion of AVP [18]. There is also evidence that the secretion of several steroid hormones is sex-dependent [19,20].

It is possible that during chronic stress glucocorticoids and mineralocorticoids may act oppositely on AVP's release. Hence, the expressions of CRH hnRNA and AVP hnRNA in the parvocellular neurons of PVN are significantly elevated in rats exposed to forced-swim stress and their increase is abolished by the administration of dexamethasone [21]. On the other hand, the application of eplerenone, which is a mineralocorticoid receptor antagonist, alleviated anxiety-like behavior and reduced vasopressin and corticosterone concentrations in the posterior pituitary [22].

Thus far, multiple gaps exist in understanding the mutual interactions between vasopressin and steroid hormones in health and cardiovascular diseases. Therefore, the principal aim of the present review is to analyze the cooperation of vasopressin and steroid hormones at the cellular and systematic levels in the context of their influence on cardiovascular regulation, tissue metabolism, and oxygenation. Specifically, we discuss the positive and negative consequences of the interaction of AVP and steroid hormones in the development of hypertension and heart failure.

## 2. Interactions of the Hypothalamo–Pituitary–Adrenal System with Vasopressin at the Cellular Level

### 2.1. Genomic and Nongenomic Actions of Steroid Hormones

Multiple studies show that steroid hormones exert both rapid and delayed cellular effects that are mediated by specific types of non-genomic and genomic receptors located either in the cellular membrane or intracellularly [23–26]. The receptors of specific steroids are encoded by different genes and their activation may result either in stimulatory or in inhibitory effects, depending on the cell type.

In the individual cell, steroid ligands cooperate with local transcription factors and other regulatory compounds [27,28]. The essential role of binding to receptors and mobilizing posttranslational modifications is played coactivators, which are involved in the integration of cellular processes and the adjustment of cellular responses to current needs. An especially important role is attributed to steroid receptor coactivators (SRCs, namely SRC1, SRC2, and SRC3), which are known as Nuclear Receptor Coactivators (NCOASs), and to Coactivator Binding Inhibitors (CBIs) [29,30].

Mineralocorticoid receptors (MRs, MCRs), glucocorticoid receptors (GRs), estrogen receptors (ERs, ESRs), androgen receptors, and progesterone receptors (PGRs) belong to a subfamily three of nuclear receptors possessing the ligand binding domain (LBD), the DNA-binding domain (DBD), and the N-terminal domain (NTD). In their inactive state, the receptors are present in the cytoplasm in a multiprotein chaperon complex, which contains the ligand-binding cleft identifying and binding the ligand. Binding elicits conformational transformations within the complex that permit the subcellular trafficking of the ligand to the target within the cell and its interaction with DNA [31,32].

*Glucocorticoid and Mineralocorticoid Receptors.* In humans, the function of glucocorticoids is served by cortisol, whereas in the rat it is served by corticosterone, which acts also as a mineralocorticoid. Glucocorticoid and mineralocorticoid receptors are members of the nuclear receptor superfamily of transcription factors (TFs) that modulate processes of transcription through direct binding to the glucocorticoid response element (GRE) or mineralocorticoid response element (MRE) in DNA. A DNA-binding domain is 96% identical in GRs and MRs. The receptors possess also a C-terminal ligand-binding domain (CT-LBD) and an amino-terminus domain (NTD). GR is a 97 kDA protein encoded by the *NR3C1/Nr3c1* gene (in humans located in chromosome 5) and cooperates with several co-regulators [32–34]. An amino-terminus contains AF-1 and AF-2 regions which interact with CT-LBD and can stimulate transcription in the absence of a ligand [35–39].

The binding of the ligand by a GR initiates a cascade of events enabling the translocation of the ligand-bound receptor to the nucleus. In the nucleus, the receptors bind to specific DNA sequences which are known as glucocorticoid response elements (GREs) and negative glucocorticoid response elements (nGREs) [40,41]. The regulation of GREs appears to play a dominant role in the cellular processes of neurons [23,40]. The direct occupancy of nGRE results in the repression of the target gene. Steroid receptor coactivators (SRCs) appear to participate in the repression of CRH expression by GRs in the hypothalamus. In the nucleus, MRs and GRs can also interact with some other active proteins (MAZ—myc-associated zinc finger protein, AP-1—activator protein 1, NF-κB—nuclear factor κB, and SRC-1/2/3) that operate as ligand-selective co-regulators. They can induce a remodeling of gene conformation and may initiate the formation of transcription-initiation complexes. GRE-DNA interactions are modulated by chromatin configurations. In the cardiovascular system, the expression of MRs is higher in males than in females [42].

Glucocorticoid receptors have also been identified in the mitochondria, where they regulate mitochondrial gene transcription. In neuronal mitochondria, GRs interact with Bcl-2 protein and form GR/Bcl-2 complexes. Interestingly, a short action exerted at a low density intensifies the formation of complexes, whereas, in high doses, cortisol exerts opposing effects [43]. The interaction of Bcl-2 with other regulatory factors determines the specificity of the actions of glucocorticoids in various organs [37,44,45].

Glucocorticoids also modulate the process of transcription indirectly by physical interaction (tethering), which does not require direct contact with DNA but engages the activation of transcription factors. In various types of cells, these factors may act either as co-activators or as co-repressors. The tissue–cell-dependent expression of co-regulators causes specific the tissue–cell action of steroid molecules [30,37,46–48]. It is likely that protein–protein interactions mediate the rapid effects of steroid hormones and play an essential role in the trans-repression of genes by glucocorticoids in hypoxia and inflammatory processes. For instance, it has been found that they interact with hypoxia-induced factors (HIFs) at the level of the promoter region of the inflammatory genes and can either enhance or inhibit the activation of the HIF pathway [47,49]. It has been postulated that, during inflammatory processes, the co-activating function of steroids determines collagen synthesis, the generation of reactive oxygen species, and the engagement of peroxisome proliferator-receptor gamma co-activator 1-alpha (PGC-1$\alpha$), as well as the activation of p38 mitogen-activated protein kinase and nicotinamide adenine dinucleotide phosphate oxidases (NOX) 2 and 4 [50,51]. On the other hand, glucocorticoids can inhibit inflammation through the repression of genes engaged in the synthesis of pro-inflammatory proteins (AP-1, NF$\kappa$B) and through the enhancement of the expression of genes involved in the generation of anti-inflammatory compounds [47,52].

There is evidence of reciprocal interactions between glucocorticoid receptors' pathways. For instance, it has been shown that GRs can induce the expression of genes that promote or inhibit the p38 MAP kinase pathway (MAPK) [51]. Furthermore, the expression of GRs and their responsiveness to glucocorticoids is regulated by microRNAs, whereas the expression of microRNA is regulated by glucocorticoids [53].

The mineralocorticoid receptor is also known as the nuclear receptor of subfamily 3, group C, member 2 (NR3C2). MRs bind mainly aldosterone, but they also show high affinity to cortisol and androgens. The importance of cortisol and aldosterone in specific cell types largely depends on the availability of 11-$\beta$-hydroxysteroid dehydrogenase type 2 (11$\beta$-HSD2), which converts active cortisol into inactive cortisone. The opposite action is exerted by 11$\beta$-HSD1, which transforms cortisone into cortisol. The availability of 11$\beta$-HSD2 in several regions of the brain causes aldosterone to have good access to brain MRs and be able exert potent regulatory effects in spite of the fact that its concentration in plasma is hundreds of times lower than the concentration of cortisol [54–56]. In the heart, cardiomyocytes and macrophages do not express 11$\beta$-HSD2 and both cortisol and aldosterone participate in MR stimulation. Moreover, in the heart, aldosterone exerts some effects through cross-talk with cardiac G-protein-coupled receptors (GPCRs) [57]. Activated MRs can form homodimers or can associate with GRs and form heterodimers.

MRs are present in the kidney, heart, and vessels, where mineralocorticoids participate in the regulation of hypertrophy, fibrosis, inflammation, and apoptosis. Mineralocorticoids can act either directly on NR3C2 receptors or their action can be mediated by the formation of other active molecules, such as interleukin-1 (IL-1), tumor necrosis factor $\alpha$ (TNF-$\alpha$), cardiotrophin-1 (CT-1), and Toll-like receptor 4 (TLR-4). During the inflammatory process, inflammatory cytokines (IL-1, IL-6, TNF-$\alpha$) act synergistically with mineralocorticoids and can act jointly through the inhibition of ACTH secretion in the hypothalamic–pituitary–adrenal axis [58]. In the rat's mesangial cells, aldosterone was found to stimulate NF-$\kappa$B and glucocorticoid-inducible protein kinase-1 (SGK1) activities. It also elevates promoter activities and the protein expression of intercellular adhesion molecule–1 (ICAM-1) and connective tissue growth factor (CTGF). There is evidence that these factors are involved in aldosterone-mediated mesangial fibrosis and inflammation [59].

In cardiomyocytes, the genomic action of aldosterone mediated by MRs participates in the regulation of chronotropic and hypertrophic actions. It has been shown that aldosterone enhances the expression of mRNA which codes for the $\alpha$1H protein, and the latter is a constituent of CaV3.2 channel, which is one of the two T channels of cardiomyocytes [60–62]. The action of aldosterone on the T channels is presumably indirect, because the gene *CACNA1h*, which codes for the CaV3.2 T-type channel, does not possess an MRE. The increased formation of reactive oxygen species (ROS) and their involvement in the regulation of the affinity of steroid hormones to MRs should also be taken into consideration [63–66]. In the cardiovascular system, the genomic and non-genomic effects of aldosterone are also modified by angiotensin II (Ang II) [39].

In the brain, MRs have been identified mainly in the hippocampus, septum, and other limbic structures, whereas GRs are expressed in the septum, hippocampus, brain stem, and the prefrontal cortex [44,65–67]. The affinity of corticosterone to the MRs in neurons is 10-fold higher than to their GRs [65–67]. MRs are associated with the cellular membrane and, after their activation, they are translocated with the help of β-arrestin to the cells where they can exert their action through non-genomic GPCR processes [25,57].

*Androgen, Estrogen, and Progesterone.* Both in males and females, androgens are synthesized in the adrenal glands (mainly in the *zona fascicularis* and the *zona reticularis*), in the brain (mainly in the hippocampus), and in the liver [68]. In males, testosterone is produced chiefly by the Leydig cells of the testes, while, in females, testosterone and its metabolites are produced primarily in the adrenal glands and ovaries. The cells of the adrenal cortex also synthesize dehydroepiandrosterone (DHEA), androstenedione, androstendione, androstenediol, and 11-β-hydroxyandrostenedione. Testosterone can be converted to dihydrotestosterone by 5α-reductase, while deoxycorticosterone is converted into dihydrodeoxycorticosterone. Both DHEA and testosterone are able to stimulate androgen receptors; however, DHEA has significantly greater androgenic activity than testosterone. Aromatase-producing dihydrotestosterone is also involved in the formation of estradiol, which is engaged in the stimulation of estrogen receptors. This gene, which encodes aromatase (*CYP19A1*), is located on chromatosome 15 and has been identified in the lungs, vessels, and multiple brain regions [69,70]. The process of the aromatization of testosterone to estradiol occurs in several peripheral tissues and in the brain [68].

Androgens, estrogens, and progesterone interact with receptors in the brain, heart, and vessels and participate in the regulation of the cardiovascular system by means of classic genomic and non-classic pathways [71–78]. Androgen receptors were identified in cells of the reproductive system, bones, vessels, and brain [78,79]. In the brain, ARs are present in the cortex, midbrain, brain stem, and spinal cord. Specifically, a high density of AR immunoreactivity was found in the olfactory bulb, the nucleus accumbens, the medial amygdala, the bed nucleus of the stria terminalis, the medial preoptic area, the septum, the mesencephalic periaqueductal gray (PAG), the dorsal raphe nucleus, the substantia nigra, the area postrema, the dorsal motor vagus nucleus, and in the preganglionic cells of the autonomic nervous system [75,77].

Androgen signaling engages several molecular pathways. The primary androgen receptor is a nuclear transcription factor that is activated mainly by testosterone and dihydrotestosterone. Non-stimulated ARs are present mainly in the cytoplasm and are associated with heat shock proteins (HSPs). The association of androgen with ARs allows the dissociation of these receptors from chaperone proteins and the translocation of the androgen–AR complex to the nucleus, where it binds to the androgen-response element (ARE) and regulates gene transcription. Androgens also regulate rapid non-genomic processes, engaging G-protein-coupled receptor family C (GPRC6A), zinc transporter ZIP9 membrane-receptor, and oxoeicosanoid receptor (OXER). Consequentially, through their activation of the genomic-dependent and non-genomic dependent signaling pathways, androgens initiate transcription processes and activate canonical pathways associated with the activation of ionotropic receptors, G-protein-coupled receptors activating PLC, calcium transporters, and endothelial nitric oxide synthase (eNOS). Testosterone can also stimulate membrane ARs, which bind to Src and activate the MAPK pathway. The transactivation of membrane ARs by other ligands has also been reported [78,79]. Most likely, the activation of rapid non-genomic processes is essential for the fast action of androgens, such as their cell migration, mitosis, and inflammatory processes [80–82].

Estrogens easily penetrate the cellular membrane, reaching their highest concentration within the nucleus compartment [83–85]. In cells, they regulate long-lasting processes by means of nuclear receptors and rapid non-genomic processes. Nuclear estrogen receptors ESR1 (ERα) and ESR2 (ERβ) are codified by the *ESR1* and *ESR2* genes. *ESR1* is located on chromosome 6 (6q25.1) and *ESR2* on chromosome 14 (14q23.2). Both receptors act as transcription factors that mediate the transcriptional activity of estrogens with reference to specific genes. In the absence of the ligand, ESRs are associated with HSP and do not express transcriptional activity. After activation by the ligand, ESRs interact with the estrogen response element (ERE) and operate either as monomers or as dimers (ESR1-ESR1; ESR2-ESR2 or ESR1-ESR2). In the nucleus, estrogens enhance the transcription of specific target genes [74,83–85] Estrogens can also modulate the expression of other genes acting indirectly via the activation of PI3K/Akt and MAPK/ERK pathways, as well as through the inhibition of the JNK pathway [86].

Multiple essential actions of estrogens, such as the inhibition of ROS production, the regulation of mitochondrial ATP levels, and the formation of mitochondrial structural conglomerations, occur in the mitochondria [87,88]. Recently, attention has been drawn to the essential role of estrogens in the regulation of mitochondrial bioenergetics in human subjects [89]. Estrogens also interact with the membrane-associated G-protein-coupled receptor (GPER, named GPR30) which is present in the endoplasmic nucleus, Golgi apparatus, and cellular membrane. GPERs are involved in the rapid non-genomic actions of estrogens that are mediated by extracellularly activated kinase (ERK), cyclic adenosine monophosphate (cAMP), and $Ca^2$. With regard to the cardiovascular system, it is essential to note that the stimulation of estrogen receptors results in the activation of several rapid and long-lasting cellular processes which are essential for the function of the heart and vessels. Functionally effective ESRs are present in the cardiac and vascular smooth muscle cells and modulate the function of the perivascular unit [90,91]. GPERs, ERα, Erβ, and GPRE1 are widely represented in most cell types of the cardiovascular system and in the adipose tissue [86,92,93]. The processes of the stimulation of ERs involve the activation of the phosphoinoisitide 3-kinase-serin/threonine-specific kinase B (PI3K/Akt/eNOS) and mitogen-activated protein kinase MAPK/eNOS pathways, which are engaged in the production of NO and play an essential role in vasodilation. Moreover, it has been shown that, in cardiomyocytes, estrogens regulate the activity of calcium-handling proteins, including L-type $Ca^{2+}$ channel (LTCC), ryanodine channel (RYR), sarcoplasmic reticulum $Ca^{2+}$ ATPase, and sodium–calcium exchanger (NCX). Thus, it has been suggested that complex reciprocal interactions between the activation of estrogen receptors and calcium signaling pathways may play an essential role in the regulation of cardiomyocytes' activities [73,94]. Furthermore, it has been reported that estrogens exert an antioxidant action and regulate cell contractility through effects exerted on the calcium-dependent signaling pathways operating in the cardiac mitochondria and sarcoplasmic/endoplasmic reticulum, where they modulate $Ca^{2+}$-ATPase 2a (SERCA2a) activity and the function of calcium ion channels [73,95–98]. Altogether, it is likely that, in the heart, a deficiency of estrogen may result in disturbances in calcium homeostasis. It has also been reported that the interaction of ERα with peroxisome proliferator-activator receptors (PPARs) causes a repression of the transactivation of the PPAR in the heart and vessels [86].

It is likely that estrogens may also influence the function of the cardiovascular system through actions exerted in the brain as their receptors are widely expressed in multiple brain regions involved in cardiovascular regulation, such as the frontal cortex, the sensorimotor cortex, the thalamus, the hypothalamus, the amygdala, the ventral tegmental area, the hippocampus, the dorsal raphe nucleus, and the cerebellum [72,99]. Finally, it should be noted that the activation of ERα and GPER1 plays a significant role in the modulation of the immune processes that are activated in cardiovascular diseases [74,84,91,100,101].

The action of progesterone (P4) is mediated through genomic signaling, engaging two subtypes of nuclear receptors (PGRA, PGRB), and through non-genomic G-protein-associated membrane-progestin receptors (mPRs). Some actions of PG can also be exerted by GRs [102–104].

*Neurosteroids.* Some steroids, which are known as neurosteroids, have been identified in the nervous system and act preferably on neuronal membrane receptors. Among them are steroid sulfates, such as DHEAS, which is a product of the sulfation of DHEA [105]. It is suggested that neurosteroids play a significant role in the modulation of the action of other steroids and classical neurotransmitters. Their action is not a subject of discussion in the present survey (see [103–106] for a further review of this topic).

### 2.2. Genomic and Non-Genomic Effects of Vasopressin

Vasopressin, which is a principal vasopressin peptide in mammals, is synthesized mainly in the neurons of the hypothalamic supraoptic, paraventricular, and suprachiasmatic nuclei. The majority of the axons of these neurons reach the posterior pituitary, where AVP is released into the blood and can be distributed to peripheral organs. Some of the axons reach the median eminence and release AVP into the hypophyseal portal system and the anterior pituitary, where vasopressin contributes to the regulation of ACTH. Vasopressin-expressing cells have also been identified in the brain regions engaged in the regulation of blood pressure, metabolism, pain, stress, and anxiety. Among them are neurons of the brain cortex, olfactory bulb, BNST, dorsomedial hypothalamic nucleus, nucleus of the diagonal band of Broca, circumventricular organs, brain stem, and spinal cord [107–109]. In addition, AVP mRNA has been detected in the heart and vessels and in pancreatic tissue [110,111].

The vasopressin gene is located in chromosome 20 and consists of three exons (A, B, and C). The exon A codes for a signal peptide, vasopressin peptide (Cys-Tyr-phe-Gln-Asn-Cys-Pro-Arg-Gly-NH2), a three-amino acid spacer, and the first nine ($NH_2$ terminal) aminoacids of neurophysin (NP). Exon B codes for the mid-portion of NP, whereas exon C codes for the terminal portion of NP, a cleavage site, and the COOH-terminal glycopeptide (GP, known as copeptin). Copeptin consists of 39 aminoacids and is released from neurons in equimolar quantities with AVP [112,113]. Because the copeptin molecule is more stable than the vasopressin molecule, measurements of GP concentrations are frequently used as a marker of AVP levels. Measurements of GP levels have been included in ESC guidelines on myocardial infarction and as a biomarker of inflammation [114–116].

The expression of the vasopressin gene is influenced by changes in body fluid's osmolality and blood volume, as well as by constituents of the hypothalamo–hypophysial–adrenal axis, pain, cytokines, and inflammation factors, especially those associated with COVID-19 [113,117–121]. Vasopressin gene expression and AVP mRNA abundance are enhanced by chronic osmotic simulation and are decreased by hypoosmolality. The osmotically induced increase in AVP mRNA and the release of AVP by neurons of the hypothalamo–neurohypophyseal axis are potentiated by the administration of lipopolysaccharide and the enhanced release of IL-1β and IL-6 in the posterior pituitary [122]. The expression of the AVP gene in the hypothalamus is potentiated by IL-1 and IL-2. Moreover, IL-1β stimulates the release of CRH, AVP, and melanocortin-stimulating hormone (α-MSH) [123,124].

It has been shown that hypoosmolality induces GRs' expression and that this is related to corticosterone-negative feedback on AVP transcriptions. The above data support the hypothesis that the AVP gene is directly inhibited by glucocorticoids and that the induction of GRs in the hypothalamic cells suppresses AVP expression during prolonged hypoosmolality [112]. However, it should be noted that during prolonged increases in corticosteroids' concentrations, such as the one takes place during autoimmune inflammation, vasopressin neurons can escape from glucocorticoids' inhibition, presumably due to the increased engagement of inflammatory cytokines [125].

Osmotic stimulation induces the rapid upregulation of CRH in vasopressinergic neurons of the hypothalamic magnocellular nuclei [126]. There is evidence of the coordinated regulation of AVP and CRH genes by glucocorticoids. It has been shown that adrenalectomy causes the enhancement of CRH and AVP immunoreactivity in the hypothalamus and elevated CRH immunoreactivity in the cerebral cortex, the amygdala, and the BNST.

Since the stimulatory effect of adrenalectomy on the expression of AVP and CRH in the hypothalamus could be reduced by the administration of dexamethasone, it was concluded that glucocorticoids produced in the adrenal glands play a primarily inhibitory role in the regulation of AVP and CRH secretions [127]. The PVN neurons express glucocorticoid receptors and glucocorticoids reduce AVP gene expression in the parvocellular neurosecretory neurons of the PVN [128]. The AVP gene and CRH gene possess cAMP response elements (CREs), that are activated by intracellular cAMP, and AP1 and AP2 transcription factors can be repressed by glucocorticoids. In the PVN neurons, the expression of the CRH and AVP genes is regulated by nGREs and serum response elements [113,129,130]. Vasopressin neurons also express MRs and there is evidence that aldosterone and corticosterone increase the sodium channel (ENaC) leak current through an action exerted at the promoter region of the $\gamma$ ENaC gene [131]. The magnocellular neurosecretory neurons of the PVN and SON, as well as of other brain regions, express 11$\beta$-HSD2, and this increases their sensitivity for aldosterone [56,132].

It appears that, during restraint stress, the inhibitory effect of glucocorticoids on CRH and AVP mRNA expressions in the PVN neurons can be modulated by the concomitant release of testosterone [133]. During chronic stress, glutamergic, gamma-aminobutyric acid (GABA), and noradrenergic terminals exert a number of convergent actions that jointly regulate the activity of the CRF and AVP neurons of the PVN [134]. Growing evidence indicates that androgens play an essential role in the regulation of neurons expressing CRH, AVP, dopamine, and serotonin during stress-related behavior [135].

*Vasopressin receptors.* Arginine vasopressin stimulates two subtypes of V1 receptors (V1R), known as V1aR and V1bR, and one type of V2 receptor (V2R). The AVP receptors belong to a family of G-protein-coupled receptors. In high concentrations, AVP also interacts with oxytocin receptors [109,136–139]. The AVPR1a gene has been mapped to the 12q14.2 locus and AVPR1b to the 1q32.1 locus. The AVP2R gene is located on the long arm of the X-chromosome (Xq28). The mutation of the V2R gene is inherited in an X-linked manner and results in congenital nephrogenic diabetes insipidus, which is characterized by strong polydipsia and polyuria [140,141]. It has been shown that vasopressin receptors can act as homodimers and as heterodimers and it is likely that their dimerization influences the effectiveness of the stimulation of the target cells. Specifically, the formation of V1aR and OTR and V1bR and corticotropin-releasing hormone receptor (CRHR) heterodimers has been well documented [142,143]. The expression of vasopressin receptors is regulated by corticosteroids [144].

Vasopressin V1aR mRNA and protein have been detected in multiple organs and tissues, including the heart and vessels, and their expression is altered in pathological processes [113,138–145]. In cardiac ventricular sarcolemma, vasopressin was found to open $K_{ATP}$ channels through an action exerted on V1R [146]. AVP was also found to reduce $Ca^{2+}$ influx through L-type $Ca^{2+}$ channels, and this effect can be abolished by a blockade of V1aR [147]. There is evidence that the stimulation of cardiac V1aR decreases cardiac beta receptors' responsiveness [148].

In vitro experiments on H9c2 rat ventricle cardiomyocytes exposed to hypoxia revealed that AVP acting on V1aR in a V1aR/GRK2/$\beta$- arrestin1/ERK1/2-dependent manner enhances cell survival. It has been suggested that, during heart failure, when the levels of circulating AVP are elevated, the inhibition of G protein-coupled receptor kinase 2 (GRK2) can potentially exacerbate negative V1aR-mediated effects by preventing receptor desensitization and augmenting G$\alpha$q protein-dependent signaling [149,150]. In the heart, the stimulation of V1aR is also engaged in the generation of pro-inflammatory cytokines and in the development of inflammation and fibrosis. AVP increases IL-6 mRNA and protein levels in cardiac fibroblasts and this effect requires the activation of GRK2 and NF-$\kappa$B [151]. In addition, it has been shown that endotoxemia induced by lipopolysaccharides and the concomitant increase in IL-1$\beta$, TNF-$\alpha$, and interferon gamma causes a downregulation of V1aR gene expression in the heart, vessels, liver, and lungs, as well as a reduction in the responsiveness of vascular smooth muscle cells. It is likely that the diminished responsiveness of V1aR to vasopressin accounts for its inadequate stimulation during the circulatory shock that can occur during endotoxemia [152].

A growing number of studies provide evidence of the interaction of AVP with corticosteroids in other organs, whose proper action is necessary for the appropriate regulation of cardiovascular functions, such as the brain and the gastrointestinal system. The administration of corticosterone was found to decrease the expression of V1aR in the lateral septum and the hippocampus [144]. The stimulation of V1aR in cortical astrocytes was found to exert a neuroprotective action and this effect was associated with the activation of the nuclear transcription factor cAMP response element-binding protein (CREB) and with a prominent decrease in IL-1β and TNF-α gene expressions [146]. Vasopressin also plays an essential role in the regulation of the gut microbiome [151]. The enhanced stimulation of V1aR may participate in the habituation of the release of ACTH, corticosterone, and testosterone to repeated stress in Sprague Dawley rats [153].

V1b receptor mRNA is expressed in the corticotropic cells of the brain and in the anterior pituitary, the adrenal gland, the heart, the kidney, the thymus, the lung, the spleen, the pancreas, the uterus, and in white adipose tissue [154–156]. In the brain, V1bR mRNA was detected in the olfactory bulb, the cortex, the hippocampus, the hypothalamus, the septum, and the cerebellum; however, its expression was lower than the expression of V1aR mRNA [156]. In the pancreas, stimulation of V1bR potentiates the secretion of insulin from β cells where AVP acts synergistically with CRH [156]. Thus far, there is no convincing evidence of the presence of V2R in the heart and vessels and of their direct involvement in the regulation of cardiovascular functions via vasopressin. Nevertheless, through the stimulation of V2R in the kidneys and through the regulation of urine concentrations and body fluid natremia, AVP may modulate the effects of the stimulation of V1R in the cardiovascular system [153–158].

## 3. Role of the Hypothalamo–Pituitary–Adrenal System and Vasopressin in the Regulation of Energy Balance and Water–Electrolyte Balance at Rest and during Neurogenic Stress

### *3.1. Regulation of Steroids Secretion*

The synthesis of glucocorticoids in the adrenal cortex is regulated mainly by ACTH, which is released from the pituitary gland. The secretion of ACTH is closely regulated by CRH and AVP, which are supplied principally by neurons of the PVN [158–160]. Glucocorticoids released from the adrenal cortex and ACTH secreted from the corticotropic cells of the pituitary gland are able to inhibit the reciprocal secretion of CRH from the hypothalamic neurons [113,129,161] (Figure 2).

The secretion of glucocorticoids from the adrenal cortex is regulated by the circadian rhythm; however, the daily rhythm of glucocorticoid release may be influenced by other factors, such as gender, aging, early life stress, inflammatory diseases, pregnancy, and lactation [159,162]. ACTH binds on the surface of cells of the fascicular layer of the adrenal cortex with the G protein-coupled melanocortin type 2 receptor (MC2R), which activates the 3′,5′-cyclic AMP signaling cascade, enabling the penetration of cholesterol into the inner mitochondrial membrane. Cholesterol is converted by cytochrome P450 into pregnenolone, which is converted in several steps to progesterone, 11-deoxycortisol, and cortisol (the main glucocorticoid in humans), or corticosterone (the main glucocorticoid in rodents) (Figure 3) [159].

ACTH also stimulates the synthesis of aldosterone in the glomerular layer of the adrenal cortex. Aldosterone is synthesized from cholesterol in the inner mitochondrial membrane with the participation of aldosterone synthase (CYP11B2). The process of its synthesis includes the formation of pregnenolone and the conversion of pregnenolone to 11-deoxycorticosterone (DOC) in the smooth endoplasmic reticulum (Figure 3). Further processing to aldosterone takes place again in the mitochondria [163,164]. The synthesis of aldosterone in the adrenal cortex can also be stimulated by the renin-angiotensin system (RAS) and by extracellular potassium concentrations [163].

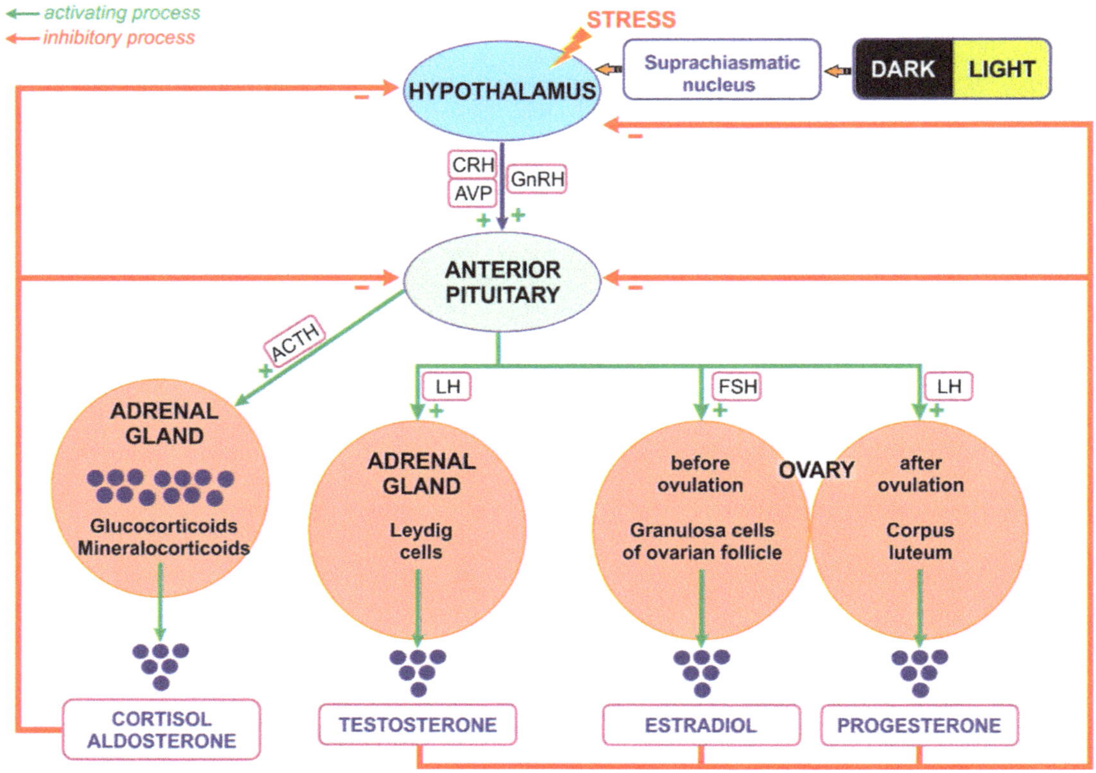


**Figure 2.** The regulation of the secretion of hormones of the hypothalamo–pituitary–adrenal axis. ACTH—adrenocorticotropic hormone, AVP—arginine vasopressin, CRH—corticotropine-releasing hormone, FSH—follicle-stimulating hormone; GnRH—gonadotropine-releasing hormone, LH—luteinizing hormone, + stimulation, - inhibition. Other explanations are in the text.

Androgens, as well as estrogen and progesterone, are released during the activation of the HPA, which plays a crucial role in the regulation of the ovarian cycle. The hypothalamic cells synthesize GnRH, which stimulates the release of LH and FSH from the anterior pituitary gland. In the woman, FSH stimulates follicular maturation and the release of estrogen in the follicular phase, while LH triggers ovulation as well as the release of progesterone from the corpus luteum *(ruptured follicle)* in the luteal phase. Estrogens released from the ovaries and acting through reciprocal inhibition are able to slow down the activity of both the pituitary gland (the synthesis and secretion of FSH and LH) and the hypothalamic neurons (the synthesis and secretion of GnRH), thereby generating the menstrual cycle [164,165]. Estrogens are synthesized as a result of the action of the aromatase, which is responsible for the aromatization of androgens into estrogens (Figure 3) [166]. The main site of the expression of aromatase in premenopausal women is located in the ovarian granulosa cells. In pregnant women it is also present in the placental syncytiotrophoblast. Both in men and in postmenopausal women, fibroblasts of adipose tissue and skin contribute to estrogen synthesis [167]. During reproductive life, women show a characteristic periodicity in their secretion of ovarian hormones, which corresponds to the monthly menstrual cycle.

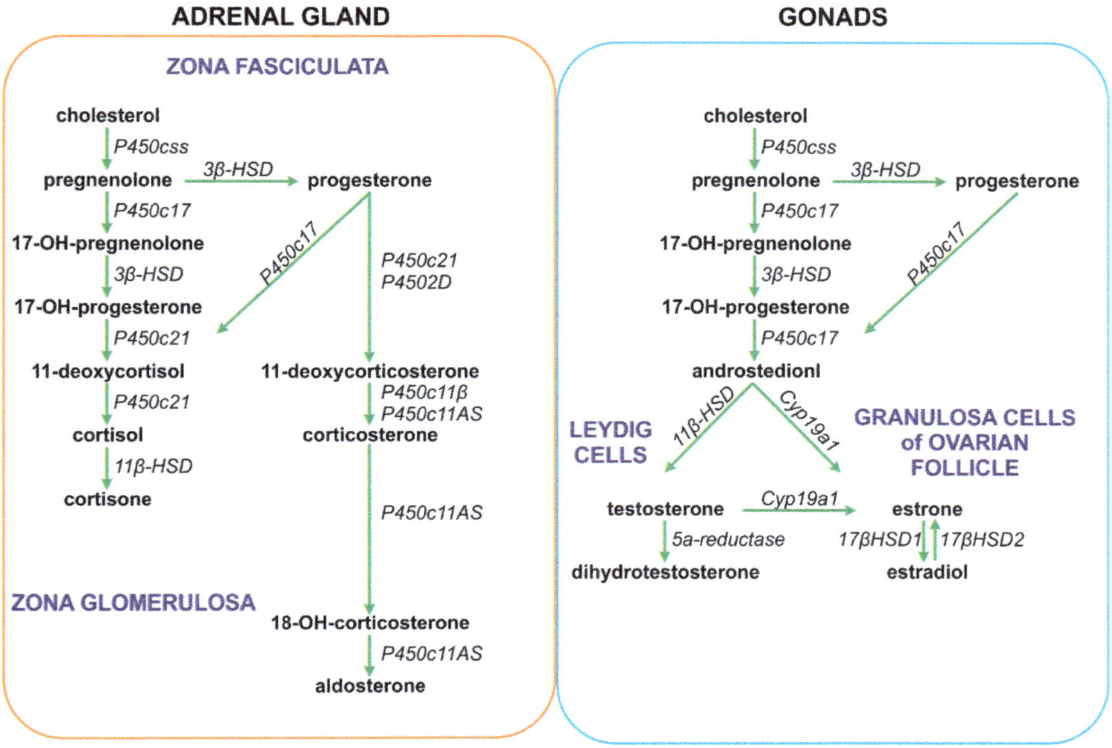


**Figure 3.** Steps of the synthesis of steroid hormones in the adrenal glands and gonads. Other explanations are in the text.

As already mentioned, androgens are synthesized mainly in adult female ovaries and in male testes. In addition, their synthesis is stimulated by ACTH in the reticular zone of the adrenal glands (mainly dehydroepiandrosterone sulfate, DHEA/-S; 4-androstenedione, A4; and 11β-hydroxyandrostenedione). Androgens are metabolized in the liver and subsequently excreted by the kidneys [168]. In men, the dominant androgen is testosterone; its concentration is 15 times higher than in women [169,170]. Testosterone synthesis in men takes place in the Leydig cells of the testes (Figures 2 and 3). The activation of the Leydig cell LH receptors by LH initiates an intracellular signaling cascade, which involves the stimulation of adenylate cyclase, the increased production of cAMP, and the activation of adenosine-dependent kinase [171]. As a consequence of the activation of cAMP-dependent intracellular signal transduction, cholesterol can be transferred to the mitochondria by the steroidogenic acute regulatory protein (STAR), the translocator protein (18 kDa; TSPO), and other transducosome proteins [172]. In the mitochondria of Leydig cells, cholesterol is converted to pregnenolone by the cytochrome P450 enzyme located on the matrix side of the inner mitochondrial membrane, which cleaves the C27 cholesterol side chain (CYP11A1). Subsequently, after transfer to the smooth endoplasmic reticulum, pregnenolone is metabolized to testosterone by 3β-hydroxysteroid dehydrogenase (3β-HSD; HSD3B), 17α-hydroxylase/17,20 lyase (CYP17A1), and 17β-hydroxysteroid dehydrogenase type 3 (17β-HSD3, HSD17B) (Figure 3) [173–175].

### 3.2. Interaction between Steroids and AVP in the Regulation of Energy Balance

Multiple studies indicate that both steroid hormones and vasopressin play an important role in the regulation of energy balance. Glucocorticoids promote glucose production in the liver and reduce peripheral glucose uptake by the muscle and adipose tissue. They also increase the breakdown of the constituents of fat and muscle, thereby providing additional substrates for glucose and free fatty acids' production [176].

Vasopressin is involved in the regulation of food intake and energy balance via a wide range of central and peripheral actions exerted on cellular growth and proliferation, protein turnover, lipid metabolism, and glucose homeostasis [177].

#### 3.2.1. Animal Studies

It has been shown that endogenous nucleobindin-2 (NUCB2)/nesfatin-1 regulates AVP and oxytocin in the PVN of male mice, and this is associated with a resetting of fluid balance and food intake, and with an increase in body weight without a change in energy expenditure [178]. Experiments on mice showed that the administration of an AVP analog (Ac3IV) twice a day for 22 days not only reduces energy consumption, body weight, and fat content, but also decreases blood glucose concentration, increases insulin sensitivity, and significantly improves glucose tolerance and glucose-induced insulin secretion. Moreover, it lowers total cholesterol and low-density lipoprotein (LDL) cholesterol and triglycerides and increases high-density lipoprotein (HDL) cholesterol [179]. Several hormones affecting energy balance play a significant role in the regulation of AVP secretion. Ghrelin, an orexigenic hormone, has been shown to stimulate vasopressinergic neurons in the hypothalamic paraventricular nucleus and AVP secretion in a nutritional-status-dependent manner in in vivo experiments. It also activated excitatory GABAergic synaptic input via a retrograde neuron–glial circuit in hypothalamic slices obtained from fasted rats [180]. On the other hand, the cerebroventricular administration of adiponectin significantly reduced basal plasma AVP concentrations in conscious rats in a dose-dependent manner, with the maximum effect achieved 10 min after administration [181].

Additionally, it appears that AVP may exert thermogenic functions, as the exposure of brown adipose tissue (BAT) adipocytes to AVP was found to increase uncoupling protein-1 (UCP-1) protein expression, induce a time- and dose-dependent increase in p38 MAP kinase phosphorylation, and increase monocyte chemoattractant protein-1 (MCP-1) mRNA and IL-6 expressions. In contrast, adiponectin mRNA expression was reduced [182]. Former studies have shown that acting, in the adipose tissue, on V1aR AVP may exert both lipolytic and antilipolytic effects [183,184]. Its action on lipid metabolism through V1bR is different. It has been shown that V1bR-deficient mice have a lower body weight and greater epididymal adipose tissue mass in comparison to wild-type mice. It is believed that AVP acting on V1bR receptors is able to influence lipid metabolism through the modulation of insulin signaling. Currently, it appears that the stimulation of V1aRs results in the impairment of glucose tolerance and in lipolytic action, whereas the activation of V1bRs improves glucose tolerance and exerts an antilipolytic effect [185,186].

#### 3.2.2. Human Studies

Elevated plasma copeptin (AVP substitute) concentrations have been demonstrated in patients with type 2 diabetes and with type 1 diabetes [187]. Moreover, recent studies indicate that higher concentrations of AVP or copeptin predispose patients to the development of type 2 diabetes and metabolic syndrome and that the consumption of larger amounts of water by people with high plasma copeptin concentrations results in a reduction in fasting glucose or glucagon levels [188]. Studies on human populations revealed an association of polymorphisms of the gene encoding AVP with metabolic disorders. A significant relationship was found between high copeptin levels and reduced insulin sensitivity, as well as between AVP gene tagSNPs (CC genotype rs6084264, TT genotype rs2282018, C allele rs2770381, and CC genotype rs1410713) and the incidence of hyperglycemia [189]. Similarly, a specific polymorphism of the V1aR gene (T allele of rs1042615) has been associated with an increased incidence of type 2 diabetes in people who consume large amounts of fat or are overweight [190]. In another study, the major A allele of rs35810727, a tagSNP of the V1bR, has been associated with an increased body mass index (BMI) and type 2 diabetes [191].

*3.3. Interaction of AVP and Steroids in the Regulation of Water–Electrolyte Balance*

It is well known that the vasopressin system (VS) plays a pivotal role in the regulation of water and electrolytes' balance and that a lack of AVP results in the excretion of large amounts of free water in the urine (polyuria) and in subsequent polydipsia [108,192–194]. The release of AVP is regulated by a variety of internal and external cooperating factors, acting as anticipatory or consecutive signals [192,194–199]. The regulation of water–electrolyte balance via AVP is a complex process in which the osmolality of the extracellular fluid (ECF), including the plasma and cerebrospinal fluid, plays a key role. ECF tonicity is sensed by osmoreceptive neurons located mainly in the subfornical organ (SFO) and the organum vasculosum of the lamina terminalis (OVLT), which are able to sense osmolarity using the aquaporin (AQP) receptor and have direct connections with the PVN and SON [200,201]. The osmotically induced shrinkage of osmosensitive cells during dehydration causes the activation of TRPV1 delta-N channels (a family of transient receptor potential cation channels) that sense mechanical stretch and allow the influx of cations into the cell [200,202]. In addition, the brain osmoreceptors' activity may be regulated by anticipatory signals generated in receptors in the oropharyngeal region, esophagus, or stomach, which participate in the regulation of fluid intake. Specifically, information about the taste of water is transmitted centripetally by the tympanic cord belonging to the facial nerve, information about dry mouth is received via the trigeminal nerve, and information about the stretching of the esophagus and stomach and the volume of water ingested are transmitted by the glossopharyngeal nerve and by the vagus nerve [203].

In the kidneys, AVP regulates the activity of the aquaporin-2 (AQ2) water channel in the collecting duct and participates in the process of urine concentration [204]. Moreover, it stimulates sodium reabsorption, acting on the luminal sodium channel ENaC in the cortical and outer medullary parts of the collecting tubules, and activates urea transporters UT-A1 and UT-A3 in the inner terminal medullary, increasing urea reabsorption in the collecting tubules [205,206]. Additionally, at higher concentrations, AVP also increases sodium reabsorption through the activation of Na-K-2Cl cotransporters (NKCC2) in the thick ascending limb of the nephron loop (Henle's loop) [207].

There is strong evidence that the regulation of water–electrolyte balance by vasopressin is significantly influenced by steroid hormones.

Studies on Animals and Human Subjects

Experimental studies have shown that glucocorticoids' deficiency facilitates the activation of V2 receptors in the renal collecting tubule [208]. It has also been shown that the application of dexamethasone improves cardiac functions in rats with congestive heart failure elicited by coronary artery ligation and that this effect is associated with a reduced expression of V2R and AQP2 and AQP3 and with a reduced number of ENaC and Na-K-2Cl cotransporters 3 in the renal collecting tubule [205]. In rats, the effects of dexamethasone were abolished by the use of the glucocorticoid receptor inhibitor RU486 [205]. However, clinical studies could not confirm the significant involvement of glucocorticoids applied alone in fluid retention in critically ill patients. Therefore, it has been postulated that the effects of steroid hormones on water–electrolyte balance are mediated mainly by the activation of mineralocorticoid receptors [206].

It is well known that mineralocorticoids have a significant impact on electrolyte balance, especially on sodium and potassium turnover, as well as on the clearance of free water [207,208].

Currently, it also appears that gonadal steroids have a significant effect on the vasopressinergic system [209]. Experimental studies have revealed that the antidiuretic function of AVP is more effective in male rats than in female rats [209]. Furthermore, its effect is more potent in females during estrus, when the level of circulating estrogen is low [210]. In addition, ovariectomy was found to increase their antidiuretic response to AVP to a level comparable to that observed in males, whereas estradiol substitution reduced the antidiuretic effect of AVP to the level observed in non-estrus females [210]. It appears that estrogens may also influence the hypothalamic vasopressinergic neurons, acting both through the ERα and ERβ receptors. In this context, the presence of ERα on osmoreceptive neurons of the SFO and OVLT, which send projections to the SON, has been demonstrated [211–213]. Moreover, both ERα and ERβ have been detected in the kidney, where the expression of ERα is prevailing [214]. Interestingly, GPER is a newly discovered aldosterone receptor that mediates nongenomic aldosterone pathways. GPER activation by aldosterone mediates water and sodium retention in the body and contributes to vasoconstriction [215]. Recently, it has been shown that the production of aldosterone may be influenced by progesterone [216].

It should be noted that testosterone, in addition to estrogen, can directly inhibit AVP secretion [217]. However, it is frequently emphasized that the effect of testosterone, and the action of some of its metabolites (α-androstane-3β), on the release of AVP may be mediated by ERβ [214].

### *3.4. Interaction of AVP and Steroids in Neurogenic Stress*

A number of studies provide evidence that neurogenically mediated stress causes s joint stimulation of vasopressin-secreting neurons and CRH-secreting neurons, which suggests that a coordinated regulation of AVP and CRH's release and action may play an important role in the modulation of neurogenic stress. Moreover, it has been found that, in chronic stress and depression, the effects of CRH on ACTH release are strongly enhanced by vasopressin, which is produced in increasing amounts when the hypothalamic PVN and SON neurons are chronically activated [218,219]. It has been found that AVP is engaged in stress-induced tachycardia and baroreceptor reflex (BRR) desensitization [220]. Clinical studies have shown that healthy adults exposed to a social stress test respond with an increase in serum copeptin concentration [218]. Experiments on mice showed that chronic unpredictable stress (UCS) and ovariectomy influence AVP immunoreactivity in the lateral magnocellular and the medial magnocellular subdivisions of the PVN [221]. The role of vasopressin in the response to stress appears to differ in female and male C57BL/6 mice. Namely, a six-week chronic variable stress (CVS) paradigm has been found to increase sociability in female mice and to decrease AVP mRNA in the PVN, whereas, in male mice, CVS had no effect on social behavior or AVP expression [222,223]. Studies conducted on male Sprague Dawley rats subjected to 4 weeks of unpredictable chronic mild stress (UCMS) showed that injections of small interfering RNA (siRNA) directed against V1aR into the PVN prevent an increase in blood pressure and the elevation of renal sympathetic nerve activity (RSNA), whereas the administration of scrambled RNA (scrRNA) into the PVN elicits an increase in blood pressure. The results suggested that V1aR signaling in the PVN contributes to the generation of neuro-cardiovascular responses to stress [224]. Research performed by Komnenov et al. (2021) showed that rats subjected to UCMS have higher plasma AVP levels and a greater abundance of V1aR and V1bR transcripts in their PVN. The rats also manifested a higher resting MAPK, heart rate, and RSNA, and these effects could be eliminated by the combined inhibition of V1aR and V1bR [225].

More recently, human studies have shown that, during chronic stress, the hypothalamic activation of pituitary changes from the dominant CRH phenotype to the dominant AVP phenotype; however, the cortisol level remains elevated because its metabolism is reduced [222]. In addition, recent studies have provided evidence that the rs10877969 polymorphism of the V1aR gene is associated with the elevation of symptoms of stress and acute pain [226].

Sex Differences

It is well known that there are sex differences in the tolerance of stress and in the predisposition to anxiety and depression that may be partly associated with different regulations of the HPA in males and females [227,228]. Experimental studies have demonstrated that female rats respond with larger increases in ACTH to neurogenic stress than male rats and suggested that, during neurogenic stress, the gonadal steroids involved in the regulation of sexual behavior and reproduction may have a potential impact on the pituitary secretion of ACTH [229,230]. Subsequently, it has been shown that the secretion of ACTH is regulated by a testosterone-dependent effect on the synthesis of AVP and by a corticosterone-dependent effect on the synthesis of CRH in the PVN [231]. It is likely that one of the reasons for differences in sex-related susceptibility to stress is the functional heterogeneity of GRs and MRs in the brains of females and males, because a higher expression of MRs was found in the brain of the male mice than in the brain of female mice manifesting depressive behavior [232]. Research performed by Woodward et al. (2023) on female and male mice subjected to the UCMS procedure showed that female mice show anxiety and depressive behavior associated with FosB activation in the neurons of their medial prefrontal cortex (mPFC), expressing parvalbumin after 4 weeks of exposure, while in male mice behavioral and biochemical alterations are observed not earlier than after 8 weeks of UCMS. Furthermore, the use of patch-clamp electrophysiology allowed the researchers to demonstrate that there were time-corresponding sex-specific differences in the altered neuronal excitability of mPFC cells after 4 and 8 weeks of exposure [228]. Rosinger et al. (2019), using a corticotropin-releasing factor 1 receptor (CRFR1) reporter mouse line, demonstrated that male mice showed a significantly higher distribution of cells expressing CRFR1 in their PVN compared to females; however, it should be noted that this relationship was age-related, because it was observed only in older (20–24-month) mice and not in mice during early post-natal life. Gonadectomy in adult six-week-old mice resulted in a significant decrease in the number of CRFR1-immunoreactive cells in the PVN in males but not in females. In addition, their CRFR1 cells showed moderate co-expression with estrogen receptor alpha and high co-expression with androgen receptors. The use of restraint stress resulted in a greater activation of CRFR1 cells in the PVN of male mice than in the PVN of female mice [232]. Cox et al. (2015) examined the association of X chromosome genes with behavioral disorders in an experimental model (fragile X syndrome, autism) using a mouse model with an atypical sex chromosome configuration resembling Turner (45, XO) and Klinefelter (47, XXY) syndromes. The researchers showed higher AVP expression in the amygdala of female mice with one copy of the X-chromosome gene. A reduced level of plasma AVP was found in girls with Turner syndrome [233].

The recent clinical studies of Cohen et al. (2023) have revealed that gender differences in the reaction to stress already occur in young people aged 18–25, and they may be related to the different activation of their prefrontal cortexes. Using GABA in magnetic resonance spectroscopy, the authors were able to demonstrate significant differences in neuronal activity in the responses to stress among the sexes. It should be noted that the differences were especially evident in the ventromedial prefrontal cortex, which plays an essential role in the regulation of the hypothalamic–pituitary, hypothalamic–pituitary–adrenal, and hypothalamic–pituitary–gonadal axes [234].

## 4. Altered Interactions of Vasopressin with the Hypothalamo–Pituitary–Adrenal System in Cardiovascular and Metabolic Diseases

### 4.1. Cardiovascular Diseases

Substantial evidence indicates that the central and peripheral interactions of steroid hormones and vasopressin are reprogrammed in hypertension and that these modifications may initiate and/or potentiate cardiovascular complications. Pietranera et al. (2004) revealed that subcutaneous injections of deoxycorticosterone acetate cause significantly greater increases in AVP and V1aR mRNA in the magnocellular divisions of the PVN in spontaneously hypertensive (SHR) rats than in control SHR rats receiving oil vehicle [235].

It is possible that AVP contributes to the elevation of blood pressure in DOCA-induced hypertension through the augmentation of the neurogenic component of vascular resistance. Studies conducted on DOC-treated and saline-treated rats revealed a significant increase in vascular resistance that was associated with amplification of the central sympathetic tone only in DOC–salt rats. These effects were intensified by vasopressin. Moreover, a reduction in vascular resistance was observed in DOC–salt rats treated with AVP antagonist (I-deaminopenicillamine, 4-valine, 8-D-arginine vasopressin, dPVDAVP) and in those who underwent lumbar sympathectomy [236]. Experiments on dogs maintained on a normal-salt diet showed that a subcutaneous administration of DOC elicited an increase in blood pressure, which was accompanied by an increase in cardiac output, hypernatremia, and an elevation of vasopressin concentration in plasma and in the cerebrospinal fluid (CSF) at the early stage of hypertension [237].

Some evidence suggests that an inappropriate interaction of steroid hormones and vasopressin may play a role in the development of congestive heart failure. For instance, it has been reported that the treatment of sheep with paced-induced heart failure with urocortin 2 (Ucn2—a group of peptides sharing structural similarities with CRH combined with canrenoic acid), which is an antagonist of mineralocorticoid receptors, led to better hemodynamics than the application of canrenoic acid alone. The combined treatment also contributed to reductions in PRA, Ang II, and AVP concentrations, as well as to an improvement of the kidneys' function [238].

### *4.2. Metabolic Diseases*

A growing number of studies address the question of whether interactions between AVP and steroids are altered in metabolic diseases. Clinical data, based on measurements of blood copeptin concentrations, suggest that AVP may play a role in the pathogenesis of the metabolic syndrome (MetS). In this line, obese men with elevated fat free mass and total fat mass show higher fasting glucose and insulin concentrations, as well as an enhanced pituitary response to combined CRH/AVP stimulation. These results suggest that the joint stimulation of CRH and AVP, which is associated with an excessive stimulation of ACTH release, may promote the development of body overweight, adiposity, and hyperinsulinemia [239]. There is also evidence that obesity causes a significant rearrangement of the activation of the HPA by the noradrenergic system. For instance, noradrenergic transporter activity (NAT) assessed in positron emission tomography (PET) was found to correlate differently in obese and non-obese patients. In the non-obese patients, a positive correlation was found between NAT and HPA activity, whereas in the obese patients the correlation was negative. The study suggested that, in the obese patients, the regulation of HPA by the noradrenergic system mainly activates the hypothalamic neurons, whereas in non-obese subjects it engages the prefrontal–limbic cortex more intensively [240]. It has also been postulated that the altered association of copeptin and insulin resistance is related to the elevation of hepatic glycogenolysis; increased insulin, glucagon, and ACTH secretions; and the enhanced activation of 11β-HSD2 [241]. Links between specific steroid hormones and vasopressin in hypoglycemia are less evident. It has been reported that insulin-induced hypoglycemia increases the secretion of cortisol, but it does not affect plasma AVP levels [242]. In another study, more drastic hypoglycemia (1.6 mmol/L), probably producing hypoglycemic stress, caused a rapid significant increase in serum copeptin concentration, which was positively correlated with ACTH and cortisol concentrations. Interestingly, these effects were observed only in women. In men, there was no correlation between copeptin and blood ACTH levels during hypoglycemia and only a poor correlation was found between copeptin and blood cortisol levels during hypoglycemia [243]. Studies conducted on AVP-deficient Brattleboro rats that were also suffering from diabetes mellitus showed that AVP does not play a significant role in the regulation of the release of HPA hormones during acute stress [244].

Balapattabi et al. (2021) examined the associations of liver failure with AVP secretion and hyponatremia in terms of sex differences. They showed that male rats with a bile duct ligation (BDL) were hyponatremic and manifested significantly higher concentrations of plasma copeptin and higher FosB expression in their supraoptic AVP neurons than the sham males, whereas similar associations were not observed in female BDL rats [245]. A study conducted in women with polycystic ovary syndrome revealed that their AVP peak responses to hypoglycemia were negatively correlated with testosterone, androstenedione, and endogenous insulin levels; however, there was no correlation between AVP and basal and hypoglycemia-induced peak cortisol concentrations [246].

Overweight/obese women manifested higher ACTH and cortisol responses to AVP tests and significantly greater hormone inhibition after alprazolam (benzodiazepine used for the treatment of stress and depression) than controls. In both groups, AVP-induced delta-peak cortisol values before and after alprazolam pre-treatment were significantly correlated. Body fat distribution had no effect on the HPA's response to AVP either before or after alprazolam [247].

## 5. Impact of Therapeutic Interventions on Interactions of Vasopressin with the Hypothalamo–Pituitary–Adrenal System in Health and in Cardiovascular and Metabolic Diseases

As discussed in the previous parts of this review, there are several interconnections between steroids and vasopressin that are significantly affected in cardiovascular, metabolic, and psychogenic disorders [157,168,239]. Thus, it is justified to assume that various therapeutic interventions which have an impact on the vasopressinergic system and the hypothalamo–pituitary–adrenal axis may potently influence the effectiveness of AVP-HPA interactions.

Systemically applied corticosteroids are widely used in the therapy of inflammatory and autoimmune disorders [163,236,248]. Their influence on the HPA and vasopressin secretion resembles the role of endogenous corticosteroids in a negative feedback loop. A post-mortem study by Erkut et al. (1998) on nine patients treated with a corticosteroid and on eight control subjects demonstrated a significant decrease in CRH- and AVP-releasing neurons in the hypothalamus. The study showed that the corticosteroid-treated patients retained only 3.3% of their CRH-releasing cells and 33% of their AVP-releasing cells in comparison to the control group [249]. Thus far, the clinical implications of these changes have not been fully recognized. A study on premature infants with bronchopulmonary dysplasia revealed that treatment with dexamethasone improved their pulmonary functions but did not modify their AVP levels [250]. On the other hand, the administration of corticosteroids in supraphysiological doses in a clinical trial on children with acute lymphoblastic leukemia (ALL) resulted in the suppression of the HPA for a few weeks in the majority of patients [251]. A decrease in AVP mRNA was also found in the SChN of patients treated with corticosteroid (2-fold lower level in the SChN of patients exposed to corticosteroid in comparison to control subjects). Accordingly, it has been proposed that corticosteroid-induced changes in AVP synthesis may account for the circadian rhythm abnormalities and sleep disturbances in steroid-treated patients [252].

It has been postulated that vasopressin synthesized in the magnocellular cells of the hypothalamus participates in a phenomenon known as "corticosteroid escape". According to this hypothesis, inflammatory mediators (for instance IL-6) increase AVP release from hypothalamic magnocellular PVN cells above physiological concentrations, which results in a strong activation of adenylyl cyclases (AC2 and AC7) and an elevation of cAMP to a level at which it can inhibit the negative feedback mediated by endogenous corticosteroids in the HPA. Most likely, the rearrangement of mutual relations between CRH, AVP, interleukins, and corticosteroids accounts for the "corticosteroid escape" that is observed during the activation of the host immune system and that maintains corticosteroid release in spite of the high dose of exogenous corticosteroid imposed by the therapy (Figure 4) [253]. However, the role of "corticosteroid escape" in the achievement of the optimum results from systemic corticosteroid therapy is not yet fully determined.

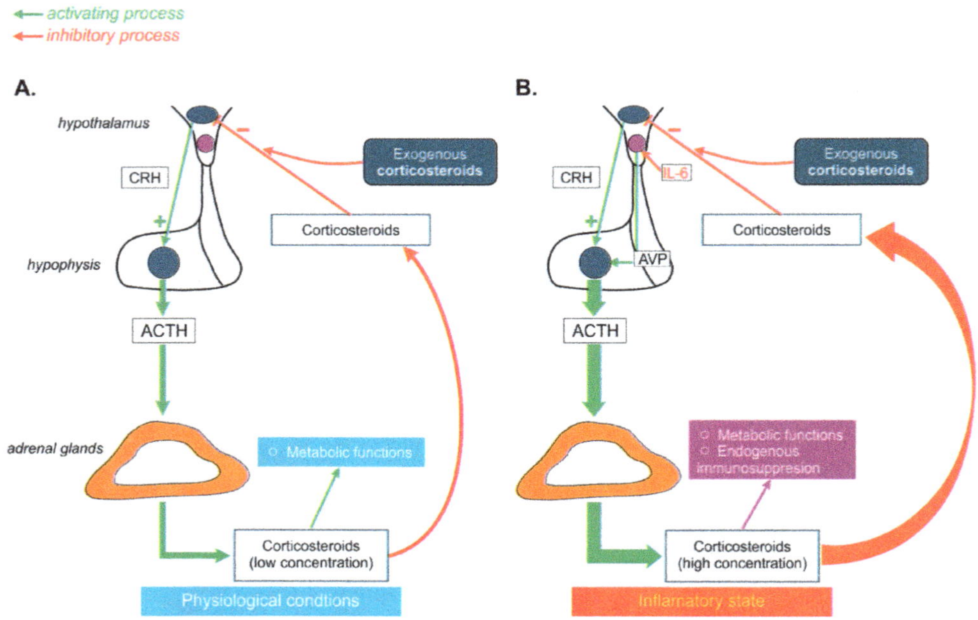


**Figure 4.** The phenomenon of "corticosteroid escape". (**A**)—under normal conditions, corticosteroids inhibit the release of corticotrpine-releasing hormone (CRH), (**B**)—during inflammatory states, the activation of autoimmune processes' and inhibition of CRH secretion via negative feedback are impaired and the activation of the HPA is maintained in spite of the high concentration of corticosteroids. ACTH—adrenocorticotropic hormone, IL-6—interleukin 6, green arrows—stimulation, purple (red) arrows—inhibition.

### 5.1. Steroids and Vasopressin Treatments in Cardiovascular Diseases

It is likely that the resetting of the vasopressin-HPA system by corticosteroids plays a beneficial role in the treatment of severe cardiovascular disorders such as myocardial infarction, shock, and cardiac arrest. A study on the rat model of myocardial infarction revealed that a hydrocortisone administration during the early reperfusion period results in decreased infarct size and in reduced oxidative stress [254]. The potentially beneficial role of corticosteroids may result from their anti-inflammatory activity. There are some concerns regarding the use of corticosteroids in acute myocardial infarction because of their negative effects on wound healing and scar formation. The meta-analysis performed by Giugliano et al. [255] indicated a slight reduction in cardiovascular mortality in patients with acute myocardial infarction treated with corticosteroids. However, it should be noted that this survey was based mainly on studies including small number of patients (less than 100) and it did not equivocally indicate the benefit of treatment with corticosteroids [255]. In another study, the best survival rate (91%) was found in patients with a low baseline cortisol level and with an appropriate adrenal response to ACTH analogs (a difference between minimum and maximum cortisol levels higher than 9 µg/dL) [256]. A study on a cohort of 159 patients with septic shock revealed that the survival rate was significantly greater when vasopressin was used together with hydrocortisone [257,258]. This finding suggested a potential benefit of the joint application of corticosteroids and vasopressin in critically ill patients. More recently, a randomized clinical trial on 512 patients from 10 hospitals in Denmark showed a significant increase in the probability of the return of spontaneous circulation (ROSC) in patients receiving a combined treatment of methylprednisolone and vasopressin in comparison to patients receiving a placebo. However, there was no difference in the 30-day mortality and neurological outcomes of the patients [259].

Studies on overweight/obese women have shown that that obesity is associated with higher ACTH and cortisol responses to AVP tests than those of women with normal body weight, the control, which suggests that obesity may cause the disarrangement of interactions of vasopressin with the HPA [247].

*5.2. Impact of Anti-Depressive and Neuroleptic Treatments on Vasopressin–HPA Interactions*

As was mentioned above, the vasopressin–HPA system is controlled by monoamine neurotransmitters, and the administration of anti-depressive or other neuroleptic compounds, which interfer with monoaminergic transmission, significantly influences AVP-HPA interactions (see Section 3.4).

Moreover, it has been reported that the activation of the HPA during stress seems to be altered in patients with depression and that vasopressin plays a more important role than CRH in the regulation of ACTH release in this group of patients [259,260]. The altered action of vasopressin in patients with depressive and stress-related disorders opens a discussion on the usefulness of selective AVP antagonists as anti-depressive or anxiolytic drugs [261,262]. Undoubtedly, further studies are needed to explain whether the down-regulation of the vasopressinergic pathways would play a beneficial role in these disorders.

A study upon an animal model of depression (olfactory bulbectomy in mice, OB mice) revealed that OB mice demonstrated higher vasopressin and ACTH plasma levels. Moreover, it was possible to reverse theses elevations by chronic treatment with anti-depressive drugs, namely fluoxetine (SSRI) and venlafaxine (SNRI) [261]. Another study on healthy rats showed that antidepressant compounds, such as SSRI and desipramine (tricyclic antidepressant) influence the function of the HPA by means of vasopressin V1bRs [263]. Studies on Wistar rats have shown that the AVP release from magnocellular and parvocellular cells may be affected by neuroleptic drugs and that clozapine and olanzapine are more effective than haloperidol [264].

A study on six patients with atrial fibrillation who qualified for electrical cardioversion and six patients with depression who qualified for electroconvulsive therapy revealed a significantly enhanced activation of their HPA after these interventions. In particular, their AVP concentration increased by 7 times after electrical cardioversion and 2 times after electroconvulsive therapy, whereas their plasma ACTH levels were elevated 3 times and 4 times, respectively [265].

## 6. Summary

The present review analyses the complex mechanisms underlying the cooperation of vasopressin with hormones of the hypothalamo–pituitary–adrenal axis. As shown in Figure 5, AVP cooperates with specific components of the HPA through interactions occurring both in the central nervous system and in the peripheral organs.

This cooperation plays a significant role in the regulation of blood circulation, metabolism, water–electrolyte balance, and behavioral adaptations to stress challenges. Vasopressin and components of the HPA closely interact and can be considered a functionally united AVP-HPA system. Growing evidence shows that cardiovascular and metabolic diseases, as well as inflammation and stress, are associated with an inappropriate functioning of the AVP-HPA system, and this question should be taken into consideration when pharmacological treatment is planned. The final action of a specific steroid depends on the concentration of the hormone, the type of stimulated receptors, the number of receptors, and the presence of specific enzymatic pathways in the targeted cell. In the cardiovascular system and in the organs regulating energy balance, stimulations of vasopressin and steroid receptors initiate a wide range of actions whose final effect may be either beneficial or detrimental (Figure 5).

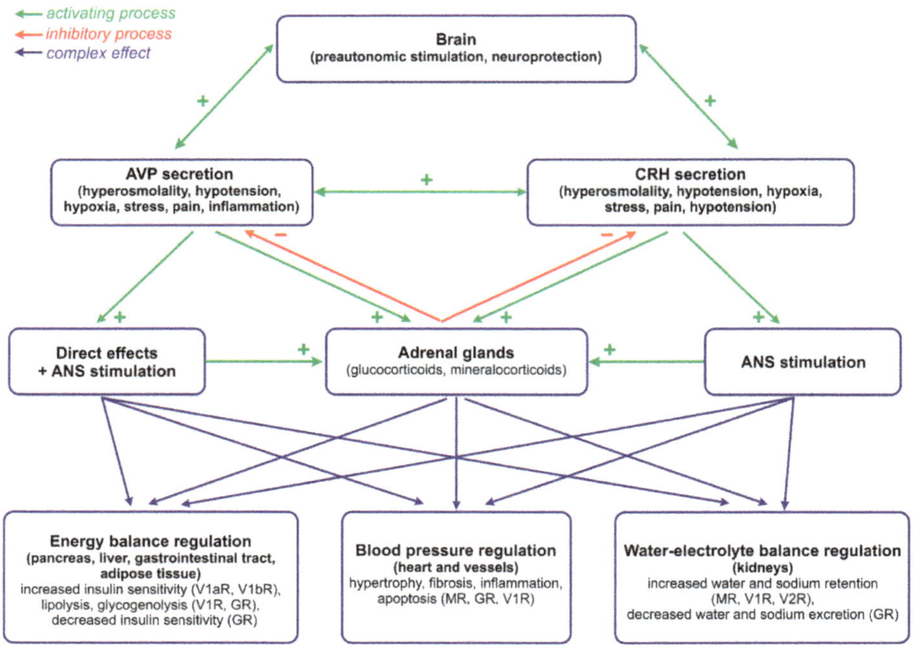


**Figure 5.** The stimulatory and inhibitory interactions between vasopressin and hormones secreted by the hypothalamo–pituitary–adrenal system that play essential roles in the regulation of energy balance, water–electrolyte balance, and blood pressure. ANS—autonomic nervous system, AVP—arginine vasopressin, GR—glucocorticoid receptor, MR—mineralocorticoids receptor, V1R, V1aR, and V1bR—vasopressin receptors. See text for further explanations.

## 7. Future Directions

Our abundant knowledge of the molecular processes initiated by vasopressin and steroid hormones in specific cells and organs is in contrast to our sparse knowledge of their actions in whole organisms. Experimental studies provide strong evidence that the secretion and action of vasopressin and steroid hormones is significantly altered in cardiovascular and metabolic diseases but it is not yet sufficiently understood which mechanisms are responsible for these changes and to what extent the alterations depend on age, sex, the time of application, the presence of other challenges, such as stress and other pathogenic factors, and treatment with specific pharmaceuticals. These denote essential directions for further research in this area.

Moreover, it should be noted that the prevailing number of studies showing the significant interaction of vasopressin and steroids in the regulation of metabolism, cardiovascular functions, and water–electrolyte balance comes from experiments performed on animals, whereas these questions remain largely unexplored in human beings. Undoubtedly, the clinical importance of the findings provided by the animal experiments should be confirmed in studies on large cohorts of human populations, including patients suffering from hypertension, obesity, diabetes mellitus, and atherosclerosis.

## 8. Conclusions

1. Vasopressin (AVP) and steroid hormones are frequently released together and closely cooperate in the regulation of blood pressure, metabolism, water–electrolyte balance, and behavior.

2. Vasopressin interacts with specific components of the hypothalamo–pituitary–adrenal axis in the brain and in several peripheral organs and tissues, including the heart, vessels, kidneys, and adipose tissue.
3. Appropriate interactions of AVP with the HPA are essential for the efficient regulation of water–electrolyte balance, blood pressure, and energy balance, and it is justified to consider vasopressin and the hypothalamo–pituitary axis as a highly coordinated, functional AVP-HPA system.
4. Interactions between AVP and HPA are significantly altered in cardiovascular, respiratory, and metabolic diseases and during inflammation and neurogenic stress.
5. Inappropriate interactions of AVP and steroids may initiate or intensify cardiovascular complications in metabolic diseases.
6. The interplay of vasopressin and steroid hormones is not yet fully recognized and further studies are needed to determine the potentially beneficial or harmful consequences of interference with these factors in the treatment of specific pathological states.

**Author Contributions:** E.S.-S. proposed the design and the final version of the review; E.S.-S., K.C., W.B.-R. and M.K. searched the literature; K.C. and M.K. designed the figures. All authors wrote the first draft of the manuscript. All authors have read and agreed to the published version of the manuscript.

**Funding:** This study was supported by the Medical University of Warsaw Scientific Projects 1MA/N. The work did not receive any specific grant from funding agencies in the public, commercial, or not-for-profit sectors.

**Institutional Review Board Statement:** Not applicable.

**Data Availability Statement:** Not applicable.

**Acknowledgments:** The authors wish to thank to Marcin Kumosa from the Department of Experimental and Clinical Physiology of the Medical University of Warsaw for the technical preparation of the figures.

**Conflicts of Interest:** The authors declare no conflicts of interest.

## Abbreviations

ACh—acetylcholine, ACTH—adrenocorticotropic hormone, Ang—angiotensin, AR—androgen receptor, AQP—aquaporin, AVP—arginine vasopressin, BDL—bile duct ligation, BNST—bed nucleus of the stria terminalis, cAMP—cyclic adenosine monophosphate, CBI—Coactivator Binding Inhibitor, COVID-19—coronavirus disease 2019, CRF—cortictropin-releasing factor, CRH—corticotropin-releasing hormone, CT—cardiotrophin, CTGF—connective tissue growth factor, DBD—DNA-binding domain, DHEA—dehydroepiandrosterone, DOC—deoxycorticosterone, ERK—extracellular signal-regulated kinase, ENaC—epithelial sodium channel, ES—estrogen, ESR—estrogen receptor, FSH—follicle-stimulating hormone, GABA—gamma aminobutyric acid, GC—glucocorticoid, GnRH—gonadotropine-releasing hormone, GP—glycopeptin, copeptine, GPER—G-protein-coupled receptor, GR—glucocorticoid receptor, GRK2—G-protein-coupled receptor kinase 2, HDL—high-density lipoprotein, HIF—hypoxia-inducible factor, HSD—hydroxysteroid dehydrogenase, ICAM—intercellular adhesion molecule, IL—interleukin, JNK—Jun N-terminal kinase, LBD—ligand-binding domain, LDL—low-density lipoprotein, LH—luteinizing hormone, MAPK—mitogen-activated protein kinase, MC—mineralocorticoid, MCP-1—monocyte chemoattractant protein-1, MCR, MR—mineralocorticoid receptor, MRE—mineralocorticoid response element, NADPH—dinicotinamide adenine dinucleotide phosphate, NAT—noradrenergic transporter activity, NCOAS—nuclear receptor co-activator, NCX—sodium–calcium exchanger, NFκB—nuclear factor kappa B, NO—nitric oxide, NOX—nicotinamide adenine dinucleotide phosphate oxidase, NP—neurophysin, NR3C2—nuclear receptor of subfamily 3, group C, member 2, NTD—N-terminal domain, NST—nucleus of the solitary tract, OVLT—organum vasculosum of the lamina terminalis, PET—positron emission tomography, PG—progesterone, PGR—progesterone receptor, PKA—protein kinase A, PKC—ptotein kinase C,

PLC—phospholipase C, PPAR—peroxisome proliferator-activator receptor, PVN—parventricular nucleus, RAS—renin angiotensin system, ROS—reactive oxygen species, ROSC—return of spontaneous circulation, SFO—subfornical organ, SHR—spontaneously hypertensive rat, SON—supraoptic nucleus, SNP—single nucleotide polymorphism, SRC—steroid receptor co-activator, TNF-α—tumor necrosis factor alpha, UCMS—unpredictable chronic mild stress, UCP—uncoupling protein, UT—urea transporter, V1aR—vasopressin receptor of type V1a, V1bR—vasopressin receptor of type V1b, V2R—V2 vasopressin receptor of type 2, VS—vasopressin system.

## References


1. Hilton, J.G.; Scian, L.F.; Westermann, S.D.; Nakano, J.; Kruesi, O.R. Vasopressin stimulation of the isolated adrenal glands: Nature and mechanism of hydrocortisone secretion. *Endocrinology* **1960**, *67*, 298–310. [CrossRef] [PubMed]
2. Larsen, P.J.; Vrang, N.; Møller, M.; Jessop, D.S.; Lightman, S.L.; Chowdrey, H.S.; Mikkelsen, J.D. The diurnal expression of genes encoding vasopressin and vasoactive intestinal peptide within the rat suprachiasmatic nucleus is influenced by circulating glucocorticoids. *Brain Res. Mol. Brain Res.* **1994**, *27*, 342–346. [CrossRef] [PubMed]
3. Mouri, T.; Itoi, K.; Takahashi, K.; Suda, T.; Murakami, O.; Yoshinaga, K.; Andoh, N.; Ohtani, H.; Masuda, T.; Sasano, N. Colocalization of corticotropin-releasing factor and vasopressin in the paraventricular nucleus of the human hypothalamus. *Neuroendocrinology* **1993**, *57*, 34–39. [CrossRef] [PubMed]
4. Otubo, A.; Kawakami, N.; Maejima, S.; Ueda, Y.; Morris, J.F.; Sakamoto, T.; Sakamoto, H. Vasopressin gene products are colocalised with corticotrophin-releasing factor within neurosecretory vesicles in the external zone of the median eminence of the Japanese macaque monkey (Macaca fuscata). *J. Neuroendocr.* **2020**, *32*, e12875. [CrossRef] [PubMed]
5. Engler, D.; Pham, T.; Fullerton, M.J.; Ooi, G.; Funder, J.W.; Clarke, I.J. Studies of the secretion of corticotropin-releasing factor and arginine vasopressin into the hypophysial-portal circulation of the conscious sheep. I. Effect of an audiovisual stimulus and insulin-induced hypoglycemia. *Neuroendocrinology* **1989**, *49*, 367–381. [CrossRef] [PubMed]
6. Familari, M.; Smith, A.I.; Smith, R.; Funder, J.W. Arginine vasopressin is a much more potent stimulus to ACTH release from ovine anterior pituitary cells than ovine corticotropin-releasing factor. 1. In vitro studies. *Neuroendocrinology* **1989**, *50*, 152–157. [CrossRef] [PubMed]
7. Labrie, F.; Giguere, V.; Proulx, L.; Lefevre, G. Interactions between CRF, epinephrine, vasopressin and glucocorticoids in the control of ACTH secretion. *J. Steroid Biochem.* **1984**, *20*, 153–160. [CrossRef] [PubMed]
8. Veldhuis, H.D.; de Kloet, E.R. Vasopressin-related peptides increase the hippocampal corticosterone receptor capacity of diabetes insipidus (Brattleboro) rat. *Endocrinology* **1982**, *110*, 153–157. [CrossRef] [PubMed]
9. Papanek, P.E.; Sladek, C.D.; Raff, H. Corticosterone inhibition of osmotically stimulated vasopressin from hypothalamic-neurohypophysial explants. *Am. J. Physiol.* **1997**, *272 Pt 2*, R158–R162. [CrossRef] [PubMed]
10. Liu, X.; Wang, C.A.; Chen, Y.Z. Nongenomic effect of glucocorticoid on the release of arginine vasopressin from hypothalamic slices in rats. *Neuroendocrinology* **1995**, *62*, 628–633. [CrossRef] [PubMed]
11. Calogero, A.E.; Liapi, C.; Chrousos, G.P. Hypothalamic and suprahypothalamic effects of prolonged treatment with dexamethasone in the rat. *J. Endocrinol. Investig.* **1991**, *14*, 277–286. [CrossRef] [PubMed]
12. Woodcock, E.A.; Mcleod, J.K.; Johnston, C.I. Vasopressin stimulates phosphatidylinositol turnover and aldosterone synthesis in rat adrenal glomerulosa cells: Comparison with angiotensin II. *Endocrinology* **1986**, *118*, 2432–2436. [CrossRef] [PubMed]
13. Saito, R.; Ishiharada, N.; Ban, Y.; Honda, K.; Takano, Y.; Kamiya, H. Vasopressin V1 receptor in rat hippocampus is regulated by adrenocortical functions. *Brain Res.* **1994**, *646*, 170–174. [CrossRef] [PubMed]
14. Watters, J.J.; Poulin, P.; Dorsa, D.M. Steroid hormone regulation of vasopressinergic neurotransmission in the central nervous system. *Prog. Brain Res.* **1998**, *119*, 247–261. [CrossRef] [PubMed]
15. Watters, J.J.; Wilkinson, C.W.; Dorsa, D.M. Glucocorticoid regulation of vasopressin V1a receptors in rat forebrain. *Brain Res. Mol. Brain Res.* **1996**, *38*, 276–284. [CrossRef] [PubMed]
16. Rabadan-Diehl, C.; Makara, G.; Kiss, A.; Lolait, S.; Zelena, D.; Ochedalski, T.; Aguilera, G. Regulation of pituitary V1b vasopressin receptor messenger ribonucleic acid by adrenalectomy and glucocorticoid administration. *Endocrinology* **1997**, *138*, 5189–5194. [CrossRef] [PubMed]
17. Hu, S.B.; Tannahill, L.A.; Biswas, S.; Lightman, S.L. Release of corticotrophin-releasing factor-41, arginine vasopressin and oxytocin from rat fetal hypothalamic cells in culture: Response to activation of intracellular second messengers and to corticosteroids. *J. Endocrinol.* **1992**, *132*, 57–65. [CrossRef] [PubMed]
18. Currie, I.S.; Gillies, G.; Brooks, A.N. Modulation of arginine vasopressin secretion from cultured ovine hypothalamic cells by glucocorticoids and opioid peptides. *Neuroendocrinology* **1994**, *60*, 360–367. [CrossRef] [PubMed]
19. Kokras, N.; Hodes, G.E.; Bangasser, D.A.; Dalla, C. Sex differences in the hypothalamic-pituitary-adrenal axis: An obstacle to antidepressant drug development? *Br. J. Pharmacol.* **2019**, *176*, 4090–4106. [CrossRef] [PubMed]
20. Orshal, J.M.; Khalil, R.A. Gender, sex hormones, and vascular tone. *Am. J. Physiol. Regul. Integr. Comp. Physiol.* **2004**, *286*, R233–R249. [CrossRef] [PubMed]

21. Jiang, Y.Q.; Kawashima, H.; Iwasaki, Y.; Uchida, K.; Sugimoto, K.; Itoi, K. Differential effects of forced swim-stress on the corticotropin-releasing hormone and vasopressin gene transcription in the parvocellular division of the paraventricular nucleus of rat hypothalamus. *Neurosci. Lett.* **2004**, *358*, 201–204. [CrossRef] [PubMed]
22. Hlavacova, N.; Bakos, J.; Jezova, D. Eplerenone, a selective mineralocorticoid receptor blocker, exerts anxiolytic effects accompanied by changes in stress hormone release. *J. Psychopharmacol.* **2010**, *24*, 779–786. [CrossRef] [PubMed]
23. de Kloet, E.R.; Meijer, O.C.; de Nicola, A.F.; de Rijk, R.H.; Joëls, M. Importance of the brain corticosteroid receptor balance in metaplasticity, cognitive performance and neuro-inflammation. *Front. Neuroendocr.* **2018**, *49*, 124–145. [CrossRef] [PubMed]
24. Karst, H.; Berger, S.; Turiault, M.; Tronche, F.; Schütz, G.; Joëls, M. Mineralocorticoid receptors are indispensable for nongenomic modulation of hippocampal glutamate transmission by corticosterone. *Proc. Natl. Acad. Sci. USA* **2005**, *102*, 19204–19207. [CrossRef] [PubMed]
25. Karst, H.; den Boon, F.S.; Vervoort, N.; Adrian, M.; Kapitein, L.C.; Joëls, M. Non-genomic steroid signaling through the mineralocorticoid receptor: Involvement of a membrane-associated receptor? *Mol. Cell. Endocrinol.* **2022**, *541*, 111501. [CrossRef] [PubMed]
26. Vasudevan, N.; Pfaff, D.W. Non-genomic actions of estrogens and their interaction with genomic actions in the brain. *Front. Neuroendocr.* **2008**, *29*, 238–257. [CrossRef] [PubMed]
27. Lachize, S.; Apostolakis, E.M.; van der Laan, S.; Tijssen, A.M.; Xu, J.; de Kloet, E.R.; Meijer, O.C. Steroid receptor coactivator-1 is necessary for regulation of corticotropin-releasing hormone by chronic stress and glucocorticoids. *Proc. Natl. Acad. Sci. USA* **2009**, *106*, 8038–8042. [CrossRef] [PubMed]
28. van Weert, L.T.C.M.; Buurstede, J.C.; Mahfouz, A.; Braakhuis, P.S.M.; Polman, J.A.E.; Sips, H.C.M.; Roozendaal, B.; Balog, J.; de Kloet, E.R.; Datson, N.A.; et al. NeuroD Factors Discriminate Mineralocorticoid From Glucocorticoid Receptor DNA Binding in the Male Rat Brain. *Endocrinology* **2017**, *158*, 1511–1522. [CrossRef] [PubMed]
29. Skowron, K.J.; Booker, K.; Cheng, C.; Creed, S.; David, B.P.; Lazzara, P.R.; Lian, A.; Siddiqui, Z.; Speltz, T.E.; Moore, T.W. Steroid receptor/coactivator binding inhibitors: An update. *Mol. Cell. Endocrinol.* **2019**, *493*, 110471. [CrossRef] [PubMed]
30. Stashi, E.; York, B.; O'Malley, B.W. Steroid receptor coactivators: Servants and masters for control of systems metabolism. *Trends Endocrinol. Metab.* **2014**, *25*, 337–347. [CrossRef] [PubMed]
31. Grossmann, C.; Almeida-Prieto, B.; Nolze, A.; Alvarez de la Rosa, D. Structural and molecular determinants of mineralocorticoid receptor signalling. *Br. J. Pharmacol.* **2022**, *179*, 3103–3118. [CrossRef] [PubMed]
32. Vanderhaeghen, T.; Beyaert, R.; Libert, C. Bidirectional Crosstalk Between Hypoxia Inducible Factors and Glucocorticoid Signalling in Health and Disease. *Front. Immunol.* **2021**, *12*, 684085. [CrossRef] [PubMed]
33. Fadel, L.; Dacic, M.; Fonda, V.; Sokolsky, B.A.; Quagliarini, F.; Rogatsky, I.; Uhlenhaut, N.H. Modulating glucocorticoid receptor actions in physiology and pathology: Insights from coregulators. *Pharmacol. Ther.* **2023**, *251*, 108531. [CrossRef] [PubMed]
34. Vandevyver, S.; Dejager, L.; Libert, C. Comprehensive overview of the structure and regulation of the glucocorticoid receptor. *Endocr. Rev.* **2014**, *35*, 671–693. [CrossRef] [PubMed]
35. Clayton, S.A.; Jones, S.W.; Kurowska-Stolarska, M.; Clark, A.R. The role of microRNAs in glucocorticoid action. *J. Biol. Chem.* **2018**, *293*, 1865–1874. [CrossRef] [PubMed]
36. Knutti, D.; Kaul, A.; Kralli, A. A tissue-specific coactivator of steroid receptors, identified in a functional genetic screen. *Mol. Cell. Biol.* **2000**, *20*, 2411–2422. [CrossRef] [PubMed]
37. Meijer, O.C.; Buurstede, J.C.; Schaaf, M.J.M. Corticosteroid Receptors in the Brain: Transcriptional Mechanisms for Specificity and Context-Dependent Effects. *Cell Mol. Neurobiol.* **2019**, *39*, 539–549. [CrossRef] [PubMed]
38. Oakley, R.H.; Cidlowski, J.A. The biology of the glucocorticoid receptor: New signaling mechanisms in health and disease. *J. Allergy Clin. Immunol.* **2013**, *132*, 1033–1044. [CrossRef] [PubMed]
39. Oakley, R.H.; Cruz-Topete, D.; He, B.; Foley, J.F.; Myers, P.H.; Xu, X.; Gomez-Sanchez, C.E.; Chambon, P.; Willis, M.S.; Cidlowski, J.A. Cardiomyocyte glucocorticoid and mineralocorticoid receptors directly and antagonistically regulate heart disease in mice. *Sci. Signal.* **2019**, *12*, eaau9685. [CrossRef] [PubMed]
40. Koning, A.C.A.M.; Buurstede, J.C.; van Weert, L.T.C.M.; Meijer, O.C. Glucocorticoid and Mineralocorticoid Receptors in the Brain: A Transcriptional Perspective. *J. Endocr. Soc.* **2019**, *3*, 1917–1930. [CrossRef] [PubMed]
41. Sacta, M.A.; Chinenov, Y.; Rogatsky, I. Glucocorticoid Signaling: An Update from a Genomic Perspective. *Annu. Rev. Physiol.* **2016**, *78*, 155–180. [CrossRef] [PubMed]
42. Faulkner, J.L.; Belin de Chantemèle, E.J. Mineralocorticoid Receptor and Endothelial Dysfunction in Hypertension. *Curr. Hypertens. Rep.* **2019**, *21*, 78. [CrossRef] [PubMed]
43. Kokkinopoulou, I.; Moutsatsou, P. Mitochondrial Glucocorticoid Receptors and Their Actions. *Int. J. Mol. Sci.* **2021**, *22*, 6054. [CrossRef] [PubMed]
44. Viho, E.M.G.; Buurstede, J.C.; Mahfouz, A.; Koorneef, L.L.; van Weert, L.T.C.M.; Houtman, R.; Hunt, H.J.; Kroon, J.; Meijer, O.C. Corticosteroid Action in the Brain: The Potential of Selective Receptor Modulation. *Neuroendocrinology* **2019**, *109*, 266–276. [CrossRef] [PubMed]
45. Winnay, J.N.; Xu, J.; O'Malley, B.W.; Hammer, G.D. Steroid receptor coactivator-1-deficient mice exhibit altered hypothalamic-pituitary-adrenal axis function. *Endocrinology* **2006**, *147*, 1322–1332. [CrossRef] [PubMed]
46. Yi, P.; Yu, X.; Wang, Z.; O'Malley, B.W. Steroid receptor-coregulator transcriptional complexes: New insights from CryoEM. *Essays Biochem.* **2021**, *65*, 857–866. [CrossRef] [PubMed]

47. Marchi, D.; van Eeden, F.J.M. Homeostatic Regulation of Glucocorticoid Receptor Activity by Hypoxia-Inducible Factor 1: From Physiology to Clinic. *Cells* **2021**, *10*, 3441. [CrossRef] [PubMed]
48. Syed, A.P.; Greulich, F.; Ansari, S.A.; Uhlenhaut, N.H. Anti-inflammatory glucocorticoid action: Genomic insights and emerging concepts. *Curr. Opin. Pharmacol.* **2020**, *53*, 35–44. [CrossRef] [PubMed]
49. Kodama, T.; Shimizu, N.; Yoshikawa, N.; Makino, Y.; Ouchida, R.; Okamoto, K.; Hisada, T.; Nakamura, H.; Morimoto, C.; Tanaka, H. Role of the glucocorticoid receptor for regulation of hypoxia-dependent gene expression. *J. Biol. Chem.* **2003**, *278*, 33384–33391. [CrossRef] [PubMed]
50. Callera, G.E.; Touyz, R.M.; Tostes, R.C.; Yogi, A.; He, Y.; Malkinson, S.; Schiffrin, E.L. Aldosterone activates vascular p38MAP kinase and NADPH oxidase via c-Src. *Hypertension* **2005**, *45*, 773–779. [CrossRef] [PubMed]
51. Zeyen, L.; Seternes, O.M.; Mikkola, I. Crosstalk between p38 MAPK and GR Signaling. *Int. J. Mol. Sci.* **2022**, *23*, 3322. [CrossRef] [PubMed]
52. Quatrini, L.; Ugolini, S. New insights into the cell-and tissue-specificity of glucocorticoid actions. *Cell. Mol. Immunol.* **2021**, *18*, 269–278. [CrossRef] [PubMed]
53. Davel, A.P.; Anwar, I.J.; Jaffe, I.Z. The endothelial mineralocorticoid receptor: Mediator of the switch from vascular health to disease. *Curr. Opin. Nephrol. Hypertens.* **2017**, *26*, 97–104. [CrossRef] [PubMed]
54. Geerling, J.C.; Loewy, A.D. Aldosterone in the brain. *Am. J. Physiol. Ren. Physiol.* **2009**, *297*, F559–F576. [CrossRef] [PubMed]
55. Haque, M.; Wilson, R.; Sharma, K.; Mills, N.J.; Teruyama, R. Localisation of 11β-Hydroxysteroid Dehydrogenase Type 2 in Mineralocorticoid Receptor Expressing Magnocellular Neurosecretory Neurones of the Rat Supraoptic and Paraventricular Nuclei. *J. Neuroendocrinol.* **2015**, *27*, 835–849. [CrossRef] [PubMed]
56. Vassiliou, A.G.; Athanasiou, N.; Vassiliadi, D.A.; Jahaj, E.; Keskinidou, C.; Kotanidou, A.; Dimopoulou, I. Glucocorticoid and mineralocorticoid receptor expression in critical illness: A narrative review. *World J. Crit. Care Med.* **2021**, *10*, 102–111. [CrossRef] [PubMed]
57. Parker, B.M.; Wertz, S.L.; Pollard, C.M.; Desimine, V.L.; Maning, J.; McCrink, K.A.; Lymperopoulos, A. Novel Insights into the Crosstalk between Mineralocorticoid Receptor and G Protein-Coupled Receptors in Heart Adverse Remodeling and Disease. *Int. J. Mol. Sci.* **2018**, *19*, 3764. [CrossRef] [PubMed]
58. Perlstein, R.S.; Whitnall, M.H.; Abrams, J.S.; Mougey, E.H.; Neta, R. Synergistic roles of interleukin-6, interleukin-1, and tumor necrosis factor in the adrenocorticotropin response to bacterial lipopolysaccharide in vivo. *Endocrinology* **1993**, *132*, 946–952. [CrossRef] [PubMed]
59. Terada, Y.; Ueda, S.; Hamada, K.; Shimamura, Y.; Ogata, K.; Inoue, K.; Taniguchi, Y.; Kagawa, T.; Horino, T.; Takao, T. Aldosterone stimulates nuclear factor-kappa B activity and transcription of intercellular adhesion molecule-1 and connective tissue growth factor in rat mesangial cells via serum- and glucocorticoid-inducible protein kinase-1. *Clin. Exp. Nephrol.* **2012**, *16*, 81–88. [CrossRef] [PubMed]
60. Maturana, A.; Lenglet, S.; Python, M.; Kuroda, S.; Rossier, M.F. Role of the T-type calcium channel CaV3.2 in the chronotropic action of corticosteroids in isolated rat ventricular myocytes. *Endocrinology* **2009**, *150*, 3726–3734. [CrossRef] [PubMed]
61. Rossier, M.F. The Cardiac Mineralocorticoid Receptor (MR): A Therapeutic Target Against Ventricular Arrhythmias. *Front. Endocrinol.* **2021**, *12*, 694758. [CrossRef] [PubMed]
62. Funder, J.W. Aldosterone and Mineralocorticoid Receptors-Physiology and Pathophysiology. *Int. J. Mol. Sci.* **2017**, *18*, 1032. [CrossRef] [PubMed]
63. Igbekele, A.E.; Jia, G.; Hill, M.A.; Sowers, J.R.; Jia, G. Mineralocorticoid Receptor Activation in Vascular Insulin Resistance and Dysfunction. *Int. J. Mol. Sci.* **2022**, *23*, 8954. [CrossRef] [PubMed]
64. Vanderhaeghen, T.; Timmermans, S.; Watts, D.; Paakinaho, V.; Eggermont, M.; Vandewalle, J.; Wallaeys, C.; Van Wyngene, L.; Van Looveren, K.; Nuyttens, L.; et al. Reprogramming of glucocorticoid receptor function by hypoxia. *EMBO Rep.* **2022**, *23*, e53083. [CrossRef] [PubMed]
65. McEown, K.; Treit, D. Mineralocorticoid receptors in the medial prefrontal cortex and hippocampus mediate rats' unconditioned fear behaviour. *Horm. Behav.* **2011**, *60*, 581–588. [CrossRef] [PubMed]
66. Reul, J.M.; de Kloet, E.R. Two receptor systems for corticosterone in rat brain: Microdistribution and differential occupation. *Endocrinology* **1985**, *117*, 2505–2511. [CrossRef] [PubMed]
67. Reul, J.M.; de Kloet, E.R.; van Sluijs, F.J.; Rijnberk, A.; Rothuizen, J. Binding characteristics of mineralocorticoid and glucocorticoid receptors in dog brain and pituitary. *Endocrinology* **1990**, *127*, 907–915. [CrossRef] [PubMed]
68. Ahmadpour, D.; Grange-Messent, V. Involvement of Testosterone Signaling in the Integrity of the Neurovascular Unit in the Male: Review of Evidence, Contradictions, and Hypothesis. *Neuroendocrinology* **2021**, *111*, 403–420. [CrossRef] [PubMed]
69. Castelli, M.P.; Casti, A.; Casu, A.; Frau, R.; Bortolato, M.; Spiga, S.; Ennas, M.G. Regional distribution of 5α-reductase type 2 in the adult rat brain: An immunohistochemical analysis. *Psychoneuroendocrinology* **2013**, *38*, 281–293. [CrossRef] [PubMed]
70. Takahashi, K.; Hosoya, T.; Onoe, K.; Takashima, T.; Tanaka, M.; Ishii, A.; Nakatomi, Y.; Tazawa, S.; Takahashi, K.; Doi, H.; et al. Association between aromatase in human brains and personality traits. *Sci. Rep.* **2018**, *8*, 16841. [CrossRef] [PubMed]
71. Ghoumari, A.M.; Abi Ghanem, C.; Asbelaoui, N.; Schumacher, M.; Hussain, R. Roles of Progesterone, Testosterone and Their Nuclear Receptors in Central Nervous System Myelination and Remyelination. *Int. J. Mol. Sci.* **2020**, *21*, 3163. [CrossRef] [PubMed]

72. Hwang, W.J.; Lee, T.Y.; Kim, N.S.; Kwon, J.S. The Role of Estrogen Receptors and Their Signaling across Psychiatric Disorders. *Int. J. Mol. Sci.* **2020**, *22*, 373. [CrossRef] [PubMed]
73. Jiao, L.; Machuki, J.O.; Wu, Q.; Shi, M.; Fu, L.; Adekunle, A.O.; Tao, X.; Xu, C.; Hu, X.; Yin, Z.; et al. Estrogen and calcium handling proteins: New discoveries and mechanisms in cardiovascular diseases. *Am. J. Physiol. Heart Circ. Physiol.* **2020**, *318*, H820–H829. [CrossRef] [PubMed]
74. Trenti, A.; Tedesco, S.; Boscaro, C.; Trevisi, L.; Bolego, C.; Cignarella, A. Estrogen, Angiogenesis, Immunity and Cell Metabolism: Solving the Puzzle. *Int. J. Mol. Sci.* **2018**, *19*, 859. [CrossRef] [PubMed]
75. Coolen, R.L.; Cambier, J.C.; Spantidea, P.I.; van Asselt, E.; Blok, B.F.M. Androgen receptors in areas of the spinal cord and brainstem: A study in adult male cats. *J. Anat.* **2021**, *239*, 125–135. [CrossRef] [PubMed]
76. Bernstein, S.R.; Kelleher, C.; Khalil, R.A. Gender-based research underscores sex differences in biological processes, clinical disorders and pharmacological interventions. *Biochem. Pharmacol.* **2023**, *215*, 115737. [CrossRef] [PubMed]
77. Cunningham, R.L.; Lumia, A.R.; McGinnis, M.Y. Androgen receptors, sex behavior, and aggression. *Neuroendocrinology* **2012**, *96*, 131–140. [CrossRef] [PubMed]
78. Cheng, J.; Watkins, S.C.; Walker, W.H. Testosterone activates mitogen-activated protein kinase via Src kinase and the epidermal growth factor receptor in sertoli cells. *Endocrinology* **2007**, *148*, 2066–2074. [CrossRef] [PubMed]
79. Davey, R.A.; Grossmann, M. Androgen Receptor Structure, Function and Biology: From Bench to Bedside. *Clin. Biochem. Rev.* **2016**, *37*, 3–15. [PubMed]
80. Thomas, P. Membrane Androgen Receptors Unrelated to Nuclear Steroid Receptors. *Endocrinology* **2019**, *160*, 772–781. [CrossRef] [PubMed]
81. Venkatesh, V.S.; Grossmann, M.; Zajac, J.D.; Davey, R.A. The role of the androgen receptor in the pathogenesis of obesity and its utility as a target for obesity treatments. *Obes. Rev.* **2022**, *23*, e13429. [CrossRef] [PubMed]
82. Lucas-Herald, A.K.; Touyz, R.M. Androgens and Androgen Receptors as Determinants of Vascular Sex Differences Across the Lifespan. *Can. J. Cardiol.* **2022**, *38*, 1854–1864. [CrossRef] [PubMed]
83. Arterburn, J.B.; Prossnitz, E.R. G Protein-Coupled Estrogen Receptor GPER: Molecular Pharmacology and Therapeutic Applications. *Annu. Rev. Pharmacol. Toxicol.* **2023**, *63*, 295–320. [CrossRef] [PubMed]
84. Chakraborty, B.; Byemerwa, J.; Krebs, T.; Lim, F.; Chang, C.Y.; McDonnell, D.P. Estrogen Receptor Signaling in the Immune System. *Endocr. Rev.* **2023**, *44*, 117–141. [CrossRef] [PubMed]
85. Gregorio, K.C.R.; Laurindo, C.P.; Machado, U.F. Estrogen and Glycemic Homeostasis: The Fundamental Role of Nuclear Estrogen Receptors ESR1/ESR2 in Glucose Transporter GLUT4 Regulation. *Cells* **2021**, *10*, 99. [CrossRef] [PubMed]
86. Rzemieniec, J.; Castiglioni, L.; Gelosa, P.; Muluhie, M.; Mercuriali, B.; Sironi, L. Nuclear Receptors in Myocardial and Cerebral Ischemia-Mechanisms of Action and Therapeutic Strategies. *Int. J. Mol. Sci.* **2021**, *22*, 12326. [CrossRef] [PubMed]
87. Guajardo-Correa, E.; Silva-Agüero, J.F.; Calle, X.; Chiong, M.; Henríquez, M.; García-Rivas, G.; Latorre, M.; Parra, V. Estrogen signaling as a bridge between the nucleus and mitochondria in cardiovascular diseases. *Front. Cell Dev. Biol.* **2022**, *10*, 968373. [CrossRef] [PubMed]
88. Wang, H.; Zhao, Z.; Lin, M.; Groban, L. Activation of GPR30 inhibits cardiac fibroblast proliferation. *Mol. Cell. Biochem.* **2015**, *405*, 135–148. [CrossRef] [PubMed]
89. Klinge, C.M. Estrogenic control of mitochondrial function. *Redox Biol.* **2020**, *31*, 101435. [CrossRef] [PubMed]
90. Kurmann, L.; Okoniewski, M.; Dubey, R.K. Estradiol Inhibits Human Brain Vascular Pericyte Migration Activity: A Functional and Transcriptomic Analysis. *Cells* **2021**, *10*, 2314. [CrossRef] [PubMed]
91. Machuki, J.O.; Zhang, H.Y.; Harding, S.E.; Sun, H. Molecular pathways of oestrogen receptors and β-adrenergic receptors in cardiac cells: Recognition of their similarities, interactions and therapeutic value. *Acta Physiol.* **2018**, *222*, e12978. [CrossRef] [PubMed]
92. da Silva, J.S.; Montagnoli, T.L.; Rocha, B.S.; Tacco, M.L.C.A.; Marinho, S.C.P.; Zapata-Sudo, G. Estrogen Receptors: Therapeutic Perspectives for the Treatment of Cardiac Dysfunction after Myocardial Infarction. *Int. J. Mol. Sci.* **2021**, *22*, 525. [CrossRef] [PubMed]
93. Fuentes, N.; Silveyra, P. Estrogen receptor signaling mechanisms. *Adv. Protein Chem. Struct. Biol.* **2019**, *116*, 135–170. [CrossRef] [PubMed]
94. Tran, Q.K. Reciprocality Between Estrogen Biology and Calcium Signaling in the Cardiovascular System. *Front. Endocrinol.* **2020**, *11*, 568203. [CrossRef] [PubMed]
95. Menazza, S.; Murphy, E. The Expanding Complexity of Estrogen Receptor Signaling in the Cardiovascular System. *Circ. Res.* **2016**, *118*, 994–1007. [CrossRef] [PubMed]
96. Nilsson, S.; Mäkelä, S.; Treuter, E.; Tujague, M.; Thomsen, J.; Andersson, G.; Enmark, E.; Pettersson, K.; Warner, M.; Gustafsson, J.A. Mechanisms of estrogen action. *Physiol. Rev.* **2001**, *81*, 1535–1565. [CrossRef] [PubMed]
97. Sun, J.; Aponte, A.M.; Menazza, S.; Gucek, M.; Steenbergen, C.; Murphy, E. Additive cardioprotection by pharmacological postconditioning with hydrogen sulfide and nitric oxide donors in mouse heart: S-sulfhydration vs. S-nitrosylation. *Cardiovasc. Res.* **2016**, *110*, 96–106. [CrossRef] [PubMed]
98. Zimmerman, M.A.; Budish, R.A.; Kashyap, S.; Lindsey, S.H. GPER-novel membrane oestrogen receptor. *Clin. Sci.* **2016**, *130*, 1005–1016. [CrossRef] [PubMed]

99. Shughrue, P.J.; Lane, M.V.; Merchenthaler, I. Comparative distribution of estrogen receptor-alpha and -beta mRNA in the rat central nervous system. *J. Comp. Neurol.* **1997**, *388*, 507–525. [CrossRef] [PubMed]
100. Prossnitz, E.R.; Barton, M. The G-protein-coupled estrogen receptor GPER in health and disease. *Nat. Rev. Endocrinol.* **2011**, *7*, 715–726. [CrossRef] [PubMed]
101. Brann, D.W.; Lu, Y.; Wang, J.; Zhang, Q.; Thakkar, R.; Sareddy, G.R.; Pratap, U.P.; Tekmal, R.R.; Vadlamudi, R.K. Brain-derived estrogen and neural function. *Neurosci. Biobehav. Rev.* **2022**, *132*, 793–817. [CrossRef] [PubMed]
102. Aickareth, J.; Hawwar, M.; Sanchez, N.; Gnanasekaran, R.; Zhang, J. Membrane Progesterone Receptors (mPRs/PAQRs) Are Going beyond Its Initial Definitions. *Membranes* **2023**, *13*, 260. [CrossRef] [PubMed]
103. Shah, N.M.; Imami, N.; Johnson, M.R. Progesterone Modulation of Pregnancy-Related Immune Responses. *Front. Immunol.* **2018**, *9*, 1293. [CrossRef] [PubMed]
104. Singh, M.; Su, C.; Ng, S. Non-genomic mechanisms of progesterone action in the brain. *Front. Neurosci.* **2013**, *7*, 159. [CrossRef] [PubMed]
105. Vítků, J.; Hampl, R. Steroid Conjugates and Their Physiological Role. *Physiol. Res.* **2023**, *72*, S317–S322. [CrossRef] [PubMed]
106. Pinna, G. Allopregnanolone (1938–2019): A trajectory of 80 years of outstanding scientific achievements. *Neurobiol. Stress* **2020**, *13*, 100246. [CrossRef] [PubMed]
107. Buijs, R.M.; Hermes, M.H.; Kalsbeek, A. The suprachiasmatic nucleus-paraventricular nucleus interactions: A bridge to the neuroendocrine and autonomic nervous system. *Prog. Brain Res.* **1998**, *119*, 365–382. [CrossRef] [PubMed]
108. Szczepanska-Sadowska, E.; Czarzasta, K.; Cudnoch-Jedrzejewska, A. Dysregulation of the Renin-Angiotensin System and the Vasopressinergic System Interactions in Cardiovascular Disorders. *Curr. Hypertens. Rep.* **2018**, *20*, 19. [CrossRef] [PubMed]
109. Szczepanska-Sadowska, E.; Zera, T.; Sosnowski, P.; Cudnoch-Jedrzejewska, A.; Puszko, A.; Misicka, A. Vasopressin and Related Peptides; Potential Value in Diagnosis, Prognosis and Treatment of Clinical Disorders. *Curr. Drug Metab.* **2017**, *18*, 306–345. [CrossRef] [PubMed]
110. Amico, J.A.; Finn, F.M.; Haldar, J. Oxytocin and vasopressin are present in human and rat pancreas. *Am. J. Med. Sci.* **1988**, *296*, 303–307. [CrossRef] [PubMed]
111. Hupf, H.; Grimm, D.; Riegger, G.A.; Schunkert, H. Evidence for a vasopressin system in the rat heart. *Circ. Res.* **1999**, *84*, 365–370. [CrossRef] [PubMed]
112. Burbach, J.P.; Luckman, S.M.; Murphy, D.; Gainer, H. Gene regulation in the magnocellular hypothalamo-neurohypophysial system. *Physiol. Rev.* **2001**, *81*, 1197–1267. [CrossRef] [PubMed]
113. Sparapani, S.; Millet-Boureima, C.; Oliver, J.; Mu, K.; Hadavi, P.; Kalostian, T.; Ali, N.; Avelar, C.M.; Bardies, M.; Barrow, B.; et al. The Biology of Vasopressin. *Biomedicines* **2021**, *9*, 89. [CrossRef] [PubMed]
114. Morgenthaler, N.G.; Struck, J.; Alonso, C.; Bergmann, A. Assay for the measurement of copeptin, a stable peptide derived from the precursor of vasopressin. *Clin. Chem.* **2006**, *52*, 112–119. [CrossRef] [PubMed]
115. Rechlin, T.; Hochholzer, W.; Stelzi, C.; Laule, K.; Freidank, H.; Morgenthaler, N.G.; Bergmann, A.; Potocki, M.; Noveanu, M.; Breidthardt, T.; et al. Incremental value of Copeptin for rapid rule out of acute myocardial infarction. *J. Am. Coll. Cardiol.* **2009**, *54*, 60–68. [CrossRef] [PubMed]
116. Roffi, M.; Patrono, C. CardioPulse: 'Ten Commandments' of 2015 European Society of Cardiology Guidelines for the management of acute coronary syndromes in patients presenting without persistent ST-segment elevation (NSTE-ACS). *Eur. Heart J.* **2016**, *37*, 208. [CrossRef] [PubMed]
117. Aikins, A.O.; Nguyen, D.H.; Paundralingga, O.; Farmer, G.E.; Shimoura, C.G.; Brock, C.; Cunningham, J.T. Cardiovascular Neuroendocrinology: Emerging Role for Neurohypophyseal Hormones in Pathophysiology. *Endocrinology* **2021**, *162*, bqab082. [CrossRef] [PubMed]
118. Costello, H.M.; Krilis, G.; Grenier, C.; Severs, D.; Czopek, A.; Ivy, J.R.; Nixon, M.; Holmes, M.C.; Livingstone, D.E.W.; Hoorn, E.J.; et al. High salt intake activates the hypothalamic-pituitary-adrenal axis, amplifies the stress response, and alters tissue glucocorticoid exposure in mice. *Cardiovasc. Res.* **2023**, *119*, 1740–1750. [CrossRef] [PubMed]
119. Goldstein, D.S. The extended autonomic system, dyshomeostasis, and COVID-19. *Clin. Auton. Res.* **2020**, *30*, 299–315. [CrossRef] [PubMed]
120. Iwasaki, Y.; Oiso, Y.; Saito, H.; Majzoub, J.A. Positive and negative regulation of the rat vasopressin gene promoter. *Endocrinology* **1997**, *138*, 5266–5274. [CrossRef] [PubMed]
121. Lee, S.; Rivier, C. Hypophysiotropic role and hypothalamic gene expression of corticotropin-releasing factor and vasopressin in rats injected with interleukin-1 beta systemically or into the brain ventricles. *J. Neuroendocr.* **1994**, *6*, 217–224. [CrossRef] [PubMed]
122. Grinevich, V.; Ma, X.M.; Jirikowski, G.; Verbalis, J.; Aguilera, G. Lipopolysaccharide endotoxin potentiates the effect of osmotic stimulation on vasopressin synthesis and secretion in the rat hypothalamus. *J. Neuroendocr.* **2003**, *15*, 141–149. [CrossRef] [PubMed]
123. Pardy, K.; Murphy, D.; Carter, D.; Hui, K.M. The influence of interleukin-2 on vasopressin and oxytocin gene expression in the rodent hypothalamus. *J. Neuroimmunol.* **1993**, *42*, 131–138. [CrossRef] [PubMed]
124. Zelazowski, P.; Patchev, V.K.; Zelazowska, E.B.; Chrousos, G.P.; Gold, P.W.; Sternberg, E.M. Release of hypothalamic corticotropin-releasing hormone and arginine-vasopressin by interleukin 1 beta and alpha MSH: Studies in rats with different susceptibility to inflammatory disease. *Brain Res.* **1993**, *631*, 22–26. [CrossRef] [PubMed]

125. Antoni, F.A. Magnocellular Vasopressin and the Mechanism of "Glucocorticoid Escape". *Front. Endocrinol.* **2019**, *10*, 422. [CrossRef] [PubMed]
126. Luo X, Kiss A, Makara G, Lolait SJ, Aguilera G Stress-specific regulation of corticotropin releasing hormone receptor expression in the paraventricular and supraoptic nuclei of the hypothalamus in the rat. *J. Neuroendocr.* **1994**, *6*, 689–696. [CrossRef] [PubMed]
127. Sawchenko, P.E. Adrenalectomy-induced enhancement of CRF and vasopressin immunoreactivity in parvocellular neurosecretory neurons: Anatomic, peptide, and steroid specificity. *J. Neurosci.* **1987**, *7*, 1093–1106. [CrossRef] [PubMed]
128. Kovács, K.J.; Földes, A.; Sawchenko, P.E. Glucocorticoid negative feedback selectively targets vasopressin transcription in parvocellular neurosecretory neurons. *J. Neurosci.* **2000**, *20*, 3843–3852. [CrossRef] [PubMed]
129. Kageyama, K.; Suda, T. Regulatory mechanisms underlying corticotropin-releasing factor gene expression in the hypothalamus. *Endocr. J.* **2009**, *56*, 335–344. [CrossRef] [PubMed]
130. Yoshida, M. Gene regulation system of vasopressin and corticotropin-releasing hormone. *Gene Regul. Syst. Bio.* **2008**, *2*, 71–88. [CrossRef] [PubMed]
131. Mills, N.J.; Sharma, K.; Haque, M.; Moore, M.; Teruyama, R. Aldosterone Mediated Regulation of Epithelial Sodium Channel (ENaC) Subunits in the Rat Hypothalamus. *Neuroscience* **2018**, *390*, 278–292. [CrossRef] [PubMed]
132. Aguilera, G.; Liu, Y. The molecular physiology of CRH neurons. *Front. Neuroendocrinol.* **2012**, *33*, 67–84. [CrossRef] [PubMed]
133. Bous, J.; Fouillen, A.; Orcel, H.; Granier, S.; Bron, P.; Mouillac, B. Structures of the arginine-vasopressin and oxytocin receptor signaling complexes. *Vitam. Horm.* **2023**, *123*, 67–107. [CrossRef] [PubMed]
134. Holmes, C.L.; Landry, D.W.; Granton, J.T. Science Review: Vasopressin and the cardiovascular system part 2—Clinical physiology. *Crit. Care* **2004**, *8*, 15–23. [CrossRef] [PubMed]
135. Bao, A.M.; Meynen, G.; Swaab, D.F. The stress system in depression and neurodegeneration: Focus on the human hypothalamus. *Brain Res. Rev.* **2008**, *57*, 531–553. [CrossRef] [PubMed]
136. Viau, V.; Chu, A.; Soriano, L.; Dallman, M.F. Independent and overlapping effects of corticosterone and testosterone on corticotropin-releasing hormone and arginine vasopressin mRNA expression in the paraventricular nucleus of the hypothalamus and stress-induced adrenocorticotropic hormone release. *J. Neurosci.* **1999**, *19*, 6684–6693. [CrossRef] [PubMed]
137. Herman, J.P.; Tasker, J.G. Paraventricular Hypothalamic Mechanisms of Chronic Stress Adaptation. *Front. Endocrinol.* **2016**, *7*, 137. [CrossRef] [PubMed]
138. Sheng, J.A.; Tan, S.M.L.; Hale, T.M.; Handa, R.J. Androgens and Their Role in Regulating Sex Differences in the Hypothalamic/Pituitary/Adrenal Axis Stress Response and Stress-Related Behaviors. *Androg. Clin. Res. Ther.* **2021**, *2*, 261–274. [CrossRef] [PubMed]
139. Thai, B.S.; Chia, L.Y.; Nguyen, A.T.N.; Qin, C.; Ritchie, R.H.; Hutchinson, D.S.; Kompa, A.; White, P.J.; May, L.T. Targeting G protein-coupled receptors for heart failure treatment. *Br. J. Pharmacol.* **2024**, *181*, 2270–2286. [CrossRef] [PubMed]
140. Morello, J.P.; Bichet, D.G. Nephrogenic diabetes insipidus. *Annu. Rev. Physiol.* **2001**, *63*, 607–630. [CrossRef] [PubMed]
141. Li, Q.; Tian, D.; Cen, J.; Duan, L.; Xia, W. Novel *AVPR2* mutations and clinical characteristics in 28 Chinese families with congenital nephrogenic diabetes insipidus. *J. Endocrinol. Investig.* **2021**, *44*, 2777–2783. [CrossRef] [PubMed]
142. Dekan, Z.; Kremsmayr, T.; Keov, P.; Godin, M.; Teakle, N.; Dürrauer, L.; Xiang, H.; Gharib, D.; Bergmayr, C.; Hellinger, R.; et al. Nature-inspired dimerization as a strategy to modulate neuropeptide pharmacology exemplified with vasopressin and oxytocin. *Chem. Sci.* **2021**, *12*, 4057–4062. [CrossRef] [PubMed]
143. Murat, B.; Devost, D.; Andrés, M.; Mion, J.; Boulay, V.; Corbani, M.; Zingg, H.H.; Guillon, G. V1b and CRHR1 receptor heterodimerization mediates synergistic biological actions of vasopressin and CRH. *Mol. Endocrinol.* **2012**, *26*, 502–520. [CrossRef] [PubMed]
144. Patchev, V.K.; Almeida, O.F. Corticosteroid regulation of gene expression and binding characteristics of vasopressin receptors in the rat brain. *Eur. J. Neurosci.* **1995**, *7*, 1579–1583. [CrossRef] [PubMed]
145. Wasilewski, M.A.; Grisanti, L.A.; Song, J.; Carter, R.L.; Repas, A.A.; Myers, V.D.; Gao, E.; Koch, W.J.; Cheung, J.Y.; Feldman, A.M.; et al. Vasopressin type 1A receptor deletion enhances cardiac contractility, β-adrenergic receptor sensitivity and acute cardiac injury-induced dysfunction. *Clin. Sci.* **2016**, *130*, 2017–2027. [CrossRef] [PubMed]
146. Tsuchiya, M.; Tsuchiya, K.; Maruyama, R.; Takemura, G.; Minatoguchi, S.; Fujiwara, H. Vasopressin inhibits sarcolemmal ATP-sensitive K+ channels via V1 receptors activation in the guinea pig heart. *Circ. J.* **2002**, *66*, 277–282. [CrossRef] [PubMed]
147. Hantash, B.M.; Thomas, A.P.; Reeves, J.P. Regulation of the cardiac L-type calcium channel in L6 cells by arginine-vasopressin. *Biochem. J.* **2006**, *400*, 411–419. [CrossRef] [PubMed]
148. Tilley, D.G.; Zhu, W.; Myers, V.D.; Barr, L.A.; Gao, E.; Li, X.; Song, J.; Carter, R.L.; Makarewich, C.A.; Yu, D.; et al. β-adrenergic receptor-mediated cardiac contractility is inhibited via vasopressin type 1A-receptor-dependent signaling. *Circulation* **2014**, *130*, 1800–1811. [CrossRef] [PubMed]
149. Zhu, W.; Tilley, D.G.; Myers, V.D.; Coleman, R.C.; Feldman, A.M. Arginine vasopressin enhances cell survival via a G protein-coupled receptor kinase 2/β-arrestin1/extracellular-regulated kinase 1/2-dependent pathway in H9c2 cells. *Mol. Pharmacol.* **2013**, *84*, 227–235. [CrossRef] [PubMed]
150. Zhu, W.; Tilley, D.G.; Myers, V.D.; Tsai, E.J.; Feldman, A.M. Increased vasopressin 1A receptor expression in failing human hearts. *J. Am. Coll. Cardiol.* **2014**, *63*, 375–376. [CrossRef] [PubMed]
151. Xu, F.; Sun, S.; Wang, X.; Ni, E.; Zhao, L.; Zhu, W. GRK2 Mediates Arginine Vasopressin-Induced Interleukin-6 Production via Nuclear Factor-kappaB Signaling Neonatal Rat Cardiac Fibroblast. *Mol. Pharmacol.* **2017**, *92*, 278–284. [CrossRef] [PubMed]

152. Bucher, M.; Hobbhahn, J.; Taeger, K.; Kurtz, A. Cytokine-mediated downregulation of vasopressin V(1A) receptors during acute endotoxemia in rats. *Am. J. Physiol. Regul. Integr. Comp. Physiol.* **2002**, *282*, R979–R984. [CrossRef] [PubMed]
153. Gray, M.; Innala, L.; Viau, V. Central vasopressin V1A receptor blockade impedes hypothalamic-pituitary-adrenal habituation to repeated restraint stress exposure in adult male rats. *Neuropsychopharmacology* **2012**, *37*, 2712–2719. [CrossRef] [PubMed]
154. Roper, J.; O'Carroll, A.M.; Young, W., 3rd; Lolait, S. The vasopressin Avpr1b receptor: Molecular and pharmacological studies. *Stress* **2011**, *14*, 98–115. [CrossRef] [PubMed]
155. Lolait, S.J.; O'Carroll, A.M.; Mahan, L.C.; Felder, C.C.; Button, D.C.; Young, W.S., 3rd; Mezey, E.; Brownstein, M.J. Extrapituitary expression of the rat V1b vasopressin receptor gene. *Proc. Natl. Acad. Sci. USA* **1995**, *92*, 6783–6787. [CrossRef] [PubMed]
156. O'Carroll, A.M.; Howell, G.M.; Roberts, E.M.; Lolait, S.J. Vasopressin potentiates corticotropin-releasing hormone-induced insulin release from mouse pancreatic beta-cells. *J. Endocrinol.* **2008**, *197*, 231–239. [CrossRef] [PubMed]
157. Khan, S.; Raghuram, V.; Chen, L.; Chou, C.L.; Yang, C.R.; Khundmiri, S.J.; Knepper, M.A. Vasopressin V2 receptor, tolvaptan, and ERK1/2 phosphorylation in the renal collecting duct. *Am. J. Physiol. Ren. Physiol.* **2024**, *326*, F57–F68. [CrossRef] [PubMed]
158. Schrier, R.W. Molecular mechanisms of clinical concentrating and diluting disorders. *Prog. Brain Res.* **2008**, *170*, 539–550. [CrossRef] [PubMed]
159. Lightman, S.L.; Birnie, M.T.; Conway-Campbell, B.L. Dynamics of ACTH and Cortisol Secretion and Implications for Disease. *Endocr. Rev.* **2020**, *41*, bnaa002. [CrossRef]
160. Yates, F.E.; Russell, S.M.; Dallman, M.F.; Hodge, G.A.; McCann, S.M.; Dhariwal, A.P. Potentiation by vasopressin of corticotropin release induced by corticotropin-releasing factor. *Endocrinology* **1971**, *88*, 3–15. [CrossRef]
161. Joseph, D.N.; Whirledge, S. Stress and the HPA Axis: Balancing Homeostasis and Fertility. *Int. J. Mol. Sci.* **2017**, *18*, 2224. [CrossRef]
162. Robertson-Dixon, I.; Murphy, M.J.; Crewther, S.G.; Riddell, N. The Influence of Light Wavelength on Human HPA Axis Rhythms: A Systematic Review. *Life* **2023**, *13*, 1968. [CrossRef]
163. Otsuka, H.; Abe, M.; Kobayashi, H. The Effect of Aldosterone on Cardiorenal and Metabolic Systems. *Int. J. Mol. Sci.* **2023**, *24*, 5370. [CrossRef]
164. Sztechman, D.; Czarzasta, K.; Cudnoch-Jedrzejewska, A.; Szczepanska-Sadowska, E.; Zera, T. Aldosterone and mineralocorticoid receptors in regulation of the cardiovascular system and pathological remodelling of the heart and arteries. *J. Physiol. Pharmacol.* **2018**, *69*, 829–845. [CrossRef]
165. Albert, K.M.; Newhouse, P.A. Estrogen, Stress, and Depression: Cognitive and Biological Interactions. *Annu. Rev. Clin. Psychol.* **2019**, *15*, 399–423. [CrossRef]
166. Patel, S.; Homaei, A.; Raju, A.B.; Meher, B.R. Estrogen: The necessary evil for human health, and ways to tame it. *Biomed. Pharmacother.* **2018**, *102*, 403–411. [CrossRef]
167. Nelson, L.R.; Bulun, S.E. Estrogen production and action. *J. Am. Acad. Dermatol.* **2001**, *45* (Suppl. S3), S116–S124. [CrossRef]
168. Naamneh Elzenaty, R.; du Toit, T.; Flück, C.E. Basics of androgen synthesis and action. *Best Pract. Res. Clin. Endocrinol. Metab.* **2022**, *36*, 101665. [CrossRef]
169. Fanelli, F.; Baronio, F.; Ortolano, R.; Mezzullo, M.; Cassio, A.; Pagotto, U.; Balsamo, A. Normative Basal Values of Hormones and Proteins of Gonadal and Adrenal Functions from Birth to Adulthood. *Sex. Dev.* **2018**, *12*, 50–94. [CrossRef]
170. Kulle, A.E.; Riepe, F.G.; Melchior, D.; Hiort, O.; Holterhus, P.M. A novel ultrapressure liquid chromatography tandem mass spectrometry method for the simultaneous determination of androstenedione, testosterone, and dihydrotestosterone in pediatric blood samples: Age- and sex-specific reference data. *J. Clin. Endocrinol. Metab.* **2010**, *95*, 2399–2409. [CrossRef]
171. Zirkin, B.R.; Papadopoulos, V. Leydig cells: Formation, function, and regulation. *Biol. Reprod.* **2018**, *99*, 101–111. [CrossRef]
172. Midzak, A.; Akula, N.; Lecanu, L.; Papadopoulos, V. Novel androstenetriol interacts with the mitochondrial translocator protein and controls steroidogenesis. *J. Biol. Chem.* **2011**, *286*, 9875–9887. [CrossRef]
173. Beattie, M.C.; Adekola, L.; Papadopoulos, V.; Chen, H.; Zirkin, B.R. Leydig cell aging and hypogonadism. *Exp. Gerontol.* **2015**, *68*, 87–91. [CrossRef]
174. Payne, A.H.; Hales, D.B. Overview of steroidogenic enzymes in the pathway from cholesterol to active steroid hormones. *Endocr. Rev.* **2004**, *25*, 947–970. [CrossRef]
175. Wang, Y.; Chen, F.; Ye, L.; Zirkin, B.; Chen, H. Steroidogenesis in Leydig cells: Effects of aging and environmental factors. *Reproduction* **2017**, *154*, R111–R122. [CrossRef]
176. Kuo, T.; McQueen, A.; Chen, T.C.; Wang, J.C. Regulation of Glucose Homeostasis by Glucocorticoids. *Adv. Exp. Med. Biol.* **2015**, *872*, 99–126. [CrossRef]
177. Yoshimura, M.; Conway-Campbell, B.; Ueta, Y. Arginine vasopressin: Direct and indirect action on metabolism. *Peptides* **2021**, *142*, 170555. [CrossRef]
178. Nakata, M.; Gantulga, D.; Santoso, P.; Zhang, B.; Masuda, C.; Mori, M.; Okada, T.; Yada, T. Paraventricular NUCB2/Nesfatin-1 Supports Oxytocin and Vasopressin Neurons to Control Feeding Behavior and Fluid Balance in Male Mice. *Endocrinology* **2016**, *157*, 2322–2332. [CrossRef]
179. Mohan, S.; Flatt, P.R.; Irwin, N.; Moffett, R.C. Weight-reducing, lipid-lowering and antidiabetic activities of a novel arginine vasopressin analogue acting at the V1a and V1b receptors in high-fat-fed mice. *Diabetes Obes. Metab.* **2021**, *23*, 2215–2225. [CrossRef]

180. Haam, J.; Halmos, K.C.; Di, S.; Tasker, J.G. Nutritional state-dependent ghrelin activation of vasopressin neurons via retrograde trans-neuronal-glial stimulation of excitatory GABA circuits. *J. Neurosci.* **2014**, *34*, 6201–6213. [CrossRef]
181. Iwama, S.; Sugimura, Y.; Murase, T.; Hiroi, M.; Goto, M.; Hayashi, M.; Arima, H.; Oiso, Y. Central adiponectin functions to inhibit arginine vasopressin release in conscious rats. *J. Neuroendocr.* **2009**, *21*, 753–759. [CrossRef]
182. Küchler, S.; Perwitz, N.; Schick, R.R.; Klein, J.; Westphal, S. Arginine-vasopressin directly promotes a thermogenic and pro-inflammatory adipokine expression profile in brown adipocytes. *Regul. Pept.* **2010**, *164*, 126–132. [CrossRef]
183. Rofe, A.M.; Williamson, D.H. Metabolic effects of vasopressin infusion in the starved rat. Reversal of ketonaemia. *Biochem. J.* **1983**, *212*, 231–239. [CrossRef]
184. Vaughan, M. Effect of pitressin on lipolysis and on phosphorylase activity in rat adipose tissue. *Am. J. Physiol.* **1964**, *207*, 1166–1168. [CrossRef]
185. Hiroyama, M.; Aoyagi, T.; Fujiwara, Y.; Oshikawa, S.; Sanbe, A.; Endo, F.; Tanoue, A. Hyperammonaemia in V1a vasopressin receptor knockout mice caused by the promoted proteolysis and reduced intrahepatic blood volume. *J. Physiol.* **2007**, *581 Pt. 3*, 1183–1192. [CrossRef]
186. Hiroyama, M.; Fujiwara, Y.; Nakamura, K.; Aoyagi, T.; Mizutani, R.; Sanbe, A.; Tasaki, R.; Tanoue, A. Altered lipid metabolism in vasopressin V1B receptor-deficient mice. *Eur. J. Pharmacol.* **2009**, *602*, 455–461. [CrossRef]
187. Velho, G.; El Boustany, R.; Lefèvre, G.; Mohammedi, K.; Fumeron, F.; Potier, L.; Bankir, L.; Bouby, N.; Hadjadj, S.; Marre, M.; et al. Plasma Copeptin, Kidney Outcomes, Ischemic Heart Disease, and All-Cause Mortality in People With Long-standing Type 1 Diabetes. *Diabetes Care* **2016**, *39*, 2288–2295. [CrossRef]
188. Vanhaecke, T.; Perrier, E.T.; Melander, O. A Journey through the Early Evidence Linking Hydration to Metabolic Health. *Ann. Nutr. Metab.* **2020**, *76* (Suppl. S1), 4–9. [CrossRef]
189. Roussel, R.; El Boustany, R.; Bouby, N.; Potier, L.; Fumeron, F.; Mohammedi, K.; Balkau, B.; Tichet, J.; Bankir, L.; Marre, M.; et al. Plasma Copeptin, AVP Gene Variants, and Incidence of Type 2 Diabetes in a Cohort From the Community. *J. Clin. Endocrinol. Metab.* **2016**, *101*, 2432–2439. [CrossRef]
190. Enhörning, S.; Leosdottir, M.; Wallström, P.; Gullberg, B.; Berglund, G.; Wirfält, E.; Melander, O. Relation between human vasopressin 1a gene variance, fat intake, and diabetes. *Am. J. Clin. Nutr.* **2009**, *89*, 400–406. [CrossRef] [PubMed]
191. Enhörning, S.; Sjögren, M.; Hedblad, B.; Nilsson, P.M.; Struck, J.; Melander, O. Genetic vasopressin 1b receptor variance in overweight and diabetes mellitus. *Eur. J. Endocrinol.* **2016**, *174*, 69–75. [CrossRef]
192. Hew-Butler, T.; Smith-Hale, V.; Pollard-McGrandy, A.; VanSumeren, M. Of Mice and Men-The Physiology, Psychology, and Pathology of Overhydration. *Nutrients* **2019**, *11*, 1539. [CrossRef]
193. Szczepanska-Sadowska, E.; Wsol, A.; Cudnoch-Jedrzejewska, A.; Żera, T. Complementary Role of Oxytocin and Vasopressin in Cardiovascular Regulation. *Int. J. Mol. Sci.* **2021**, *22*, 11465. [CrossRef]
194. Bichet, D.G. Regulation of Thirst and Vasopressin Release. *Annu. Rev. Physiol.* **2019**, *81*, 359–373. [CrossRef]
195. Szczepanska-Sadowska, E. Neuromodulation of Cardiac Ischemic Pain: Role of the Autonomic Nervous System and Vasopressin. *J. Integr. Neurosci.* **2024**, *23*, 49. [CrossRef] [PubMed]
196. Danziger, J.; Zeidel, M.L. Osmotic homeostasis. *Clin. J. Am. Soc. Nephrol.* **2015**, *10*, 852–862. [CrossRef]
197. Zimmerman, C.A.; Lin, Y.C.; Leib, D.E.; Guo, L.; Huey, E.L.; Daly, G.E.; Chen, Y.; Knight, Z.A. Thirst neurons anticipate the homeostatic consequences of eating and drinking. *Nature* **2016**, *537*, 680–684. [CrossRef]
198. Zaelzer, C.; Hua, P.; Prager-Khoutorsky, M.; Ciura, S.; Voisin, D.L.; Liedtke, W.; Bourque, C.W. ΔN-TRPV1: A Molecular Co-detector of Body Temperature and Osmotic Stress. *Cell Rep.* **2015**, *13*, 23–30. [CrossRef]
199. Saker, P.; Farrell, M.J.; Adib, F.R.; Egan, G.F.; McKinley, M.J.; Denton, D.A. Regional brain responses associated with drinking water during thirst and after its satiation. *Proc. Natl. Acad. Sci. USA* **2014**, *111*, 5379–5384. [CrossRef]
200. Cheung, P.W.; Bouley, R.; Brown, D. Targeting the Trafficking of Kidney Water Channels for Therapeutic Benefit. *Annu. Rev. Pharmacol. Toxicol.* **2020**, *60*, 175–194. [CrossRef]
201. Bankir, L.; Bichet, D.G.; Bouby, N. Vasopressin V2 receptors, ENaC, and sodium reabsorption: A risk factor for hypertension? *Am. J. Physiol. Ren. Physiol.* **2010**, *299*, F917–F928. [CrossRef] [PubMed]
202. Fenton, R.A. Essential role of vasopressin-regulated urea transport processes in the mammalian kidney. *Pflug. Arch.* **2009**, *458*, 169–177. [CrossRef] [PubMed]
203. Bankir, L. Antidiuretic action of vasopressin: Quantitative aspects and interaction between V1a and V2 receptor-mediated effects. *Cardiovasc. Res.* **2001**, *51*, 372–390. [CrossRef] [PubMed]
204. Wang, W.; Li, C.; Summer, S.N.; Falk, S.; Cadnapaphornchai, M.A.; Chen, Y.C.; Schrier, R.W. Molecular analysis of impaired urinary diluting capacity in glucocorticoid deficiency. *Am. J. Physiol. Ren. Physiol.* **2006**, *290*, F1135–F1142. [CrossRef] [PubMed]
205. Zhu, X.; Huang, Y.; Li, S.; Ge, N.; Li, T.; Wang, Y.; Liu, K.; Liu, C. Glucocorticoids Reverse Diluted Hyponatremia Through Inhibiting Arginine Vasopressin Pathway in Heart Failure Rats. *J. Am. Heart Assoc.* **2020**, *9*, e014950. [CrossRef] [PubMed]
206. Frenkel, A.; Abuhasira, R.; Bichovsky, Y.; Bukhin, A.; Novack, V.; Brotfain, E.; Zlotnik, A.; Klein, M. Examination of the association of steroids with fluid accumulation in critically ill patients, considering the possibility of biases. *Sci. Rep.* **2021**, *11*, 5557. [CrossRef] [PubMed]
207. Kenyon, C.J.; Saccoccio, N.A.; Morris, D.J. Aldosterone effects on water and electrolyte metabolism. *J. Endocrinol.* **1984**, *100*, 93–100. [CrossRef] [PubMed]

208. Johnston, J.G.; Welch, A.K.; Cain, B.D.; Sayeski, P.P.; Gumz, M.L.; Wingo, C.S. Aldosterone: Renal Action and Physiological Effects. *Compr. Physiol.* **2023**, *13*, 4409–4491. [CrossRef]
209. Sladek, C.D.; Somponpun, S.J. Estrogen receptors: Their roles in regulation of vasopressin release for maintenance of fluid and electrolyte homeostasis. *Front. Neuroendocr.* **2008**, *29*, 114–127. [CrossRef]
210. Wang, Y.X.; Crofton, J.T.; Liu, H.; Sato, K.; Brooks, D.P.; Share, L. Estradiol attenuates the antidiuretic action of vasopressin in ovariectomized rats. *Am. J. Physiol.* **1995**, *268 Pt. 2*, R951–R957. [CrossRef]
211. Somponpun, S.J. Neuroendocrine regulation of fluid and electrolyte balance by ovarian steroids: Contributions from central oestrogen receptors. *J. Neuroendocr.* **2007**, *19*, 809–818. [CrossRef] [PubMed]
212. Somponpun, S.J.; Johnson, A.K.; Beltz, T.; Sladek, C.D. Estrogen receptor-alpha expression in osmosensitive elements of the lamina terminalis: Regulation by hypertonicity. *Am. J. Physiol. Regul. Integr. Comp. Physiol.* **2004**, *287*, R661–R669. [CrossRef] [PubMed]
213. Voisin, D.L.; Simonian, S.X.; Herbison, A.E. Identification of estrogen receptor-containing neurons projecting to the rat supraoptic nucleus. *Neuroscience* **1997**, *78*, 215–228. [CrossRef]
214. Kuiper, G.G.; Carlsson, B.; Grandien, K.; Enmark, E.; Häggblad, J.; Nilsson, S.; Gustafsson, J.A. Comparison of the ligand binding specificity and transcript tissue distribution of estrogen receptors alpha and beta. *Endocrinology* **1997**, *138*, 863–870. [CrossRef] [PubMed]
215. Li, X.; Kuang, W.; Qiu, Z.; Zhou, Z. G protein-coupled estrogen receptor: A promising therapeutic target for aldosterone-induced hypertension. *Front. Endocrinol.* **2023**, *14*, 1226458. [CrossRef]
216. Rodriguez-Giustiniani, P.; Rodriguez-Sanchez, N.; Galloway, S.D.R. Fluid and electrolyte balance considerations for female athletes. *Eur. J. Sport. Sci.* **2022**, *22*, 697–708. [CrossRef]
217. Swenson, K.L.; Sladek, C.D. Gonadal steroid modulation of vasopressin secretion in response to osmotic stimulation. *Endocrinology* **1997**, *138*, 2089–2097. [CrossRef]
218. Siegenthaler, J.; Walti, C.; Urwyler, S.A.; Schuetz, P.; Christ-Crain, M. Copeptin concentrations during psychological stress: The PsyCo study. *Eur. J. Endocrinol.* **2014**, *171*, 737–742. [CrossRef]
219. Swaab, D.F.; Bao, A.M.; Lucassen, P.J. The stress system in the human brain in depression and neurodegeneration. *Aging Res. Rev.* **2005**, *4*, 141–194. [CrossRef]
220. Milutinović-Smiljanić, S.; Šarenac, O.; Lozić-Djurić, M.; Murphy, D.; Japundžić-Žigon, N. Evidence for involvement of central vasopressin V1b and V2 receptors in stress-induced baroreflex desensitization. *Br. J. Pharmacol.* **2013**, *169*, 900–908. [CrossRef]
221. Grassi, D.; Lagunas, N.; Calmarza-Font, I.; Diz-Chaves, Y.; Garcia-Segura, L.M.; Panzica, G.C. Chronic unpredictable stress and long-term ovariectomy affect arginine-vasopressin expression in the paraventricular nucleus of adult female mice. *Brain Res.* **2014**, *1588*, 55–62. [CrossRef] [PubMed]
222. Russell, G.; Lightman, S. The human stress response. *Nat. Rev. Endocrinol.* **2019**, *15*, 525–534. [CrossRef]
223. Borrow, A.P.; Bales, N.J.; Stover, S.A.; Handa, R.J. Chronic Variable Stress Induces Sex-Specific Alterations in Social Behavior and Neuropeptide Expression in the Mouse. *Endocrinology* **2018**, *159*, 2803–2814. [CrossRef] [PubMed]
224. Das, S.; Komnenov, D.; Newhouse, L.; Rishi, A.K.; Rossi, N.F. Paraventricular Nucleus V1a Receptor Knockdown Blunts Neurocardiovascular Responses to Acute Stress in Male Rats after Chronic Mild Unpredictable Stress. *Physiol. Behav.* **2022**, *253*, 113867. [CrossRef] [PubMed]
225. Komnenov, D.; Quaal, H.; Rossi, N.F. V1a and V1b vasopressin receptors within the paraventricular nucleus contribute to hypertension in male rats exposed to chronic mild unpredictable stress. *Am. J. Physiol. Regul. Integr. Comp. Physiol.* **2021**, *320*, R213–R225. [CrossRef] [PubMed]
226. Powell-Roach, K.L.; Yao, Y.; Jhun, E.H.; He, Y.; Suarez, M.L.; Ezenwa, M.O.; Molokie, R.E.; Wang, Z.J.; Wilkie, D.J. Vasopressin SNP pain factors and stress in sickle cell disease. *PLoS ONE* **2019**, *14*, e0224886. [CrossRef] [PubMed]
227. Pavlidi, P.; Kokras, N.; Dalla, C. Sex Differences in Depression and Anxiety. *Curr. Top. Behav. Neurosci.* **2023**, *62*, 103–132. [CrossRef] [PubMed]
228. Woodward, E.; Rangel-Barajas, C.; Ringland, A.; Logrip, M.L.; Coutellier, L. Sex-Specific Timelines for Adaptations of Prefrontal Parvalbumin Neurons in Response to Stress and Changes in Anxiety- and Depressive-Like Behaviors. *eNeuro* **2023**, *10*, ENEURO.0300-22.2023. [CrossRef] [PubMed]
229. Rivier, C. Gender, sex steroids, corticotropin-releasing factor, nitric oxide, and the HPA response to stress. *Pharmacol. Biochem. Behav.* **1999**, *64*, 739–751. [CrossRef] [PubMed]
230. Viau, V. Functional cross-talk between the hypothalamic-pituitary-gonadal and -adrenal axes. *J. Neuroendocr.* **2002**, *14*, 506–513. [CrossRef]
231. Teo, C.H.; Wong, A.C.H.; Sivakumaran, R.N.; Parhar, I.; Soga, T. Gender Differences in Cortisol and Cortisol Receptors in Depression: A Narrative Review. *Int. J. Mol. Sci.* **2023**, *24*, 7129. [CrossRef] [PubMed]
232. Rosinger, Z.J.; Jacobskind, J.S.; De Guzman, R.M.; Justice, N.J.; Zuloaga, D.G. A sexually dimorphic distribution of corticotropin-releasing factor receptor 1 in the paraventricular hypothalamus. *Neuroscience* **2019**, *409*, 195–203. [CrossRef] [PubMed]
233. Cox, K.H.; Quinnies, K.M.; Eschendroeder, A.; Didrick, P.M.; Eugster, E.A.; Rissman, E.F. Number of X-chromosome genes influences social behavior and vasopressin gene expression in mice. *Psychoneuroendocrinology* **2015**, *51*, 271–281. [CrossRef] [PubMed]

234. Cohen, J.E.; Holsen, L.M.; Ironside, M.; Moser, A.D.; Duda, J.M.; Null, K.E.; Perlo, S.; Richards, C.E.; Nascimento, N.F.; Du, F.; et al. Neural response to stress differs by sex in young adulthood. *Psychiatry Res. Neuroimaging* **2023**, *332*, 111646. [CrossRef] [PubMed]
235. Pietranera, L.; Saravia, F.; Roig, P.; Lima, A.; De Nicola, A.F. Mineralocorticoid treatment upregulates the hypothalamic vasopressinergic system of spontaneously hypertensive rats. *Neuroendocrinology* **2004**, *80*, 100–110. [CrossRef] [PubMed]
236. Matsuguchi, H.; Schmid, P.G. Acute interaction of vasopressin and neurogenic mechanisms in DOC-salt hypertension. *Am. J. Physiol.* **1982**, *242*, H37–H43. [CrossRef] [PubMed]
237. Ferrario, C.M.; Mohara, O.; Ueno, Y.; Brosnihan, K.B. Hemodynamic and neurohormonal changes in the development of DOC hypertension in the dog. *Am. J. Med. Sci.* **1988**, *295*, 352–359. [CrossRef] [PubMed]
238. Rademaker, M.T.; Charles, C.J.; Nicholls, M.G.; Richards, A.M. Interactions of enhanced urocortin 2 and mineralocorticoid receptor antagonism in experimental heart failure. *Circ. Heart Fail.* **2013**, *6*, 825–832. [CrossRef] [PubMed]
239. Pasquali, R.; Gagliardi, L.; Vicennati, V.; Gambineri, A.; Colitta, D.; Ceroni, L.; Casimirri, F. ACTH and cortisol response to combined corticotropin releasing hormone-arginine vasopressin stimulation in obese males and its relationship to body weight, fat distribution and parameters of the metabolic syndrome. *Int. J. Obes. Relat. Metab. Disord.* **1999**, *23*, 419–424. [CrossRef] [PubMed]
240. Schinke, C.; Hesse, S.; Rullmann, M.; Becker, G.A.; Luthardt, J.; Zientek, F.; Patt, M.; Stoppe, M.; Schmidt, E.; Meyer, K.; et al. Central noradrenaline transporter availability is linked with HPA axis responsiveness and copeptin in human obesity and non-obese controls. *Stress* **2019**, *22*, 93–102. [CrossRef]
241. Canivell, S.; Mohaupt, M.; Ackermann, D.; Pruijm, M.; Guessous, I.; Ehret, G.; Escher, G.; Pechère-Bertschi, A.; Vogt, B.; Devuyst, O.; et al. Copeptin and insulin resistance: Effect modification by age and 11 β-HSD2 activity in a population-based study. *J. Endocrinol. Investig.* **2018**, *41*, 799–808. [CrossRef] [PubMed]
242. Nye, E.J.; Bornstein, S.R.; Grice, J.E.; Tauchnitz, R.; Hockings, G.I.; Strakosch, C.R.; Jackson, R.V.; Torpy, D.J. Interactions between the stimulated hypothalamic-pituitary-adrenal axis and leptin in humans. *J. Neuroendocr.* **2000**, *12*, 141–145. [CrossRef] [PubMed]
243. Kacheva, S.; Kolk, K.; Morgenthaler, N.G.; Brabant, G.; Karges, W. Gender-specific co-activation of arginine vasopressin and the hypothalamic-pituitary-adrenal axis during stress. *Clin. Endocrinol.* **2015**, *82*, 570–576. [CrossRef] [PubMed]
244. Zelena, D.; Mergl, Z.; Makara, G.B. The role of vasopressin in diabetes mellitus-induced hypothalamo-pituitary-adrenal axis activation: Studies in Brattleboro rats. *Brain Res. Bull.* **2006**, *69*, 48–56. [CrossRef] [PubMed]
245. Balapattabi, K.; Little, J.T.; Bachelor, M.E.; Cunningham, R.L.; Cunningham, J.T. Sex Differences in the Regulation of Vasopressin and Oxytocin Secretion in Bile Duct-Ligated Rats. *Neuroendocrinology* **2021**, *111*, 237–248. [CrossRef]
246. Coiro, V.; Volpi, R.; Capretti, L.; Bacchi-Modena, A.; Cigarini, C.; Bianconi, L.; Rossi, G.; Gramellini, D.; Chiodera, P. Arginine vasopressin secretion in non-obese women with polycystic ovary syndrome. *Acta Endocrinol.* **1989**, *121*, 784–790. [CrossRef] [PubMed]
247. Vicennati, V.; Ceroni, L.; Gagliardi, L.; Pagotto, U.; Gambineri, A.; Genghini, S.; Pasquali, R. Response of the hypothalamic-pituitary-adrenal axis to small dose arginine-vasopressin and daily urinary free cortisol before and after alprazolam pre-treatment differs in obesity. *J. Endocrinol. Investig.* **2004**, *27*, 541–547. [CrossRef] [PubMed]
248. Zoorob, R.J.; Cender, D. A different look at corticosteroids. *Am. Fam. Physician* **1998**, *58*, 443–450. [PubMed]
249. Erkut, Z.A.; Pool, C.; Swaab, D.F. Glucocorticoids suppress corticotropin-releasing hormone and vasopressin expression in human hypothalamic neurons. *J. Clin. Endocrinol. Metab.* **1998**, *83*, 2066–2073. [CrossRef] [PubMed]
250. Zanardo, V.; Golin, R.; Chiozza, M.L.; Faggian, D. Dexamethasone does not affect vasopressin release in bronchopulmonary dysplasia. *Pediatr. Nephrol.* **2000**, *15*, 241–244. [CrossRef] [PubMed]
251. Gordijn, M.S.; Gemke, R.J.; van Dalen, E.C.; Rotteveel, J.; Kaspers, G.J. Hypothalamic-pituitary-adrenal (HPA) axis suppression after treatment with glucocorticoid therapy for childhood acute lymphoblastic leukaemia. *Cochrane Database Syst. Rev.* **2012**, CD008727, Updated in *Cochrane Database Syst. Rev.* **2015**, CD008727. [CrossRef] [PubMed]
252. Liu, R.Y.; Unmehopa, U.A.; Zhou, J.N.; Swaab, D.F. Glucocorticoids suppress vasopressin gene expression in human suprachiasmatic nucleus. *J. Steroid Biochem. Mol. Biol.* **2006**, *98*, 248–253. [CrossRef] [PubMed]
253. von Bardeleben, U.; Holsboer, F.; Stalla, G.K.; Müller, O.A. Combined administration of human corticotropin-releasing factor and lysine vasopressin induces cortisol escape from dexamethasone suppression in healthy subjects. *Life Sci.* **1985**, *37*, 1613–1618. [CrossRef]
254. Escudero, D.S.; Fantinelli, J.C.; Martínez, V.R.; González Arbeláez, L.F.; Amarillo, M.E.; Pérez, N.G.; Díaz, R.G. Hydrocortisone cardioprotection in ischaemia/reperfusion injury involves antioxidant mechanisms. *Eur. J. Clin. Investig.* **2024**, *54*, e14172. [CrossRef] [PubMed]
255. Giugliano, G.R.; Giugliano, R.P.; Gibson, C.M.; Kuntz, R.E. Meta-analysis of corticosteroid treatment in acute myocardial infarction. *Am. J. Cardiol.* **2003**, *91*, 1055–1059. [CrossRef] [PubMed]
256. Tol, M.M.; Shekar, K.; Barnett, A.G.; McGree, J.; McWhinney, B.C.; Ziegenfuss, M.; Ungerer, J.P.; Fraser, J.F. A preliminary investigation into adrenal responsiveness and outcomes in patients with cardiogenic shock after acute myocardial infarction. *J. Crit. Care* **2014**, *29*, 470.e1–470.e6. [CrossRef] [PubMed]
257. Torgersen, C.; Luckner, G.; Schröder, D.C.; Schmittinger, C.A.; Rex, C.; Ulmer, H.; Dünser, M.W. Concomitant arginine-vasopressin and hydrocortisone therapy in severe septic shock: Association with mortality. *Intensive Care Med.* **2011**, *37*, 1432–1437. [CrossRef] [PubMed]

258. Penn, J.; Douglas, W.; Curran, J.; Chaudhuri, D.; Dionne, J.C.; Fernando, S.M.; Granton, D.; Mathew, R.; Rochwerg, B. Efficacy and safety of corticosteroids in cardiac arrest: A systematic review, meta-analysis and trial sequential analysis of randomized control trials. *Crit. Care* **2023**, *27*, 12. [CrossRef] [PubMed] [PubMed Central]
259. Andersen, L.W.; Isbye, D.; Kjærgaard, J.; Kristensen, C.M.; Darling, S.; Zwisler, S.T.; Fisker, S.; Schmidt, J.C.; Kirkegaard, H.; Grejs, A.M.; et al. Effect of Vasopressin and Methylprednisolone vs Placebo on Return of Spontaneous Circulation in Patients With In-Hospital Cardiac Arrest: A Randomized Clinical Trial. *JAMA* **2021**, *326*, 1586–1594. [CrossRef] [PubMed] [PubMed Central]
260. Scott, L.V.; Dinan, T.G. Vasopressin as a target for antidepressant development: An assessment of the available evidence. *J. Affect. Disord.* **2002**, *72*, 113–124. [CrossRef] [PubMed]
261. Simon, N.G.; Guillon, C.; Fabio, K.; Heindel, N.D.; Lu, S.F.; Miller, M.; Ferris, C.F.; Brownstein, M.J.; Garripa, C.; Koppel, G.A. Vasopressin antagonists as anxiolytics and antidepressants: Recent developments. *Recent. Pat. CNS Drug Discov.* **2008**, *3*, 77–93. [CrossRef] [PubMed]
262. Poretti, M.B.; Sawant, R.S.; Rask-Andersen, M.; de Cuneo, M.F.; Schiöth, H.B.; Perez, M.F.; Carlini, V.P. Reduced vasopressin receptors activation mediates the anti-depressant effects of fluoxetine and venlafaxine in bulbectomy model of depression. *Psychopharmacology* **2016**, *233*, 1077–1086. [CrossRef] [PubMed]
263. Stewart, L.Q.; Roper, J.A.; Young, W.S., 3rd; O'Carroll, A.M.; Lolait, S.J. The role of the arginine vasopressin Avp1b receptor in the acute neuroendocrine action of antidepressants. *Psychoneuroendocrinology* **2008**, *33*, 405–415. [CrossRef] [PubMed]
264. Kiss, A.; Bundzikova, J.; Pirnik, Z.; Mikkelsen, J.D. Different antipsychotics elicit different effects on magnocellular oxytocinergic and vasopressinergic neurons as revealed by Fos immunohistochemistry. *J. Neurosci. Res.* **2010**, *88*, 677–685. [CrossRef]
265. Florkowski, C.M.; Crozier, I.G.; Nightingale, S.; Evans, M.J.; Ellis, M.J.; Joyce, P.; Donald, R.A. Plasma cortisol, PRL, ACTH, AVP and corticotrophin releasing hormone responses to direct current cardioversion and electroconvulsive therapy. *Clin. Endocrinol.* **1996**, *44*, 163–168. [CrossRef] [PubMed]

International Journal of
*Molecular Sciences*

MDPI

*Review*

# Genetic Mutations and Mitochondrial Redox Signaling as Modulating Factors in Hypertrophic Cardiomyopathy: A Scoping Review

**Antonio da Silva Menezes Junior [1,*], Ana Luísa Guedes de França-e-Silva [1], Henrique Lima de Oliveira [1], Khissya Beatryz Alves de Lima [1], Iane de Oliveira Pires Porto [2], Thays Millena Alves Pedroso [2], Daniela de Melo e Silva [1] and Aguinaldo F. Freitas, Jr. [1]**

[1] Faculdade de Medicina, Departamento de Clínica Médica, Universidade Federal de Goiás (UFG), Goiânia 74020-020, Brazil; ana.guedes@discente.ufg.br (A.L.G.d.F.-e.-S.); henrique.lima2@discente.ufg.br (H.L.d.O.); khissya_beatryz@discente.ufg.br (K.B.A.d.L.); danielamelosilva@ufg.br (D.d.M.e.S.); afreitasjr@msn.com (A.F.F.J.)

[2] Faculdade de Medicina, Universidade de Rio Verde (UniRV), Campus Aparecida, Aparecida de Goiânia 74345-030, Brazil; iane.porto@unirv.edu.br (I.d.O.P.P.); thays.millena04@gmail.com (T.M.A.P.)

* Correspondence: a.menezes.junior@uol.com.br

**Abstract:** Hypertrophic cardiomyopathy (HCM) is a heart condition characterized by cellular and metabolic dysfunction, with mitochondrial dysfunction playing a crucial role. Although the direct relationship between genetic mutations and mitochondrial dysfunction remains unclear, targeting mitochondrial dysfunction presents promising opportunities for treatment, as there are currently no effective treatments available for HCM. This review adhered to the Preferred Reporting Items for Systematic Reviews and Meta-Analysis Extension for Scoping Reviews guidelines. Searches were conducted in databases such as PubMed, Embase, and Scopus up to September 2023 using "MESH terms". Bibliographic references from pertinent articles were also included. Hypertrophic cardiomyopathy (HCM) is influenced by ionic homeostasis, cardiac tissue remodeling, metabolic balance, genetic mutations, reactive oxygen species regulation, and mitochondrial dysfunction. The latter is a common factor regardless of the cause and is linked to intracellular calcium handling, energetic and oxidative stress, and HCM-induced hypertrophy. Hypertrophic cardiomyopathy treatments focus on symptom management and complication prevention. Targeted therapeutic approaches, such as improving mitochondrial bioenergetics, are being explored. This includes coenzyme Q and elamipretide therapies and metabolic strategies like therapeutic ketosis. Understanding the biomolecular, genetic, and mitochondrial mechanisms underlying HCM is crucial for developing new therapeutic modalities.

**Keywords:** hypertrophic cardiomyopathy; metabolism; mitochondrial redox signaling; mitochondrial dysfunction

## 1. Introduction

Hypertrophic cardiomyopathy (HCM) results from microcellular changes such as myocyte remodeling, disorganization of sarcomeric proteins, and energy metabolism impairment. This condition clinically manifests itself in left ventricular hypertrophy and abnormal diastolic function. HCM is rare, with a prevalence rate of 0.2–0.5% in adults, and is a significant cause of sudden cardiac death [1–5]. By understanding the underlying causes of HCM, medical professionals can better diagnose and treat the condition, improving patient outcomes [1–3].

The causes of HCM are multifaceted. Genetic mutations in sarcomeric proteins account for 40–60% of cases, while intrinsic mutations in mitochondrial DNA, inborn metabolic errors, and other genetic and non-genetic factors also play a role [2,6,7]. Mitochondrial dysfunction is consistently present in HCM pathophysiology, regardless of the underlying cause [8]. This cascade of molecular disturbances affects intracellular calcium homeostasis, cellular metabolism, and the production of reactive oxygen species, including superoxide anions, hydroxyl, and peroxyl radicals [9–11]. It is important to understand that the illness is not a straightforward condition. The various changes collectively contribute to the complexity of the illness, making it a challenge to treat. Without a comprehensive understanding of these changes, it is not easy to develop effective treatment options. Therefore, we must continue to study and learn more about the pathophysiological complexity of the illness to provide better care for those affected by it [12–23].

Genetic variations, mitochondrial dysfunction, and oxidative stress are significant factors in the development and pathological remodeling of hypertrophic cardiomyopathy (HCM). The impact of these factors is evident in the changed structure of heart cells, mitochondrial cristae, and the functioning of oxidative phosphorylation complexes [8,12–23]. Currently, our understanding of the genetic mutations that cause HCM is limited. As a result, the relationship between these mutations, mitochondrial modulation, and HCM is not yet fully understood [19,24]. Mitochondrial dysfunction can be either primary or secondary to other etiologies [23,25].

Mitochondrial dysfunction is associated with sarcomeric mutations that can significantly impact the tertiary structure and stability of proteins [23,26]. Inefficient sarcomere performance increases ATP demand, leading to energy depletion, chronic workload increases, and oxidative stress [27]. It is essential to address the contributing factors proactively to prevent further complications down the line, as early manifestations of metabolic disorders in HCM progression are a direct result of them. It is clear that understanding and preventing these mutations is key; therefore, preventing or ameliorating mitochondrial dysfunction by targeting metabolic and oxidative stress may offer viable therapeutic strategies for halting or reversing disease progression [15,16,20,23].

HCM is a condition that can be quite debilitating and currently has no known cure; however, mavacamten, a promising ATPase inhibitor, has been approved by the FDA and can help reduce the need for invasive procedures in individuals with the obstructive form of HCM [28]. Although mavacamten cannot cure HCM, it has demonstrated a considerable potential to reduce cardiac contractility and alleviate energy depletion and oxidative stress. To develop effective treatments, it is crucial to comprehend the initial pathophysiological mechanisms of the condition, and researchers have analyzed various variants of the pathophysiology of HCM in detail [8]. Mitochondrial and genetic therapies are emerging potential methods for preventing and improving heart disease in HCM [23,29]. With further exploration of treatment options and ongoing research into the disease's pathophysiology, we can discover alternative treatments for HCM and enhance the quality of life for those with it.

## 2. Materials and Methods

This review adhered to the Preferred Reporting Items for Systematic Reviews and Meta-Analysis Extension for Scoping Reviews (PRISMA-ScR) guidelines. Searches were conducted in databases such as PubMed, EMBASE, and Scopus up to February 2024 using "MESH terms". Bibliographic references from pertinent articles were also included.

### 2.1. Protocol and Registration

This review has been registered with the Open Science Framework (OSF) and is accessible at https://doi.org/10.17605/OSF.IO/XJGVC since 16 October 2023.

### 2.2. Eligibility Criteria

Our review focused on articles that connected metabolic dysfunction, sarcomeric or DNA mitochondrial mutations, G-negatives, mitochondrial architecture, and functioning disorders with HCM. We included articles from peer-reviewed journals published until

February 2024 that discussed HCM in the context of mitochondrial dysfunction and related mutations. The scope of our review encompasses quantitative, qualitative, and mixed-methods studies addressing genetic mutations and mitochondrial dysfunction as factors in HCM. Articles exclusively focused on heart or mitochondrial diseases unrelated to HCM were excluded. However, articles elucidating the role of calcium in metabolic changes associated with HCM were included.

### 2.3. Exclusion Criteria

Excluded from consideration in the review were secondary sources such as editorials, books, expert opinion articles, dissertations, theses, and conference abstracts, except for literature reviews, which were included.

### 2.4. Data Sources

We conducted comprehensive searches of the Excerpta Medica Database (Embase), SciVerse Scopus, and PubMed databases. Our search strategy involved a combination of controlled descriptors and keywords pertinent to HCM, and no restrictions were imposed on the language or publication period. In addition, reference lists of the initially selected studies were manually searched to identify other relevant articles.

### 2.5. Study Selection Process

The identified studies were imported into the Rayyan software [30] (https://link.springer.com/article/10.1186/s13643-016-0384-4 accessed on 22 April 2024) for duplicate removal and evaluation of the eligibility criteria. This process involved three independent blinded reviewers (ASMJr, ALGFS, and HLO) who initially examined the titles and abstracts (Phase 1), followed by a full-text review of the studies selected in Phase 1 (Phase 2). A fourth reviewer resolved any discrepancy in the selection process.

### 2.6. Data Extraction Process

Data extraction was conducted by three independent blinded reviewers using a characterization table created using Microsoft Word (2021). This table includes study characteristics, including identification (citation), study design, and country; analyzed characteristics of HCM, including etiology, sample size, sex, and average age; and main outcomes, including molecular manifestations such as mitochondrial dysfunction, oxidative stress, and metabolic disorders.

### 2.7. Data Synthesis

Data from the selected studies were qualitatively synthesized, focusing on the proposed genetic mutations and mitochondrial dysfunction. This synthesis includes consideration of the etiology of HCM, whether from genetic mutations (sarcomeric or mitochondrial) and cellular metabolic dysfunction. All information was organized into a descriptive table for a comprehensive analysis.

## 3. Results and Discussion

Thirty-two articles were selected for the final analysis from an initial search of 597 (Figure 1). The primary data from these studies are presented in Supplementary Table S1. Our analysis focused on two primary categories: (1) Genetic Mutations in HCM; (2) Mitochondrial Redox Signaling.

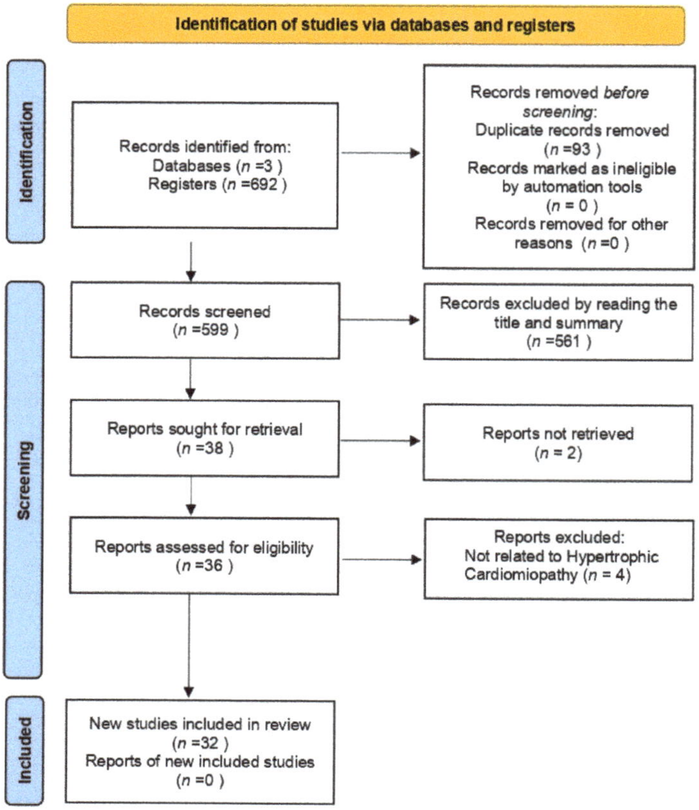


**Figure 1.** PRISMA flowchart of the reviewed articles.

### *3.1. Genetic Mutations in HCM*

HCM is a complex disorder that results from a combination of genetic and non-genetic factors. (Figure 2) [2,31,32]. Additionally, metabolic alterations and mitochondrial dysfunction can give rise to this pathology as primary disorders or secondary conditions to the primary etiology. The cardiac beta-myosin heavy chain (MYH7) and myosin-binding protein C (MYBPC) genes are responsible for roughly 70% of identifiable mutations [33]. MYH7 mutations affect myosin ATPase activity, leading to increased myocardial force. In contrast, MYBPC mutations are involved in sarcomere organization and may regulate myofibril contraction, resulting in energy overload and hypercontractility in the left ventricle [34,35]. In addition to MYH7 and MYBPC3, other genes, such as TNNT2, TNNI3, TPM1, ACTC1, MYL2, MYL3, and CSRP3, are also implicated in HCM, albeit less frequently. Some pathogenic variants exhibit high penetrance and are causal mutations, while others demonstrate incomplete penetrance influenced by genetic and environmental factors.

In HCM, allelic imbalance, particularly the characteristic autosomal dominant inheritance, can result in mosaic expression patterns of the mutant proteins, leading to cellular variability and causing differences in calcium sensitivity and contractile force in the cardiac cells [29,36].

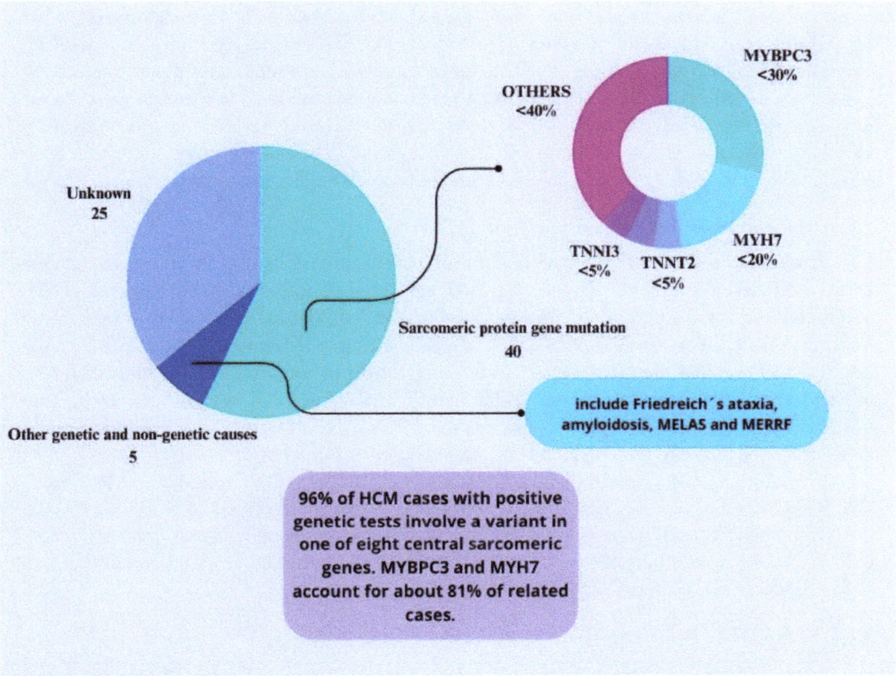


**Figure 2.** Genetic and non-genetic causes of hypertrophic cardiomyopathy. Legends: MELAS = mitochondrial encephalomyopathy, lactic acidosis, and stroke-like episodes; MERRF = myoclonic epilepsy with ragged red fibers; MYBPC3 = myosin-binding protein C, cardiac-type; MYH7 = myosin, heavy chain 7; TNNI3 = troponin I, cardiac; TNNT2 = troponin T, cardiac.

#### 3.1.1. Epigenetics and HCM

Epigenetic modifications play a significant role in HCM by influencing the levels of mutated proteins. Specifically, the DNA methylation in the promoter of the MYH7 gene is inversely correlated with beta-MyHC messenger RNA (mRNA) levels. This suggests that changes in methylation can alter the transcription of the mutated gene [26]. Moreover, the activation of chromatin-remodeling proteins, such as BRG1 and DPF3a, is associated with disease severity in patients with HCM [37]. These epigenetic changes impact the expression of mutated genes, fetal gene programs, and other hypertrophic genes [38].

These findings have significant implications for the diagnosis, treatment, and management of HCM. By understanding the role of epigenetic modifications in the expression of mutated genes, targeted therapies that address the underlying cause of the disease may be possible. Identifying specific chromatin-remodeling proteins associated with disease severity may also lead to the development of personalized treatment options that consider individual differences in genetic and epigenetic profiles.

As such, continued research into the relationship between epigenetic modifications and HCM is essential for advancing our understanding of this complex disease. By exploring the underlying mechanisms of epigenetic changes, we may be able to identify new targets for therapeutic intervention and improve the lives of individuals living with HCM.

#### 3.1.2. Myc Gene in HCM

The Myc gene encodes a vital transcription factor in regulating cellular proliferation and growth. It is also a key player in the development and progression of HCM. Animal models have consistently shown that overexpression of Myc leads to cell cycle activation and tumorigenesis [39]. Interestingly, elevated Myc mRNA levels have also been observed in mouse and patient samples with HCM [12,40]. Myc acts on mitochondrial function by modulating mitochondrial biogenesis through the PGC-1a pathway, which is critical in changing HCM and heart failure. PGC-1a, a master regulator, coordinates the expression of genes vital for mitochondrial formation and function and is involved in metabolic processes like glucose control, lipid metabolism, response to oxidative stress, and adaptation to various conditions. The regulation of PGC-1a is essential for overall metabolic health and cellular function, particularly in cardiac conditions such as cardiomyopathy [12].

#### 3.1.3. Mitochondrial DNA Mutations and HCM

Studies have shown that patients with sarcomere mutations may experience changes in heart efficiency even before hypertrophy development, indicating a link between HCM and metabolic alterations [23]. Moreover, point mutations in mitochondrial DNA (mtDNA), including m.3243A>G, m.3302A>G, m.4300A>G, and m.8344A>G, can impact ATP production and increase oxidative stress, ultimately contributing to HCM. The identification of the first mtDNA mutation related to HCM, MT-TL1, in 1991 led to significant strides in understanding the mechanisms of gene expression associated with these mutations, such as 1-deoxynojirimycin, which can aid in the recovery of mitochondrial cristae [41].

Mutations in mitochondrial genes such as MT-RNR2, ELAC2, Gtpbp3, and Mto1 have also been linked to HCM [14,16,18,20,42,43]. Research following the discovery of MT-TL1 mutations has expanded our understanding of mitochondrial gene mutations associated with HCM, revealing their impact on mitochondrial function and energy production in the pathological processes leading to HCM [41].

#### 3.1.4. MicroRNAs and Gene Therapy in HCM

Gene therapy emerges as an advanced medical therapy focusing on the genetic modification of cells for therapeutic purposes. In the context of HCM, it encompasses genome editing techniques, gene replacement therapy, and allele-specific silencing [44,45]. CRISPR-Cas9 technology presents promising potential to correct genetic mutations underlying HCM. It is at the forefront of human germline therapy, with several studies reporting efficient and successful genetic editing in these embryos [44,46]. However, it faces obstacles, including limited efficiency in homology-directed repair in somatic cardiac cells.

miRNAs represent potential targets in gene therapy to optimize metabolism and energy delivery in hypertrophied hearts. MicroRNA-146a inhibits oxidative metabolism, attenuates the hypertrophic response, and reduces cardiac erbB4 signaling, which regulates glucose metabolism [47,48]. RNA interference (RNAi) therapy and MYBPC3 cDNA replacement via gene transfer have shown therapeutic potential. However, allele-specific gene silencing with RNAi faces challenges such as off-target effects and reliance on adenoviruses [34]. Despite these hurdles, gene editing technology is emerging as a crucial future therapy [49,50]. Moreover, viral vectors, especially adeno-associated viruses (AAVs), notably AAV9, the most cardiotropic serotype, demonstrate efficacy in precisely delivering genetic material to cardiac tissue [45,51]. However, potential immune responses and ethical considerations pose considerable challenges. Furthermore, the MYBPC3 gene replacement approach is promising, particularly due to its ability to correct cMyBP-C haploinsufficiency, reduce hypertrophy, and maintain sarcomeric stoichiometry [52].

Understanding these advancements is a crucial starting point for direct interventions in HCM pathogenesis, offering promising prospects for a definitive approach to treating hereditary diseases. Ethical considerations, including germline gene editing, are thoroughly addressed to ensure the responsible application of these innovative technologies [45]. Technical challenges associated with efficient Cas9-gRNA delivery, such as immune responses to viral vectors and issues related to non-viral vector delivery, can be overcome. In that case, CRISPR-Cas9 positions itself as a significant player in treating a broad range of disorders where partial or complete gene elimination is desired [53]. This potentially revolutionary scenario instills hope for substantial clinical advancements in the coming years, notwithstanding concerns such as mosaicism, off-target alterations, and, ultimately, non-Mendelian cases of HCM, which must be meticulously considered.

### 3.2. Mitochondrial Redox Signaling

Several studies have reported reduced ATP production due to impaired mitochondrial metabolism, as shown in Figure 3 [8,12–23,29]. Adenosine triphosphate (ATP) production in a healthy human heart relies primarily on fatty acids and glucose oxidation. However, when the cardiac workload increases, the ATP generation pathway shifts toward glucose oxidation. This shift is regulated by the Randle cycle, which facilitates the maintenance of energy balance and increased energy efficiency in the heart [9,15,19].

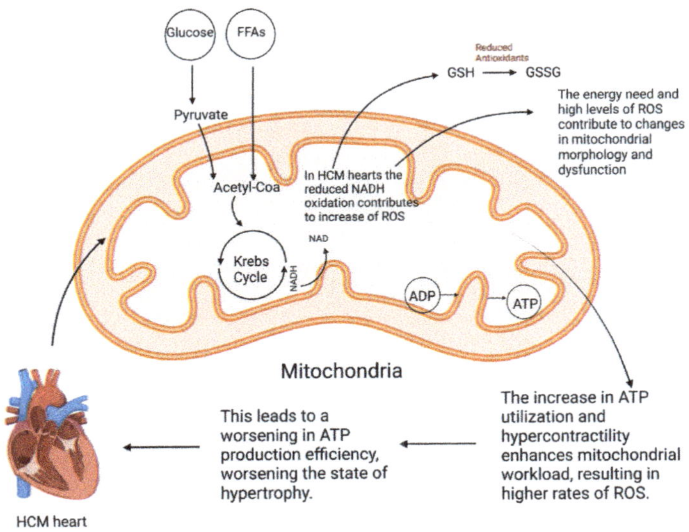


**Figure 3.** Metabolic redox integration in cardiomyocytes. Legends: HCM = Hypertrophic Cardiomyopathy, FFAs = Free Fatty Acids, NAD = Nicotinamide Adenine Dinucleotide, NADH = Nicotinamide Adenine Dinucleotide Hidrogenado, ADP = Adenosine Diphosphate, ATP = Adenosine Triphosphate, GSH = Glutathione, GSSG = Oxidized Glutathione, ROS = Reactive Oxygen Species.

In hearts affected by HCM, however, the increased energy demand during heart contraction can lead to metabolic disorders characterized by reduced ATP levels and increased intracellular ADP [26,54], with a preference for mitochondrial substrates over glucose [45]. This metabolic shift is also observed in patients with chronic myocardial infarction (CMI). It leads to cardiac hypertrophy due to increased ATP uptake, resulting in the accumulation of fatty acids within cells, causing lipotoxicity rather than increased glucose oxidation [3,9,29,55]. This metabolic deviation is a distinct characteristic of HCM, occurring independently of the sarcomeric protein genotype. HCM also reduces capillary density in cardiac tissue and hypoxia, prompting a metabolic shift from aerobic glucose metabolism to anaerobic glycolysis. This shift reduces energy efficiency and contributes to heart hypertrophy in a self-perpetuating cycle [15]. Metabolomic analyses have shown reduced cardiac bioenergetics and substrate utilization, indicating disturbances in energy metabolism that may contribute to the development of HCM-associated pathology [19].

A recent multi-omics study reported decreased metabolic supplementation through the TCA/glutamine pathway and changes in the pentose phosphate pathway. These pathways are fundamental for NADPH production and antioxidant protection. Therefore, comprehending the alterations in energy metabolism and metabolic pathways can help identify therapeutic targets for clinical improvement in HCM [22]. The therapeutic potential of perhexylin treatment, which shifts the heart's substrate utilization from fatty acids to glucose, has been demonstrated in patients with HCM. This treatment also functions as a NOX2 inhibitor, reducing oxidative stress in the heart [29,56]. Notably, these metabolic changes are partially reversible, and early detection can offer therapeutic opportunities.

Reduced mitochondrial respiration linked to NADH oxidation is a significant functional impairment in hypertrophic cardiomyopathy (HCM) [23]. However, mutations in both sarcomeric and non-sarcomeric genes, such as those in CSRP3, can potentially disturb calcium homeostasis [8]. The Muscle LIM Protein (MLP) maintains intracellular calcium balance and contributes to its pathogenesis [8,15,22]. Furthermore, the CACNA1C gene encodes the pore-forming subunit of the L-type calcium channel CaV1.2 [57]. Increased transcriptional expression of CACNA1C enhances calcium release from the sarcoplasmic reticulum (SR) and diminishes cytoplasmic calcium transport back to the SR during diastole [58,59]. These factors hinder sarcomeric relaxation, leading to diastolic dysfunction and subsequent adverse effects, including mitochondrial dysfunction, oxidative stress, and alterations in calcium-dependent pathways [17,60].

Moreover, MLP mutations raise energy demands due to inefficient sarcomeric ATP use. This leads to increased reactive oxygen species generation and further progression of HCM and heart failure. The MT-RNR2 mutation affects mitochondrial DNA, causing dysfunctions such as a decreased ATP/ADP ratio and compromised mitochondrial membrane integrity [59]. These alterations elevate intracellular $Ca^{2+}$ concentration, thereby affecting intracellular homeostasis. Research on pluripotent stem cells has shown reduced mitochondrial calcium uniporter protein levels and increased intracellular $Ca^{2+}$ concentrations and SR $Ca^{2+}$ reserves. The lowered membrane potential reduces $Ca^{2+}$ intake, potentially disrupting calcium homeostasis and triggering calcium-dependent signaling pathways contributing to heart hypertrophy [14]. This mutation also increases the number of mitochondria, possibly as an adaptation to decreased energy generation. Mutations in the *MYBPC3* and *MYH7* genes have increased calcium sensitivity in cardiac myofilaments, resulting in compromised contractility and increased ATP consumption. As a result, this hurts HCM by disturbing the balance of calcium levels and metabolic processes. Several investigations, including those undertaken by [2,41,54,59,61], have been conducted.

The binding of calcium to myofilaments leads to a decline in the regulation of cytosolic $Ca^{2+}$, inhibiting the uptake of mitochondrial $Ca^{2+}$. This process negatively affects the Krebs cycle-mediated replenishment of reduced NADH and NADPH, reducing the overall rate of metabolic reactions and modulating the system's efficiency [62]. In situations with a high demand for ATP, ADP levels increase, and NADH and NADPH levels decrease, leading to an imbalance, as shown in Figure 4, results in oxidative stress due to heightened levels of ADP that exacerbate NADH and NADPH oxidation. This disruption of the equilibrium between NAD and its oxidized form, $NAD^+$, is crucial for cellular energy production, leading to increased ATP usage. This process is associated with the initial inefficient performance of mutation-induced sarcomeres, which leads to chronic cardiac workload, energy deficiency, and oxidative stress [15,54,63].

Elevated levels of ADP have been shown to increase the sensitivity and affinity of calcium in myofibrils, thereby contributing to heart dysfunction. In normal cardiomyocytes, the enzyme creatine kinase is essential in regenerating ATP in myofilaments using mitochondrial phosphocreatine (PCr). However, in cardiomyocytes suffering from heart failure, creatine kinase does not succeed in effectively reducing ADP concentrations. This failure may contribute to the altered calcium sensitivity and affinity observed in such cells, exacerbating heart failure [64]. Elevated ADP levels and reduced ATP impair ATP-dependent ion pumps in the heart, specifically SERCA, which recaptures calcium into SR [8,65,66]. According to Wijnker et al., the direct and indirect effects of various factors significantly impact the development of HCM [29].

Patients with sarcomeric mutations often exhibit increased septum thickness, which is connected to improved mitochondrial respiration using succinate. However, they also experience reduced cellular respiration involving NADH, as per a study conducted by Lucas et al. (2003) and Nollet et al. (2023a) [23,63]. Nollet et al. (2023) further identified defects in $NAD^+$ homeostasis in myectomy samples of patients with HCM, which led to compromised bioenergetics due to diminished NADH conversion. This conversion process, which converts NADH (nicotinamide adenine dinucleotide reduced) into NAD+ (nicotinamide adenine dinucleotide), is crucial for the transfer of electrons during cellular energy production, as explained by Farhana et al. (2023) [67]. The study also revealed no incorporation of complex I in respiratory super complexes and peroxidation of cardiolipin, destabilizing the mitochondrial membrane. The use of cardiolipin-stabilizing compounds for treatment could potentially enhance the mitochondrial respiration linked to NADH [23].

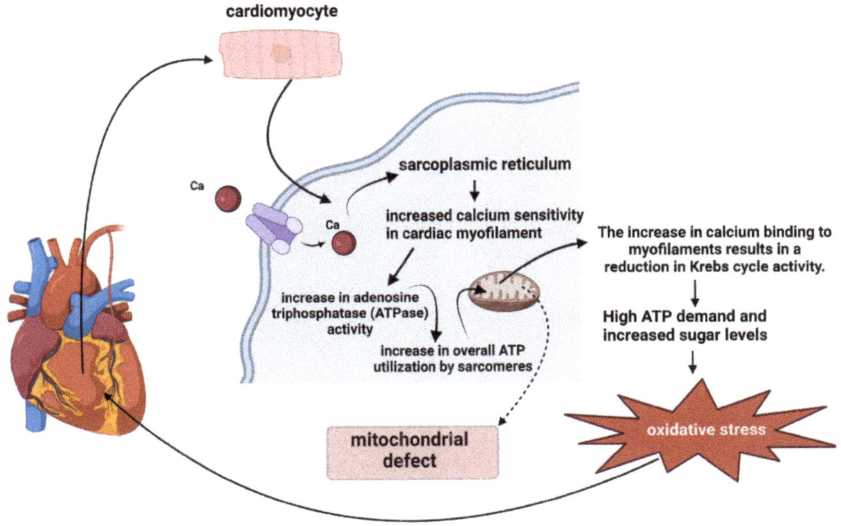


**Figure 4.** Metabolic fluxes under high ATP demand in hypertrophic cardiomyopathy. Legends: ATP = adenosine triphosphate, and ATPase = enzyme.

Mitochondrial dysfunction is a key driver of pathological remodeling in HCM, particularly in genotype-negative patients with impaired NADH-related respiration. These early metabolic changes suggest that addressing metabolic stress could be a strategic approach to treating and potentially reversing its progression [23,63,68]. Ranjbarvaziri et al. (2021) demonstrated significant energy depletion in genetically linked patients with HCM, characterized by reduced high-energy phosphate metabolites and decreased mitochondrial genes involved in creatine kinase and ATP synthesis [19]. This energy deficit activates the AMPK pathway, a cellular stress response sensor with low ATP levels [19].

Significant progress has been made in understanding the relationship between structural, sarcomeric, and mitochondrial functions in HCM. However, more research is still needed to fully comprehend the mechanisms that cause changes in cardiac energy utilization in the different HCM phenotypes. According to a study by Viola et al. (2019), specific mutations cause distinct changes in intracellular calcium homeostasis and mitochondrial metabolic function [8]. This highlights the importance of understanding the pathophysiological mechanisms of disease mutations to develop effective drug therapies. Lee et al. (2009) found a connection between mitochondrial respiration dysfunction and other disorders in transgenic mice overexpressing Myc oncogene [12]. These dysfunctions can lead to lower production of energy and an increased formation of oxygen-free radicals, which can worsen mitochondrial damage [39]. The expression of Myc oncogene in patients with terminal heart failure provides new possibilities for therapeutic approaches in the treatment and management of the condition [12].

Cardiomyocytes undergo oxidative phosphorylation, producing reactive oxygen species (ROS) in the mitochondria. The respiratory chain generates most ROS, with complexes I and III being the primary sources, as shown in Figure 5. Superoxide anion is produced in complex I through reduced flavin mononucleotide or the iron-sulfur clusters N-1a and N-1b, while in complex III, it is generated at the ubiquinol oxidation site. The respiratory chain has cellular antioxidant mechanisms that activate when an accidental leakage of electrons to oxygen occurs during energy production. These ROS, including superoxide anion, hydroxyl radical, and hydrogen peroxide, are critical signaling molecules that play crucial roles in cardiac physiology and disease [13,29,59]. Cytosolic and mitochondrial sources of ROS contribute to the intracellular pool. Under normal conditions, ROS signaling regulates cardiac development and cardiomyocyte maturation, cardiac calcium handling, excitation-contraction coupling, and vascular tone [69]. Elevated ROS levels can cause oxidative damage to mitochondria and DNA, harm proteins and lipids, activate the mitochondrial permeability transition pore, cause mitochondrial dysfunction, and lead to cellular death. Additionally, it can disrupt the ATP and ADP balance [13,29,70].

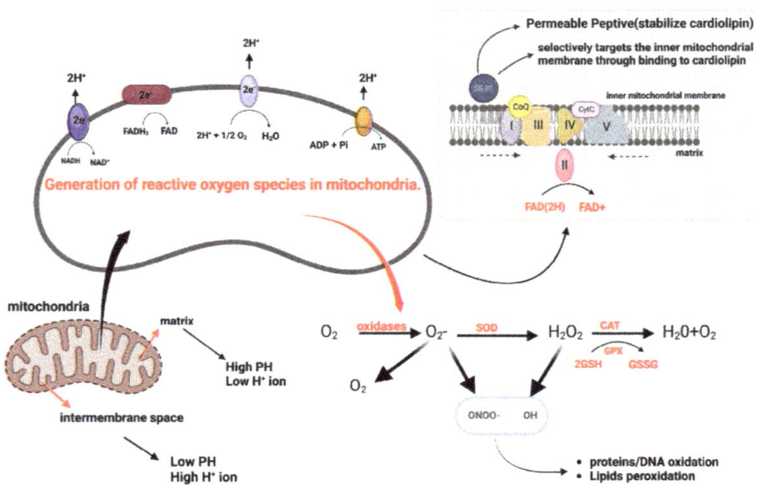


**Figure 5.** Mitochondrial electron transport chain and ROS generation. Legends: NADH = adenine dinucleotide; $NAD^+$ = nicotinamide adenine dinucleotide; PH = potential of hydrogen; SOD = superoxide dismutase; CAT = catalase; GSH = glutathione; GSSG = oxidized glutathione; ADP = adenosine diphosphate; ATP = adenosine triphosphate; Complex I = NADH dehydrogenase; Complex II = Succinate dehydrogenase; Complex III = Cytochrome b-c1; Complex IV = Cytochrome oxidase; Complex V = ATP synthase.

In addition to the respiratory chain, several proteins in the mitochondria contribute to the mitochondrial pool of ROS. P66shc, located in the mitochondrial intermembrane space, plays a significant role in oxidative stress signaling. It contributes to mitochondrial ROS production by oxidizing cytochrome c and stimulating hydrogen peroxide production. Isoforms A and B of monoamine oxidase (MAO-A and MAO-B) are crucial sources of mitochondrial hydrogen peroxide. Located in the mitochondrial outer membrane, MAOs degrade monoamines into hydrogen peroxide and aldehydes [71]. Furthermore, NOXs are a family of proteins involved in intracellular ROS production [72]. NOX4 is partially located in the mitochondrial inner membrane and is a source of mitochondrial superoxide and hydrogen peroxide, whose activity is regulated by ATP [73].

While the exact mechanisms of HCM are still not fully understood, previous non-invasive studies have shed light on certain aspects of the disease. According to these studies, compromised energy metabolism during the early stages of HCM can contribute to disease progression. In particular, mitochondrial defects have been identified as a key factor in this process [3,54,74].

Chronic mitochondrial dysfunction in myocardial heart disease (MHD) can reduce the activity of complex I, which is the primary entry point for electrons in the mitochondrial respiratory chain. This impaired function can intensify oxidative stress and increase the release of ROS, which in turn stimulates cardiac hypertrophy [75]. Metabolomic analyses have revealed a decrease in the ratio of mitochondrial DNA to genomic DNA, reduced expression of genes linked to mtDNA integrity, mitochondrial transcription, translation, and diminished cardiolipin species. These factors can affect mitochondrial respiratory activity and membrane integrity [2,3,19,54].

An insight modeling study has indicated that hearts affected by HCM undergo electrophysiological changes due to altered mitochondrial membrane potential. This implies reduced oxygen radical production in the electron transport chain and subsequent energy expenditure beyond supply. This discrepancy can be attributed to the increased oxygen consumption of cardiomyocytes [76].

Mitochondrial ROS are a significant factor in the pathogenesis of several diseases, including HCM. Superoxide dismutases (SODs) are the primary defense against mitochondrial ROS. These enzymes convert superoxide into hydrogen peroxide, and three isoforms of SOD (SOD1, SOD2, and SOD3) are present in different cellular compartments, regulating specific pools of ROS. SOD1 is mainly found in the cytosol but also localizes to the mitochondrial intermembrane space. Conversely, SOD2 is situated within the mitochondrial matrix, and SOD3 is extracellular. Precise control over the localization and activity of SOD1 and SOD2 is vital to mitigate mitochondrial ROS [69,77].

In addition to SODs, other enzymes play crucial roles in hydrogen peroxide detoxification. Catalase, predominantly found in peroxisomes and potentially in cardiac mitochondria, catalyzes the breakdown of hydrogen peroxide into water and oxygen. GSH-PXs 1 and 4, situated within the mitochondria, utilize reduced glutathione (GSH) to convert hydrogen peroxide into water. PRXs, including PRX3 and PRX5 within the mitochondria, scavenge hydrogen peroxide and peroxynitrite. These enzymes are crucial for neutralizing ROS and replenishing antioxidants [78,79].

In HCM, elevated oxidized cysteine (cystine) levels and a high oxidized glutathione (GSSG) ratio to GSH indicate increased oxidative stress. Moreover, reduced levels of essential antioxidants such as superoxide dismutase, catalase, glutathione reductase (GSR), and glutathione peroxidase (GPX) suggest potential transcriptional alterations that could compromise overall antioxidant capacity [19].

It has been observed that understanding the important role of SODs and other enzymes in regulating ROS can provide significant insights into the pathogenesis of HCM. Precise control of the localization and activity of these enzymes is crucial to reducing mitochondrial ROS and replenishing antioxidant levels. Wijnker et al. (2019) suggested that antioxidant protection through thiol-based systems and the pentose phosphate pathway occurs during the early stages of HCM [29]. However, these protective mechanisms decrease in advanced stages, resulting in chronic oxidative stress, which contributes to HCM progression and creates a cycle of mitochondrial damage and ROS generation [3,29]. Early detection of oxidative stress is essential for controlling HCM progression, enabling effective interventions, and identifying potential therapeutic targets. Primary mitochondrial diseases, although rare, often result in oxidative phosphorylation dysfunction due to mtDNA variations, with approximately 40% of affected children developing heart diseases, including heart complications [80].

Mitochondrial cristae, which are critical for cell energy production, are significantly damaged by ROS. This leads to lipid and protein oxidation, lipid peroxidation, and protein modification, affecting mitochondrial DNA, causing mutations and impaired electron transport. Oxidative stress also triggers apoptosis and permeabilization of the inner mitochondrial membrane, contributing to the pathology of HCM [2,3,13,29].

Studies have shown significant changes in mitochondrial morphology in patients with HCM, including damaged structures with disorganized and less dense cristae and an increased number of mitochondria, indicative of mitochondrial division. However, Ranjbarvaziri et al. (2019) found no evidence of increased mitochondrial fission or fusion [19]. Nollet et al. (2022 and 2023a) linked mitochondrial dysfunction in HCM to inadequate organization of interfibrillar mitochondria characterized by isolated clusters disconnected from myofilaments [23,81]. This disorganization may impair mitochondrial bioenergetics, defense against ROS, and calcium cycle disruption, leading to global energetic and metabolic cellular dysfunction [23,81]. Furthermore, prominent changes in myocardial ultrastructure have been observed, including alterations in cellular organization, a significant increase in extracellular matrix formation, and a decrease in mitochondrial mass in the sub-sarcomeric portion of cardiomyocytes [68].

Mitophagy is crucial in maintaining mitochondrial integrity and function in the myocardium by eliminating the damaged mitochondria [82]. Dysfunctional mitochondria observed in clinical models and in vitro cultures of HCM indicate potential dysfunction of overwhelming mitophagy pathways. Significant downregulation of mitophagy-related genes has been reported in patients with HCM [19]. Findings by Nollet et al. (2023a) [23] suggest that deterioration of mitochondrial quality control mechanisms contributes to the accumulation of damaged mitochondria in HCM, exacerbating the mitochondrial injury and continuing a detrimental cycle in its progression [1,22,80,83].

3.2.1. Emerging Therapeutic Strategies

This scoping review highlights the roles of genetic mutations and mitochondrial dysfunction in the development and progression of HCM. Metabolic modifications of HCM, characterized by an altered metabolic profile, abnormal mitochondrial structure, impaired respiratory function, and failure to regulate mitophagy, could guide the development of new therapeutic approaches to restore metabolic function and preserve mitochondrial integrity in the early stages of the disease. Studies have emphasized that cellular oxidative stress is a crucial factor in mitochondrial transformations, leading to structural and functional changes, alterations in cell signaling pathways, and uncontrolled ROS production, suggesting potential therapeutic benefits for various cardiovascular conditions, as shown in Figure 6 [9,80,84].

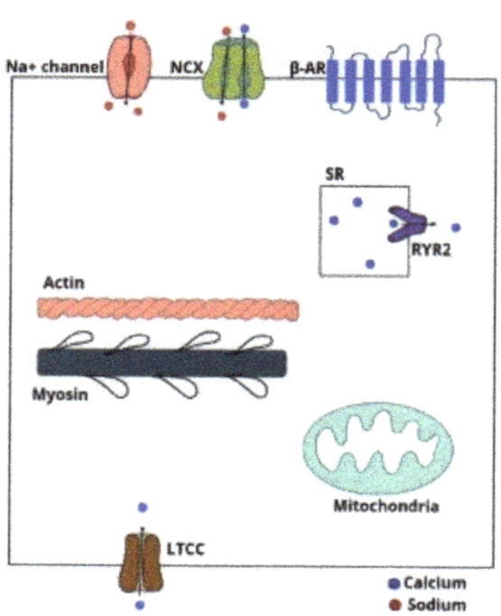


| Teraphy | Target |
|---|---|
| β-blockers | β-AR |
| Mavacamten | Myosin head |
| Ranolazine | NCX and NA+ channel |
| Dantrolene | RYR2 |
| EGCG | Troponin |
| Omecamtiv mecarbil (OM) | Myosin |
| Blebbistatin | Myosin |
| Nifedipine | LTCC |
| Verapamil | LTCC |
| Diltiazem | LTCC |
| Perhexiline | Mitochondria |
| Trimetazidine | Mitochondria |

**Figure 6.** Ongoing pharmacological strategies to treat HCM. Legends: EGCG = epigallocatechin-3-gallate, LTCC = L-type calcium channel, NCX = Sodium-Calcium Exchanger, β-AR = beta-adrenergic receptor, RYR2 = ryanodine receptor type 2, SR = sarcoplasmic reticulum.

Hypertrophic cardiomyopathy is associated with disrupted cardiac redox balance in genotype-positive patients, even without genetic alterations [56,84–86]. Factors such as wall stress, mitochondrial or microvascular dysfunction, and asymmetric septal remodeling can trigger this imbalance alongside age-related changes in cellular redox balance. Diet, antioxidants, and genetic predisposition also influence cardiac redox status. Genetic polymorphisms can affect phenotypic expression in nonfamilial HCM. Understanding the importance of these factors in other pathological mechanisms, such as myocardial energy depletion, requires further research. Current HCM therapies include β-blockers, $Ca^{2+}$ antagonists, ACE inhibitors, diuretics, antiarrhythmics, and anticoagulants, potentially impacting cardiac redox status [29].

SS-31, also known as Elamipretide, a cell-permeable peptide, selectively targets the inner mitochondrial membrane via cardiolipin binding, thereby reducing mitochondrial ROS production and restoring mitochondrial bioenergetics [87]. It stabilizes cardiolipin, promotes membrane restructuring, and facilitates the formation of mitochondrial super-complexes, thereby reducing ROS production [87–90]. Recent studies have focused on mitochondrial targets such as elamipretide and NAD+ supplementation. In HCM myectomized tissues, elamipretide treatment increased complex I incorporation and enhanced NADH-binding respiration. NADH-linked breathing impairment in HCM may stem from inefficient coupling of complex I to complexes III and IV, reduced Krebs cycle dehydrogenases, and inadequate mitochondrial calcium uptake [23,81].

Mitochondria-focused antioxidant treatments are promising for treating conditions such as HCM [91–93]. Recent studies have demonstrated the mitigating effects of these antioxidants on metabolic disorders. Traditional vitamins and antioxidant compounds such as coenzyme Q, alpha-lipoic acid, and N-acetylcysteine can combat excessive ROS production. Two mitochondrion-focused antioxidants, MitoQ, and MitoVit E, have become prominent. MitoQ targets mitochondria by incorporating the triphenylphosphonium lipophilic cation (TPP) group into coenzyme Q10 ubiquinone, reducing ROS and protecting against age-related damage. MitoVitE, a derivative of vitamin E conjugated with TPP, also targets mitochondria [9]. Coenzyme Q (CoQ) improves symptoms and quality of life in patients with HCM and reduces the interventricular septum thickness. It acts as an electron transporter in the inner mitochondrial membrane, restoring redox balance and preventing mitochondrial ROS formation. It also reduces left ventricular outflow obstruction and ventricular tachycardia [94,95].

Chen et al. (2015) discovered that 17b-estradiol improved myocardial diastolic function in HCM mice, suggesting that estrogen protects against disease progression [56]. The intervention reduced oxidative stress and prevented myocardial dysregulation.

1-Deoxynojirimycin, a treatment for morphological mitochondrial disorders, has shown benefits in recovering mitochondrial cristae and improving calcium homeostasis. Its action on optic atrophy protein 1 (OPA1), a regulator of mitochondrial cristae formation, enhances the number and width of the mitochondrial cristae, thereby improving ATP production and calcium homeostasis. Thus, 1-deoxynojirimycin is a potential therapeutic agent for treating HCM [41].

Regarding metabolic alterations, a study in rats highlighted the beneficial effects of restoring fatty acid metabolism on cardiac hypertrophy. Treatment with Tricaprylin reduced cardiomyocyte cross-sectional area, decreased interstitial fibrosis, and lowered biomarkers related to oxidative stress [96]. However, it is important to note that metabolic drugs like Perhexiline can also offer protective effects by shifting fatty acid metabolism to carbohydrates. This metabolic shift increases ATP supply and protects against catecholamine-induced cardiac damage. While fatty acid metabolism plays a crucial role in cardiac energetics, balancing its regulation with other metabolic pathways, such as carbohydrate metabolism, may be essential for optimizing cardiac function in conditions like HCM. In this regard, in HCM models, Perhexiline improved phenotypic characteristics and exercise capacity in symptomatic patients. However, chronic shift to glycolytic metabolism may be detrimental, indicating that intermittent metabolic therapy may be a more effective and innovative approach, especially compared to continuous treatment. In this context, a phase 2 clinical trial with Perhexiline showed some modest improvements in symptoms and cardiac bioenergetics compared to placebo, but the drug was discontinued due to multiorgan toxicity [97,98]. Other drugs inhibiting fatty acid β-oxidation also indicated negative outcomes, such as trimetazidine, an inhibitor of 3-ketoyl-CoA thiolase [99]. Given the negative regulation of enzymes involved in fatty acid oxidation in HCM patients, a more promising therapeutic approach would be to increase the availability of alternative fuel sources for ATP production. Therapeutic ketosis, which increases ketone bodies, has shown to be a viable strategy, considering cardiac metabolism adaptation in HCM patients. β-hydroxybutyrate, a ketone body, increased in the hearts of HCM patients, suggesting possible adaptation to reduced availability of acyl carnitines. Pharmacological oxidation of ketone bodies may have potential therapeutic applications [100–102].

B3-adrenergic receptors (b3AR) offer potential treatment avenues for metabolic changes in HCM. Activation of b3AR positively affects intracellular signaling, enhances calcium uptake, improves diastolic relaxation, and exhibits antioxidant properties [103].

Xanthine oxidase (XO) significantly contributes to ROS and uric acid generation, which are byproducts of purine metabolism. Elevated uric acid levels in patients with heart failure and HCM are correlated with an increased risk of cardiovascular events. Future research should confirm the activity of XO in HCM and investigate the blocking of ROS production by XO as a therapeutic strategy [104,105]. Restoring fatty acid metabolism in cardiac hypertrophy has beneficial effects, including reducing the cardiomyocyte cross-sectional area and oxidative stress biomarkers [96]. Perhexylin, a metabolic drug, switches metabolism from fatty acids to carbohydrates, boosting ATP delivery and protecting against catecholamine-induced cardiac damage. Intermittent metabolic therapy may be more beneficial than continuous treatment [97–99]. Therapeutic ketosis, which involves elevated levels of ketone bodies, is a viable strategy for cardiac metabolism adaptation in patients with HCM [100–102].

#### 3.2.2. Emerging Drug Therapies for HCM

Omecamtiv mecarbil (OM) is a myosin activator that has been studied as a potential treatment for systolic heart failure. Its effects may vary depending on the intracellular calcium concentration, which suggests that it could enhance contractility in cardiac failure and reduce it in diastolic failure, such as in HCM. MyoKardia, Inc. (Brisbane, CA, USA) has developed a novel oral medication called Mavacamten, which acts as an allosteric modulator of cardiac β-myosin [34]. This drug can inhibit the actin-myosin cross-bridge, reducing contractility and oxygen consumption in HCM models [106–108]. By weakening the bond between myosin heads and actin fibers, Mavacamten can prevent cardiac hypertrophy and fibrosis. It may also indirectly affect mitochondrial and energetic dysfunction by modulating contractility and cardiac function, optimizing energy balance in the heart, and enhancing efficiency in energy utilization [109]. It is worth noting that while Mavacamten shows promise as a treatment for HCM, Omecamtiv mecarbil, designed for systolic heart failure, would not be effective for HCM due to its mode of action.

Verapamil is a drug that blocks L-type calcium channels and has shown promising results in reducing hypertrophic cardiomyopathy in mice with the MYH7-Arg403Gln mutation. However, its effectiveness may vary in patients with multiple mutations. Understanding the early mechanisms involved in this disease is important to develop effective treatments. Genetic modifiers, epigenetic variations, and environmental factors can all play a role in the development of HCM, leading to various clinical outcomes [110].

The VANISH clinical trial used valsartan, an angiotensin receptor antagonist, to treat HCM with sarcomeric mutations. However, it did not significantly impact slowing subclinical HCM progression [111,112]. Sacubitril/valsartan, an angiotensin II receptor inhibitor/antagonist, has been tested in a clinical trial of patients with HCM. Lifestyle interventions such as regular exercise can activate beneficial metabolic pathways [113]. There is an unmet need for the preventive treatment of HCM. A murine model study assessed the efficacy of in vivo treatment using a derived peptide (AID-TAT) to restore mitochondrial metabolic activity and prevent HCM with the cardiac mutation Gly203Ser in troponin I [114,115], as shown in Table 1.

**Table 1.** Therapeutic possibilities and their primary effects.

| Therapeutic Possibilities | References | Primary Effects |
|---|---|---|
| Alpha-lipoic acid | [9] | Neutralizes free radicals, reduces oxidative stress, improves endothelial function, increases nitric oxide production, and influences cellular signaling pathways associated with cardiovascular health. |
| N-acetylcysteine | [9,116,117] | Has mucolytic and fluidizing activity. Reduces oxidative damage to heart tissues and increases the production of nitric oxide, which helps dilate veins and improve blood flow. |
| MitoQ | [9] | Neutralizes free radicals within the mitochondria, reduces oxidative stress, and preserves mitochondrial function. |
| MitoVit E | [9] | Has antioxidant capacity aimed at protecting mitochondria against damage caused by free radicals and reducing oxidative stress. |
| Coenzyme Q | [9,94,95] | CoQ is involved in the mitochondrial electron transport chain, where it aids in generating adenosine triphosphate (ATP), the cellular energy currency. Additionally, CoQ possesses antioxidant properties, shielding the mitochondria and the heart against damage caused by free radicals during the energy production process. |
| Elamipretide/SS-31 | [87–90] | It concentrates on mitochondrial membranes and helps regulate energy production by enhancing the efficiency of the electron transport chain. Additionally, Elamipretide can reduce oxidative stress, stabilize mitochondrial membranes, and preserve mitochondrial integrity. |
| 17b-estradiol | [56] | It influences cardiac function by regulating genes, interacting with specific receptors, modulating vascular response, and indirectly affecting metabolic and cellular signaling processes. |
| 1-Deoxynojirimycin | [41] | Inhibits metabolic pathways involved in protein glycosylation. |
| Tricaprylin | [96] | It acts as a rapid source of fuel that can be used for ATP production, thereby aiding in the heart's contractile function. This can be beneficial, especially in situations where the heart has an increased energy demand, such as certain cardiac conditions or intense physical exercise. |
| Perhexiline | [97,98,118] | Inhibits cardiac fatty acid uptake, promoting glucose utilization and reducing oxygen demand. |
| Trimetazidine | [99] | Blocks the mitochondrial protein responsible for transporting fatty acids into mitochondria. This action favors glucose metabolism, preserving cardiac function during low-oxygen-supply conditions such as angina. It improves cardiac efficiency and reduces angina pain. |
| Beta-3 adrenergic receptors | [103] | It is primarily activated by neurotransmitters like norepinephrine and adrenaline. It is involved in regulating fat metabolism and thermogenesis (heat production) and can influence cardiac function, lipolysis (fat breakdown), and muscle contraction in specific tissues, such as the smooth muscle of the bladder. |
| Omecamtiv Mecarbil | [106] | A myocardial activator that increases the sensitivity of the cardiac muscle (myocardium) to calcium concentration, resulting in greater efficiency in cardiac muscle contraction. Binds to myosin in the myocardium, altering its conformation and allowing a more effective interaction with calcium. |
| Mavacamten | [34,106–108] | An allosteric inhibitor of heart-specific myosin adenosine triphosphatase. |
| Diltiazem | [110] | A non-selective calcium channel antagonist that affects the heart and blood vessels. |
| Valsartan | [111,112] | Displaces angiotensin II from the AT1 receptor and produces its blood-pressure-lowering effects by antagonizing AT1-induced vasoconstriction, aldosterone release, catecholamine release, arginine vasopressin release, water intake, and hypertrophic responses. |
| Sacubitril | [119] | Simultaneously inhibits neprilysin (neutral endopeptidase; NEP) through sacubitrilat, the active metabolite of the prodrug sacubitril. |
| AID-TAT | [114,115] | This is related to activating the enzyme AMPK (AMP-activated protein kinase). This leads to a series of cellular events, including increased glucose uptake and increased ATP production. |

## 4. Conclusions

Hypertrophic cardiomyopathy (HCM) represents a complex cardiac condition marked by a spectrum of biological process alterations, including ionic homeostasis, structural remodeling, metabolic balance, and ROS regulation. Its current therapeutic landscape primarily targets symptom management and the prevention of complications, highlighting a significant gap in the availability of more effective treatment methodologies. The treatment options currently range from symptom alleviation strategies to more invasive surgical interventions, with the choice of treatment tailored to the evolving profiles of individual patients.

A key focus of researchers and clinicians is the investigation of modifiable risk factors encompassing genetic predispositions and mutations. This approach underscores the recognition that HCM is not just a product of unchangeable genetic factors but may also be influenced by modifiable elements, thus opening new avenues for intervention and management.

A comprehensive understanding of the biomolecular, genetic, and mitochondrial mechanisms that drive HCM pathophysiology is vital. This knowledge is crucial for the advancement of research in this field and the development of new and more effective therapeutic modalities. As our understanding deepens, it paves the way for innovative treatments that move beyond symptom management, potentially offering significant and long-lasting benefits to patients with this condition.

In summary, HCM presents a multifaceted challenge that requires a multifaceted response, encompassing advanced research, innovative therapeutic strategies, and a personalized approach to patient care. The future of HCM treatment lies at the intersection of these diverse yet interconnected domains.

**Supplementary Materials:** The following supporting information can be downloaded at https://www.mdpi.com/article/10.3390/ijms25115855/s1.

**Author Contributions:** Conceptualization, A.d.S.M.J. and A.F.F.J.; methodology, A.d.S.M.J., A.L.G.d.F.-e.-S. and H.L.d.O.; software, A.d.S.M.J. and A.L.G.d.F.-e.-S.; validation, A.d.S.M.J.; formal analysis, A.d.S.M.J.; investigation, A.d.S.M.J., A.L.G.d.F.-e.-S., H.L.d.O. and K.B.A.d.L.; data curation, A.d.S.M.J., A.L.G.d.F.-e.-S., H.L.d.O. and K.B.A.d.L.; writing—original draft preparation, A.d.S.M.J., A.L.G.d.F.-e.-S., H.L.d.O. and K.B.A.d.L.; writing—review and editing A.d.S.M.J., A.L.G.d.F.-e.-S., H.L.d.O., K.B.A.d.L., I.d.O.P.P., T.M.A.P. and D.d.M.e.S.; supervision, A.d.S.M.J., D.d.M.e.S. and A.F.F.J.; project administration, A.d.S.M.J. All authors have read and agreed to the published version of the manuscript.

**Funding:** This research received no external funding.

**Institutional Review Board Statement:** Not applicable.

**Informed Consent Statement:** Not applicable.

**Data Availability Statement:** The original contributions presented in the study are included in the article and Supplementary Materials; further inquiries can be directed to the corresponding author.

**Conflicts of Interest:** The authors declare no conflicts of interest.

## References

1. Dipchand, A.I.; Tein, I.; Robinson, B.; Benson, L.N. Maternally inherited hypertrophic cardiomyopathy: A manifestation of mitochondrial DNA mutations–clinical course in two families. *Pediatr. Cardiol.* **2001**, *22*, 14–22. [CrossRef] [PubMed]
2. Arbelo, E.; Protonotarios, A.; Gimeno, J.R.; Arbustini, E.; Barriales-Villa, R.; Basso, C.; Bezzina, C.R.; Biagini, E.; Blom, N.A.; de Boer, N.A.; et al. 2023 ESC Guidelines for the management of cardiomyopathies. *Eur Heart J.* **2023**, *44*, 3503–3626. [CrossRef] [PubMed]
3. Yokota, T.; Christiansen, L.B.; Dela, F. Impaired mitochondrial oxidative phosphorylation and fatty acid oxidation with enhanced mitochondrial oxidative stress in spontaneously occurring feline hypertrophic cardiomyopathy. *J. Card. Fail.* **2014**, *20*, S147. [CrossRef]
4. Maron, B.J.; Maron, M.S. Hypertrophic cardiomyopathy. *Lancet* **2013**, *381*, 242–255. [CrossRef] [PubMed]

5. Sequeira, V.; Waddingham, M.T.; Tsuchimochi, H.; Maack, C.; Pearson, J.T. Mechano-energetic uncoupling in hypertrophic cardiomyopathy: Pathophysiological mechanisms and therapeutic opportunities. *J. Mol. Cell. Cardiol. Plus* **2023**, *4*, 100036. [CrossRef]
6. Elliott, P. Hypertrophic cardiomyopathy. *Circulation* **2018**, *138*, 1399–1401. [CrossRef] [PubMed]
7. Hayashi, T. Hypertrophic cardiomyopathy: Diverse pathophysiology revealed by genetic research, toward future therapy. *Keio J. Med.* **2020**, *69*, 77–87. [CrossRef] [PubMed]
8. Viola, H.M.; Hool, L.C. Impaired calcium handling and mitochondrial metabolic dysfunction as early markers of hypertrophic cardiomyopathy. *Arch. Biochem. Biophys.* **2019**, *665*, 166–174. [CrossRef] [PubMed]
9. Bhatti, J.S.; Bhatti, G.K.; Reddy, P.H. Mitochondrial dysfunction, and oxidative stress in metabolic disorders—A step towards mitochondria based therapeutic strategies. *Biochim. Biophys. Acta Mol. Basis Dis.* **2017**, *1863*, 1066–1077. [CrossRef]
10. Wasmus, C.; Dudek, J. Metabolic Alterations Caused by Defective Cardiolipin Remodeling in Inherited Cardiomyopathies. *Life* **2020**, *10*, 277. [CrossRef]
11. Weissman, D.; Maack, C. Redox signaling in heart failure and therapeutic implications. *Free Radic. Biol. Med.* **2021**, *171*, 345–364. [PubMed]
12. Lee, H.G.; Chen, Q.; Wolfram, J.A.; Richardson, S.L.; Liner, A.; Siedlak, S.L.; Zhu, X.; Ziats, N.P.; Fujioka, H.; Felsher, D.W.; et al. Cell cycle re-entry and mitochondrial defects in myc-mediated hypertrophic cardiomyopathy and heart failure. *PLoS ONE* **2009**, *4*, e7172. [CrossRef] [PubMed]
13. Christiansen, L.B.; Dela, F.; Koch, J.; Hansen, C.N.; Leifsson, P.S.; Yokota, T. Impaired cardiac mitochondrial oxidative phosphorylation and enhanced mitochondrial oxidative stress in feline hypertrophic cardiomyopathy. *Am. J. Physiol. Heart Circ. Physiol.* **2015**, *308*, H1237–H1247. [CrossRef] [PubMed]
14. Li, S.; Pan, H.; Tan, C.; Sun, Y.; Song, Y.; Zhang, X.; Yang, W.; Wang, X.; Li, D.; Dai, Y.; et al. Mitochondrial Dysfunctions Contribute to Hypertrophic Cardiomyopathy in Patient iPSC-Derived Cardiomyocytes with MT-RNR2 Mutation. *Stem Cell Rep.* **2018**, *10*, 808–821. [CrossRef]
15. van der Velden, J.; Tocchetti, C.G.; Varricchi, G.; Bianco, A.; Sequeira, V.; Hilfiker-Kleiner, D.; Hamdani, N.; Leite-Moreira, A.F.; Mayr, M.; Falcão-Pires, I.; et al. Metabolic changes in hypertrophic cardiomyopathies: Scientific update from the Working Group of Myocardial Function of the European Society of Cardiology. *Cardiovasc. Res.* **2018**, *114*, 1273–1280. [CrossRef]
16. Chen, D.; Zhang, Z.; Chen, C.; Yao, S.; Yang, Q.; Li, F.; He, X.; Ai, C.; Wang, M.; Guan, M.X. Deletion of Gtpbp3 in zebrafish revealed the hypertrophic cardiomyopathy manifested by aberrant mitochondrial tRNA metabolism. *Nucleic Acids Res.* **2019**, *47*, 5341–5355. [CrossRef]
17. Li, X.; Lu, W.J.; Li, Y.N.; Wu, F.; Bai, R.; Ma, S.; Dong, T.; Zhang, H.; Lee, A.S.; Wang, Y.; et al. MLP-deficient human pluripotent stem cell-derived cardiomyocytes develop hypertrophic cardiomyopathy and heart failure phenotypes due to abnormal calcium handling. *Cell Death Dis.* **2019**, *10*, 610. [CrossRef] [PubMed]
18. Saoura, M.; Powell, C.A.; Kopajtich, R.; Alahmad, A.; AL-Balool, H.H.; Albash, B.; Alfadhel, M.; Alston, C.L.; Bertini, E.; Bonnen, P.E.; et al. Mutations in ELAC2 associated with hypertrophic cardiomyopathy impair mitochondrial tRNA 3′-end processing. *Hum. Mutat.* **2019**, *40*, 1731–1748. [CrossRef] [PubMed]
19. Ranjbarvaziri, S.; Kooiker, K.B.; Ellenberger, M.; Fajardo, G.; Zhao, M.; Roest, A.S.V.; Woldeyes, R.A.; Koyano, T.T.; Fong, R.; Ma, N.; et al. Altered cardiac energetics and mitochondrial dysfunction in hypertrophic cardiomyopathy. *Circulation* **2021**, *144*, 1714–1731. [CrossRef]
20. Zhang, Q.; He, X.; Yao, S.; Lin, T.; Zhang, L.; Chen, D.; Chen, C.; Yang, Q.; Li, F.; Zhu, Y.M.; et al. Ablation of Mto1 in zebrafish exhibited hypertrophic cardiomyopathy manifested by mitochondrion RNA maturation deficiency. *Nucleic Acids Res.* **2021**, *49*, 4689–4704. [CrossRef]
21. Previs, M.J.; O'Leary, T.S.; Morley, M.P.; Palmer, B.M.; LeWinter, M.; Yob, J.M.; Pagani, F.D.; Petucci, C.; Kim, M.S.; Margulies, K.B.; et al. Defects in the proteome and metabolome in human hypertrophic cardiomyopathy. *Circ. Heart Fail.* **2022**, *15*, e009521. [CrossRef] [PubMed]
22. Moore, J.; Ewoldt, J.; Venturini, G.; Pereira, A.C.; Padilha, K.; Lawton, M.; Lin, W.; Goel, R.; Luptak, I.; Perissi, V.; et al. Multi-omics profiling of hypertrophic cardiomyopathy reveals altered mechanisms in mitochondrial dynamics and excitation–contraction coupling. *Int. J. Mol. Sci.* **2023**, *24*, 4724. [CrossRef] [PubMed]
23. Nollet, E.E.; Duursma, I.; Rozenbaum, A.; Eggelbusch, M.; Wuest, R.C.; Schoonvelde, S.A.C.; Michels, M.; Jansen, M.; van der Wel, N.N.; Bedi, K.C.; et al. Mitochondrial dysfunction in human hypertrophic cardiomyopathy is linked to cardiomyocyte architecture disruption and corrected by improving NADH-driven mitochondrial respiration. *Eur. Heart J.* **2023**, *44*, 1170–1185. [CrossRef] [PubMed]
24. Marin-Garcia, J.; Ananthakrishnan, R.; Goldenthal, M.J.; Filiano, J.J.; Perez-Atayde, A. Cardiac mitochondrial dysfunction and DNA depletion in children with hypertrophic cardiomyopathy. *J. Inherit. Metab. Dis.* **1997**, *20*, 674–680. [CrossRef] [PubMed]
25. Meyers, D.E.; Basha, H.I.; Koenig, M.K. Mitochondrial cardiomyopathy: Pathophysiology, diagnosis, and management. *Tex. Heart Inst. J.* **2013**, *40*, 385–394.
26. Kiselev, I.; Kozin, M.; Baulina, N.; Pisklova, M.; Danilova, L.; Zotov, A.; Chumakova, O.; Zateyshchikov, D.; Favorova, O. Novel Genes Involved in Hypertrophic Cardiomyopathy: Data of Transcriptome and Methylome Profiling. *Int. J. Mol. Sci.* **2022**, *23*, 15280. [CrossRef] [PubMed]

27. Ormerod, J.O.; Frenneaux, M.P.; Sherrid, M.V. Myocardial energy depletion and dynamic systolic dysfunction in hypertrophic cardiomyopathy. *Nat. Rev. Cardiol.* **2016**, *13*, 677–687. [CrossRef] [PubMed]
28. Nag, S.; Gollapudi, S.K.; Del Rio, C.L.; Spudich, J.A.; McDowell, R. Mavacamten, a precision medicine for hypertrophic cardiomyopathy: From a motor protein to patients. *Sci. Adv.* **2023**, *9*, eabo7622. [CrossRef]
29. Wijnker, P.J.M.; Sequeira, V.; Kuster, D.W.D.; van der Velden, J. Hypertrophic cardiomyopathy: A vicious cycle triggered by sarcomere mutations and secondary disease hits. *Antioxid. Redox Signal.* **2019**, *31*, 318–358. [CrossRef]
30. Ouzzani, M.; Hammady, H.; Fedorowicz, Z.; Elmagarmid, A. Rayyan—A web and mobile app for systematic reviews. *Syst. Rev.* **2016**, *5*, 210. [CrossRef]
31. Sedaghat-Hamedani, F.; Kayvanpour, E.; Tugrul, O.F.; Lai, A.; Amr, A.; Haas, J.; Proctor, T.; Ehlermann, P.; Jensen, K.; Katus, H.A.; et al. Clinical outcomes associated with sarcomere mutations in hypertrophic cardiomyopathy: A meta-analysis on 7675 individuals. *Clin. Res. Cardiol.* **2018**, *107*, 30–41. [CrossRef]
32. Glavaški, M.; Velicki, L.; Vučinić, N. Hypertrophic cardiomyopathy: Genetic foundations, outcomes, interconnections, and their modifiers. *Medicina* **2023**, *59*, 1424. [CrossRef]
33. Chou, C.; Chin, M.T. Genetic and Molecular Mechanisms of Hypertrophic Cardiomyopathy. *Int. J. Mol. Sci.* **2023**, *24*, 2522. [CrossRef] [PubMed]
34. Tuohy, C.V.; Kaul, S.; Song, H.K.; Nazer, B.; Heitner, S.B. Hypertrophic cardiomyopathy: The future of treatment. *Eur. J. Heart Fail.* **2020**, *22*, 228–240. [CrossRef]
35. Marian, A.J.; Braunwald, E. Hypertrophic cardiomyopathy: Genetics, pathogenesis, clinical manifestations, diagnosis, and therapy. *Circ. Res.* **2017**, *121*, 749–770. [CrossRef] [PubMed]
36. Kraft, T.; Montag, J.; Radocaj, A.; Brenner, B. Hypertrophic cardiomyopathy: Cell-to-cell imbalance in gene expression and contraction force as trigger for disease phenotype development. *Circ. Res.* **2016**, *119*, 992–995. [CrossRef] [PubMed]
37. Cui, H.; Schlesinger, J.; Schoenhals, S.; Tönjes, M.; Dunkel, I.; Meierhofer, D.; Cano, E.; Schulz, K.; Berger, M.F.; Haack, T.; et al. Phosphorylation of the chromatin remodeling factor DPF3a induces cardiac hypertrophy through releasing HEY repressors from DNA. *Nucleic Acids Res.* **2016**, *44*, 2538–2553. [CrossRef]
38. Wolf, C.M. Hypertrophic cardiomyopathy: Genetics and clinical perspectives. *Cardiovasc. Diagn. Ther.* **2019**, *9* (Suppl. S2), S388–S415. [CrossRef]
39. Guan, F.; Yang, X.; Li, J.; Dong, W.; Zhang, X.; Liu, N.; Gao, S.; Wang, J.; Zhang, L.; Lu, D.; et al. New molecular mechanism underlying myc-mediated cytochrome P450 2E1 upregulation in apoptosis and energy metabolism in the myocardium. *J. Am. Heart Assoc.* **2019**, *8*, e009871. [CrossRef]
40. Wang, L.; Lu, F.; Xu, J. Identification of Potential miRNA-mRNA Regulatory Network Contributing to Hypertrophic Cardiomyopathy (HCM). *Front. Cardiovasc. Med.* **2021**, *8*, 660372. [CrossRef] [PubMed] [PubMed Central]
41. Zhuang, Q.; Guo, F.; Fu, L.; Dong, Y.; Xie, S.; Ding, X.; Hu, S.; Zhou, X.D.; Jiang, Y.; Zhou, H.; et al. 1-deoxynojirimycin promotes cardiac function and rescues mitochondrial cristae in mitochondrial hypertrophic cardiomyopathy. *J. Clin. Investig.* **2023**, *133*, e164660. [CrossRef]
42. García-Díaz, L.; Coserria, F.; Antiñolo, G. Hypertrophic cardiomyopathy due to mitochondrial disease: Prenatal diagnosis, management, and outcome. *Case Rep. Obstet. Gynecol.* **2013**, *2013*, 472356. [CrossRef] [PubMed]
43. Zhou, C.; Wang, J.; Zhang, Q.; Yang, Q.; Yi, S.; Shen, Y.; Luo, J.; Qin, Z. Clinical and genetic analysis of combined oxidative phosphorylation defificiency-10 caused by MTO1 mutation. *Clin. Chim. Acta* **2022**, *526*, 74–80. [CrossRef] [PubMed]
44. Prondzynski, M.; Mearini, G.; Carrier, L. Gene therapy strategies in the treatment of hypertrophic cardiomyopathy. *Pflug. Arch.* **2019**, *471*, 807–815. [CrossRef] [PubMed]
45. Paratz, E.D.; Mundisugih, J.; Rowe, S.J.; Kizana, E.; Semsarian, C. Gene therapy in cardiology: Is a cure for hypertrophic cardiomyopathy on the horizon? *Can. J. Cardiol.* **2023**, *40*, 777–788. [CrossRef] [PubMed]
46. Ma, H.; Marti-Gutierrez, N.; Park, S.W.; Wu, J.; Lee, Y.; Suzuki, K.; Koski, A.; Ji, D.; Hayama, T.; Ahmed, R.; et al. Correction of a pathogenic gene mutation in human embryos. *Nature* **2017**, *548*, 413–419. [CrossRef]
47. Halkein, J.; Tabruyn, S.P.; Ricke-Hoch, M.; Haghikia, A.; Nguyen, N.-Q.; Scherr, M.; Castermans, K.; Malvaux, L.; Lambert, V.; Thiry, M.; et al. MicroRNA-146a is a therapeutic target and biomarker for peripartum cardiomyopathy. *J. Clin. Investig.* **2013**, *123*, 2143–2154. [CrossRef] [PubMed]
48. Heggermont, W.A.; Papageorgiou, A.P.; Quaegebeur, A.; Deckx, S.; Carai, P.; Verhesen, W.; Eelen, G.; Schoors, S.; van Leeuwen, R.; Alekseev, S. Inhibition of microRNA-146a and overexpression of its target dihydrolipoyl succinyltransferase protect against pressure overload-Induced cardiac hypertrophy and dysfunction. *Circulation* **2017**, *136*, 747–761. [CrossRef] [PubMed]
49. Jiang, J.; Wakimoto, H.; Seidman, J.G.; Seidman, C.E. Allele-specific silencing of mutant Myh6 transcripts in mice suppresses hypertrophic cardiomyopathy. *Science* **2013**, *342*, 111–114. [CrossRef]
50. Mearini, G.; Stimpel, D.; Geertz, B.; Weinberger, F.; Krämer, E.; Schlossarek, S.; Mourot-Filiatre, J.; Stoehr, A.; Dutsch, A.; Wijnker, P.J.M.; et al. Mybpc3 gene therapy for neonatal cardiomyopathy enables long-term disease prevention in mice. *Nat. Commun.* **2014**, *5*, 5515. [CrossRef]
51. Tardiff, J.C.; Carrier, L.; Bers, D.M.; Poggesi, C.; Ferrantini, C.; Coppini, R.; Maier, L.S.; Ashrafian, H.; Huke, S.; van der Velden, J. Targets for therapy in sarcomeric cardiomyopathies. *Cardiovasc. Res.* **2015**, *105*, 457–470. [CrossRef] [PubMed]

52. Prondzynski, M.; Krämer, E.; Laufer, S.D.; Shibamiya, A.; Pless, O.; Flenner, F.; Müller, O.J.; Münch, J.; Redwood, C.; Hansen, A.; et al. Evaluation of MYBPC3 trans-Splicing and Gene Replacement as Therapeutic Options in Human iPSC-Derived Cardiomyocytes. *Mol. Ther. Nucleic Acids* **2017**, *7*, 475–486. [CrossRef] [PubMed]
53. German, D.M.; Mitalipov, S.; Mishra, A.; Kaul, S. Therapeutic Genome Editing in Cardiovascular Diseases. *JACC Basic Transl. Sci.* **2019**, *4*, 122–131. [CrossRef] [PubMed] [PubMed Central]
54. Vakrou, S.; Abraham, M.R. Hypertrophic cardiomyopathy: A heart in need of an energy bar? *Front. Physiol.* **2014**, *5*, 309. [CrossRef]
55. Nollet, E.E.; Algül, S.; Goebel, M.; Schlossarek, S.; van der Wel, N.N.; Jans, J.J.; van de Wiel, M.A.; Knol, J.C.; Pham, T.V.; Piersma, S.R.; et al. Western diet triggers cardiac dysfunction in a heterozygous Mybpc3-targeted knock-in hypertrophic cardiomyopa-thy mouse model. *Eur. J. Heart Fail.* **2023**, *25*, 31–32.
56. Chen, Y.; Zhang, Z.; Hu, F.; Yang, W.; Yuan, J.; Cui, J.; Hao, S.; Hu, J.; Zhou, Y.; Qiao, S. 17β-estradiol prevents cardiac diastolic dysfunction by stimulating mitochondrial function: A preclinical study in a mouse model of a human hypertrophic cardiomyopathy mutation. *J. Steroid Biochem. Mol. Biol.* **2015**, *147*, 92–102. [CrossRef] [PubMed]
57. Herold, K.G.; Hussey, J.W.; Dick, I.E. CACNA1C-Related Channelopathies. *Handb. Exp. Pharmacol.* **2023**, *279*, 159–181. [CrossRef] [PubMed]
58. Boczek, N.J.; Ye, D.; Jin, F.; Tester, D.J.; Huseby, A.; Bos, J.M.; Johnson, A.J.; Kanter, R.; Ackerman, M.J. Identification and functional characterization of a novel CACNA1C-mediated cardiac disorder characterized by prolonged QT intervals with hypertrophic cardiomyopathy, congenital heart defects, and sudden cardiac death. *Circ. Arrhythm. Electrophysiol.* **2015**, *8*, 1122–1132. [CrossRef] [PubMed]
59. Li, D.; Sun, Y.; Zhuang, Q.; Song, Y.R.; Wu, B.F.; Jia, Z.X.; Pan, H.Y.; Zhou, H.; Hu, S.Y.; Zhang, B.T.; et al. Mitochondrial dysfunction caused by m.2336T>C mutation with hypertrophic cardiomyopathy in cybrid cell lines. *Mitochondrion* **2019**, *46*, 313–320. [CrossRef]
60. Riaz, M.; Park, J.; Sewanan, L.R.; Ren, Y.; Schwan, J.; Das, S.K.; Pomianowski, P.T.; Huang, Y.; Ellis, M.W.; Luo, J.; et al. Muscle LIM Protein Force-Sensing Mediates Sarcomeric Biomechanical Signaling in Human Familial Hypertrophic Cardiomyopathy. *Circulation* **2022**, *145*, 1238–1253. [CrossRef]
61. Walsh, R.; Offerhaus, J.A.; Tadros, R.; Bezzina, C.R. Minor hypertrophic cardiomyopathy genes, major insights into the genetics of cardiomyopathies. *Nat. Rev. Cardiol.* **2022**, *19*, 151–167. [CrossRef] [PubMed]
62. Bertero, E.; Maack, C. Calcium Signaling and Reactive Oxygen Species in Mitochondria. *Circ. Res.* **2018**, *122*, 1460–1478. [CrossRef] [PubMed]
63. Lucas, D.T.; Aryal, P.; Szweda, L.I.; Koch, W.J.; Leinwand, L.A. Alterations in mitochondrial function in a mouse model of hypertrophic cardiomyopathy. *Am. J. Physiol. Heart Circ. Physiol.* **2003**, *284*, H575–H583. [CrossRef] [PubMed]
64. Sequeira, V.; Bertero, E.; Maack, C. Energetic drain driving hypertrophic cardiomyopathy. *FEBS Lett.* **2019**, *593*, 1616–1626. [CrossRef] [PubMed]
65. Feridooni, H.A.; Dibb, K.M.; Howlett, S.E. How cardiomyocyte excitation, calcium release and contraction become altered with age. *J. Mol. Cell. Cardiol.* **2015**, *83*, 62–72. [CrossRef] [PubMed]
66. Coppini, R.; Ferrantini, C.; Mugelli, A.; Poggesi, C.; Cerbai, E. Altered $Ca^{2+}$ and $Na^{+}$ homeostasis in human hypertrophic cardiomyopathy: Implications for arrhythmogenesis. *Front. Physiol.* **2018**, *9*, 1391. [CrossRef] [PubMed]
67. Farhana, A.; Lappin, S.L. Biochemistry, Lactate Dehydrogenase. In *StatPearls [Internet]*; StatPearls Publishing: Treasure Island, FL, USA, 2024.
68. Christiansen, L.B.; Prats, C.; Hyttel, P.; Koch, J. Ultrastructural myocardial changes in seven cats with spontaneous hypertrophic cardiomyopathy. *J. Vet. Cardiol.* **2015**, *17* (Suppl. S1), S220–S232. [CrossRef]
69. Povos, J.N.; Saraf, A.; Ghazal, N.; Pham, T.T.; Kwong, J.Q. Disfunção mitocondrial e estresse oxidativo nas cardiopatias. *Exp. Mol. Med.* **2019**, *51*, 1–13. [CrossRef]
70. Lombardi, M.; Lazzeroni, D.; Pisano, A.; Girolami, F.; Alfieri, O.; La Canna, G.; D'amati, G.; Olivotto, I.; Rimoldi, O.E.; Foglieni, C.; et al. Mitochondrial energetics and $Ca^{2+}$-activated ATPase in obstructive hypertrophic cardiomyopathy. *J. Clin. Med.* **2020**, *9*, 1799. [CrossRef]
71. Wang, Y.; Qu, H.; Liu, J. P66Shc Deletion Ameliorates Oxidative Stress and Cardiac Dysfunction in Pressure Overload-Induced Heart Failure. *J. Card. Fail.* **2020**, *26*, 243–253. [CrossRef] [PubMed]
72. Peng, M.L.; Fu, Y.; Wu, C.W.; Zhang, Y.; Ren, H.; Zhou, S.-S. Signaling Pathways Related to Oxidative Stress in Diabetic Cardiomyopathy. *Front. Endocrinol.* **2022**, *13*, 907757. [CrossRef]
73. Modanloo, M.; Shokrzadeh, M. Analyzing Mitochondrial Dysfunction, Oxidative Stress, and Apoptosis: Potential Role of L-carnitine. *Iran. J. Kidney Dis.* **2019**, *13*, 74–86. [PubMed]
74. Liu, X.; Ye, B.; Miller, S.; Yuan, H.; Zhang, H.; Tian, L.; Nie, J.; Imae, R.; Arai, H.; Li, Y.; et al. Ablation of ALCAT1 mitigates hypertrophic cardiomyopathy through effects on oxidative stress and mitophagy. *Mol. Cell. Biol.* **2012**, *32*, 4493–4504. [CrossRef]
75. Chouchani, E.T.; Methner, C.; Buonincontri, G.; Hu, C.-H.; Logan, A.; Sawiak, S.J.; Murphy, M.P.; Krieg, T. Complex I deficiency due to selective loss of Ndufs4 in the mouse heart results in severe hypertrophic cardiomyopathy. *PLoS ONE* **2014**, *9*, e94157. [CrossRef] [PubMed]

76. Adeniran, I.; MacIver, D.H.; Zhang, H. Myocardial electrophysiological, contractile, and metabolic properties of hypertrophic cardiomyopathy: Insights from modelling. In Proceedings of the Computing in Cardiology 2014, Cambridge, MA, USA, 7–10 September 2014; IEEE: Piscataway, NJ, USA, 2014.
77. Zhou, B.; Tian, R. Disfunção mitocondrial na fisiopatologia da insuficiência cardíaca. *J. Clin. Investig.* **2018**, *128*, 3716–3726. [CrossRef] [PubMed]
78. Andreyev, A.Y.; Kushnareva, Y.E.; Murphy, A.N.; Starkov, A.A. Metabolismo das ERO mitocondriais: 10 anos depois. *Bioquímica* **2015**, *80*, 517–531. [CrossRef]
79. Perkins, A.; Nelson, K.J.; Parsonage, D.; Poole, L.B.; Karplus, P.A. Peroxirredoxinas: Guardiões contra o estresse oxidativo e moduladores da sinalização de peróxidos. *Tend. Biochem. Sci.* **2015**, *40*, 435–445. [CrossRef]
80. Wang, L.; Lu, P.; Yin, J.; Xu, K.; Xiang, D.; Zhang, Z.; Zhang, H.; Zheng, B.; Zhou, W.; Wang, C.; et al. Case report: Rare novel MIPEP compound heterozygous variants presenting with hypertrophic cardiomyopathy, severe lactic acidosis and hypotonia in a Chinese infant. *Front. Cardiovasc. Med.* **2022**, *9*, 1095882. [CrossRef] [PubMed]
81. Nollet, E.; Burdzina, A.; Wüst, R.C.I.; Michels, M.; Asselbergs, F.W.; van der Wel, N.; Bedi, K.; Margulies, K.; Nirschl, J.; Kuster, D.; et al. "Disentangling" mitochondrial dysfunction in hypertrophic cardiomyopathy. *J. Mol. Cell. Cardiol.* **2022**, *173*, S113–S115. [CrossRef]
82. Li, Y.; Zhang, W.; Dai, Y.; Chen, K. Identification and verification of IGFBP3 and YTHDC1 as biomarkers associated with immune infiltration and mitophagy in hypertrophic cardiomyopathy. *Front. Genet.* **2022**, *13*, 986995. [CrossRef]
83. Cibi, D.M.; Bi-Lin, K.W.; Shekeran, S.G.; Sandireddy, R.; Tee, N.; Singh, A.; Wu, Y.; Srinivasan, D.K.; Kovalik, J.-P.; Ghosh, S.; et al. Prdm16 deficiency leads to age-dependent cardiac hypertrophy, adverse remodeling, mitochondrial dysfunction, and heart failure. *Cell Rep.* **2020**, *33*, 108288. [CrossRef]
84. Michałek, M.; Tabiś, A.; Pasławska, U.; Noszczyk-Nowak, A. Antioxidant defence and oxidative stress markers in cats with asymptomatic and symptomatic hypertrophic cardiomyopathy: A pilot study. *BMC Vet. Res.* **2020**, *16*, 26. [CrossRef]
85. Lin, C.S.; Sun, Y.L.; Liu, C.Y. Structural and biochemical evidence of mitochondrial depletion in pigs with hypertrophic cardiomyopathy. *Res. Vet. Sci.* **2003**, *74*, 219–226. [CrossRef] [PubMed]
86. Szyguła-Jurkiewicz, B.; Szczurek-Wasilewicz, W.; Osadnik, T.; Macioł-Skurk, K.; Małyszek-Tumidajewicz, J.; Skrzypek, M.; Romuk, E.; Gąsior, M.; Banach, M.; Jóźwiak, J.J. Oxidative Stress Markers in Hypertrophic Cardiomyopathy. *Medicina* **2021**, *58*, 31. [CrossRef]
87. Sabbah, H.N. Barth syndrome cardiomyopathy: Targeting the mitochondria with elamipretide. *Heart Fail. Rev.* **2021**, *26*, 237–253. [CrossRef] [PubMed]
88. Szeto, H.H. First-in-class cardiolipin-protective compound as a therapeutic agent to restore mitochondrial bioenergetics. *Br. J. Pharmacol.* **2014**, *171*, 2029–2050. [CrossRef] [PubMed]
89. Szeto, H.H.; Birk, A.V. Serendipity and the discovery of novel compounds that restore mitochondrial plasticity. *Clin. Pharmacol. Ther.* **2014**, *96*, 672–683. [CrossRef]
90. Tse, G.; Yan, B.P.; Chan, Y.W.; Tian, X.Y.; Huang, Y. Reactive oxygen species, endoplasmic reticulum stress and mitochondrial dysfunction: The link with cardiac arrhythmogenesis. *Front. Physiol.* **2016**, *7*, 313. [CrossRef]
91. Mailloux, R.J. Application of mitochondria-targeted pharmaceuticals for the treatment of heart disease. *Curr. Pharm. Des.* **2016**, *22*, 4763–4779. [CrossRef]
92. Ni, R.; Cao, T.; Xiong, S.; Ma, J.; Fan, G.-C.; Lacefield, J.C.; Lu, Y.; Le Tissier, S.; Peng, T. Therapeutic inhibition of mitochondrial reactive oxygen species with mito-TEMPO reduces diabetic cardiomyopathy. *Free Radic. Biol. Med.* **2016**, *90*, 12–23. [CrossRef]
93. Silva, F.S.; Simoes, R.F.; Couto, R.; Oliveira, P. Targeting mitochondria in cardiovascular diseases. *Curr. Pharm. Des.* **2016**, *22*, 5698–5717. [CrossRef] [PubMed]
94. Kadoya, T.; Sakakibara, A.; Kitayama, K.; Yamada, Y.; Higuchi, S.; Kawakita, R.; Kawasaki, Y.; Fujino, M.; Murakami, Y.; Shimura, M.; et al. Successful treatment of infantile-onset ACAD9-related cardiomyopathy with a combination of sodium pyruvate, beta-blocker, and coenzyme Q10. *J. Pediatr. Endocrinol. Metab.* **2019**, *32*, 1181–1185. [CrossRef] [PubMed]
95. Değerliyurt, A.; Gülleroğlu, N.B.; Kibar Gül, A.E. Primary $CoQ_{10}$ deficiency with a severe phenotype due to the c.901 C > T (p.R301W) mutation in the COQ8A gene. *Int. J. Neurosci.* **2022**, *134*, 148–152. [CrossRef] [PubMed]
96. Saifudeen, I.; Subhadra, L.; Konnottil, R.; Nair, R.R. Metabolic modulation by medium-chain triglycerides reduces oxidative stress and ameliorates CD36-mediated cardiac remodeling in spontaneously hypertensive rat in the initial and established stages of hypertrophy. *J. Card. Fail.* **2017**, *23*, 240–251. [CrossRef] [PubMed]
97. Abozguia, K.; Elliott, P.; McKenna, W.; Phan, T.T.; Nallur-Shivu, G.; Ahmed, I.; Maher, A.R.; Kaur, K.; Taylor, J.; Henning, A.; et al. Metabolic modulator perhexiline corrects energy deficiency and improves exercise capacity in symptomatic hypertrophic cardiomyopathy. *Circulation* **2010**, *122*, 1562–1569. [CrossRef] [PubMed]
98. Gehmlich, K.; Dodd, M.S.; Allwood, J.W.; Kelly, M.; Bellahcene, M.; Lad, H.V.; Stockenhuber, A.; Hooper, C.; Ashrafian, H.; Redwood, C.S.; et al. Changes in the cardiac metabolome caused by perhexiline treatment in a mouse model of hypertrophic cardiomyopathy. *Mol. Biosyst.* **2015**, *11*, 564–573. [CrossRef] [PubMed]
99. Coats, C.J.; Pavlou, M.; Watkinson, O.T.; Protonotarios, A.; Moss, L.; Hyland, R.; Rantell, K.; Pantazis, A.A.; Tome, M.; McKenna, W.J.; et al. Effect of trimetazidine dihydrochloride therapy on exercise capacity in patients with nonobstructive hypertrophic cardiomyopathy: A randomized clinical trial. *JAMA Cardiol.* **2019**, *4*, 230–235. [CrossRef] [PubMed]

100. Selvaraj, S.; Kelly, D.P.; Margulies, K.B. Implications of altered ketone metabolism and therapeutic ketosis in heart failure. *Circulation* **2020**, *141*, 1800–1812. [CrossRef] [PubMed]
101. Yurista, S.R.; Chong, C.R.; Badimon, J.J.; Kelly, D.P.; de Boer, R.A.; Westenbrink, B.D. Therapeutic potential of ketone bodies for patients with cardiovascular disease: JACC state-of-the-art review. *J. Am. Coll. Cardiol.* **2021**, *77*, 1660–1669. [CrossRef]
102. Yurista, S.R.; Matsuura, T.R.; Sillјé, H.H.W.; Nijholt, K.T.; McDaid, K.S.; Shewale, S.V.; Leone, T.C.; Newman, J.C.; Verdin, E.; van Veldhuisen, D.J.; et al. Ketone ester treatment improves cardiac function and reduces pathologic remodeling in preclinical models of heart failure. *Circ. Heart Fail.* **2021**, *14*, e007684. [CrossRef]
103. Belge, C.; Hammond, J.; Dubois-Deruy, E.; Manoury, B.; Hamelet, J.; Beauloye, C.; Markl, A.; Pouleur, A.-C.; Bertrand, L.; Esfahani, H.; et al. Enhanced expression of β3-adrenoceptors in cardiac myocytes attenuates neurohormone-induced hypertrophic remodeling through nitric oxide synthase. *Circulation* **2014**, *129*, 451–462. [CrossRef] [PubMed]
104. Zhu, L.; Wang, J.; Wang, Y.; Sun, K.; Wang, H.; Zou, Y.; Tian, T.; Liu, Y.; Zou, J.; Hui, R.; et al. Plasma uric acid as a prognostic marker in patients with hypertrophic cardiomyopathy. *Can. J. Cardiol.* **2015**, *31*, 1252–1258. [CrossRef] [PubMed]
105. Namai-Takahashi, A.; Sakuyama, A.; Nakamura, T.; Miura, T.; Takahashi, J.; Kurosawa, R.; Kohzuki, M.; Ito, O. Xanthine Oxidase Inhibitor, Febuxostat Ameliorates the High Salt Intake-Induced Cardiac Hypertrophy and Fibrosis in Dahl Salt-Sensitive Rats. *Am. J. Hypertens.* **2019**, *32*, 26–33. [CrossRef] [PubMed]
106. Bakkehaug, J.P.; Kildal, A.B.; Engstad, E.T.; Boardman, N.; Næsheim, T.; Rønning, L.; Aasum, E.; Larsen, T.S.; Myrmel, T.; How, O.-J.; et al. Myosin activator omecamtiv mecarbil increases myocardial oxygen consumption and impairs cardiac efficiency mediated by resting myosin ATPase activity. *Circ. Heart Fail.* **2015**, *8*, 766–775. [CrossRef] [PubMed]
107. Green, E.M.; Wakimoto, H.; Anderson, R.L.; Evanchik, M.J.; Gorham, J.M.; Harrison, B.C.; Henze, M.; Kawas, R.; Oslob, J.D.; Rodriguez, H.M.; et al. A small-molecule inhibitor of sarcomere contractility suppresses hypertrophic cardiomyopathy in mice. *Science* **2016**, *351*, 617–621. [CrossRef]
108. Kawas, R.F.; Anderson, R.L.; Ingle, S.R.B.; Song, Y.; Sran, A.S.; Rodriguez, H.M. A small-molecule modulator of cardiac myosin acts on multiple stages of the myosin chemomechanical cycle. *J. Biol. Chem.* **2017**, *292*, 16571–16577. [CrossRef]
109. Heitner, S.B.; Jacoby, D.; Lester, S.J.; Owens, A.; Wang, A.; Zhang, D.; Lambing, J.; Lee, J.; Semigran, M.; Sehnert, A.J. Mavacamten treatment for obstructive hypertrophic cardiomyopathy: A Clinical Trial. *Ann. Intern. Med.* **2019**, *170*, 741–748. [CrossRef] [PubMed]
110. Ho, C.Y.; Lakdawala, N.K.; Cirino, A.L.; Lipshultz, S.E.; Sparks, E.; Abbasi, S.A.; Kwong, R.Y.; Antman, E.M.; Semsarian, C.; González, A.; et al. Diltiazem treatment for pre-clinical hypertrophic cardiomyopathy sarcomere mutation carriers: A pilot randomized trial to modify disease expression. *JACC Heart Fail.* **2015**, *3*, 180–188. [CrossRef]
111. Axelsson Raja, A.; Shi, L.; Day, S.M.; Russell, M.; Zahka, K.; Lever, H.; Colan, S.D.; Margossian, R.; Hall, E.K.; Becker, J.; et al. Baseline characteristics of the VANISH cohort. *Circ. Heart Fail.* **2019**, *12*, e006231. [CrossRef]
112. Vissing, C.R.; Axelsson Raja, A.; Day, S.M.; Russell, M.W.; Zahka, K.; Lever, H.M.; Pereira, A.C.; Colan, S.D.; Margossian, R.; Murphy, A.M.; et al. Cardiac remodeling in subclinical hypertrophic cardiomyopathy: The VANISH randomized clinical trial. *JAMA Cardiol.* **2023**, *8*, 1083–1088. [CrossRef]
113. Zhuang, L.; Jia, K.; Chen, C.; Li, Z.; Zhao, J.; Hu, J.; Zhang, H.; Fan, Q.; Huang, C.; Xie, H.; et al. DYRK1B-STAT3 Drives Cardiac Hypertrophy and Heart Failure by Impairing Mitochondrial Bioenergetics. *Circulation* **2022**, *145*, 829–846. [CrossRef] [PubMed]
114. Solomon, T.; Filipovska, A.; Hool, L.; Viola, H. Preventative therapeutic approaches for hypertrophic cardiomyopathy. *J. Physiol.* **2021**, *599*, 3495–3512. [CrossRef] [PubMed]
115. Viola, H.M.; Shah, A.A.; Johnstone, V.P.A.; Szappanos, H.C.; Hodson, M.P.; Hool, L.C. Characterization and validation of a preventative therapy for hypertrophic cardiomyopathy in a murine model of the disease. *Proc. Natl. Acad. Sci. USA* **2020**, *117*, 23113–23124. [CrossRef]
116. Wilder, T.; Ryba, D.M.; Wieczorek, D.F.; Wolska, B.M.; Solaro, R.J. N-acetylcysteine reverses diastolic dysfunction and hypertrophy in familial hypertrophic cardiomyopathy. *Am. J. Physiol. Heart Circ. Physiol.* **2015**, *309*, H1720–H1730. [CrossRef] [PubMed]
117. Marian, A.J.; Tan, Y.; Li, L.; Chang, J.; Syrris, P.; Hessabi, M.; Rahbar, M.H.; Willerson, J.T.; Cheong, B.Y.; Liu, C.-Y.; et al. Hypertrophy Regression with N-Acetylcysteine in Hypertrophic Cardiomyopathy (HALT-HCM): A Randomized, Placebo-Controlled, Double-Blind Pilot Study. *Circ. Res.* **2018**, *122*, 1109–1118. [CrossRef]
118. Horowitz, J.D.; Chirkov, Y.Y. Perhexiline and hypertrophic cardiomyopathy: A new horizon for metabolic modulation. *Circulation* **2010**, *122*, 1547–1549. [CrossRef] [PubMed]
119. Jeremic, J.; Govoruskina, N.; Bradic, J.; Milosavljevic, I.; Srejovic, I.; Zivkovic, V.; Jeremic, N.; Turnic, T.N.; Tanaskovic, I.; Bolevich, S.; et al. Sacubitril/valsartan reverses cardiac structure and function in experimental model of hypertension-induced hypertrophic cardiomyopathy. *Mol. Cell. Biochem.* **2023**, *478*, 2645–2656. [CrossRef]

International Journal of
*Molecular Sciences*

MDPI

*Review*

# Therapies for Cirrhotic Cardiomyopathy: Current Perspectives and Future Possibilities

**Hongqun Liu [1], Daegon Ryu [1,2,†], Sangyoun Hwang [1,3,†] and Samuel S. Lee [1,*]**

1. Liver Unit, Cumming School of Medicine, University of Calgary, Calgary, AB T2N 4N1, Canada; hliu@ucalgary.ca (H.L.); gon22gon@naver.com (D.R.); mongmani@daum.net (S.H.)
2. Division of Gastroenterology, Yangsan Hospital, Pusan National University School of Medicine, Pusan 46033, Republic of Korea
3. Department of Internal Medicine, Dongnam Institute of Radiological and Medical Sciences, Pusan 46033, Republic of Korea

* Correspondence: samlee@ucalgary.ca

† Dr. Ryu and Dr. Hwang were recipients of sabbatical leaves from their institutions.

**Abstract:** Cirrhotic cardiomyopathy (CCM) is defined as cardiac dysfunction associated with cirrhosis in the absence of pre-existing heart disease. CCM manifests as the enlargement of cardiac chambers, attenuated systolic and diastolic contractile responses to stress stimuli, and repolarization changes. CCM significantly contributes to mortality and morbidity in patients who undergo liver transplantation and contributes to the pathogenesis of hepatorenal syndrome/acute kidney injury. There is currently no specific treatment. The traditional management for non-cirrhotic cardiomyopathies, such as vasodilators or diuretics, is not applicable because an important feature of cirrhosis is decreased systemic vascular resistance; therefore, vasodilators further worsen the peripheral vasodilatation and hypotension. Long-term diuretic use may cause electrolyte imbalances and potentially renal injury. The heart of the cirrhotic patient is insensitive to cardiac glycosides. Therefore, these types of medications are not useful in patients with CCM. Exploring the therapeutic strategies of CCM is of the utmost importance. The present review summarizes the possible treatment of CCM. We detail the current status of non-selective beta-blockers (NSBBs) in the management of cirrhotic patients and discuss the controversies surrounding NSBBs in clinical practice. Other possible therapeutic agents include drugs with antioxidant, anti-inflammatory, and anti-apoptotic functions; such effects may have potential clinical application. These drugs currently are mainly based on animal studies and include statins, taurine, spermidine, galectin inhibitors, albumin, and direct antioxidants. We conclude with speculations on the future research directions in CCM treatment.

**Keywords:** cirrhotic cardiomyopathy; treatments; beta blockers; antioxidants; anti-apoptosis; anti-inflammation

## 1. Introduction

Cirrhotic cardiomyopathy (CCM) is one of the most important complications in patients with cirrhosis. The definition includes systolic and/or diastolic dysfunction and morphological changes, such as chamber enlargement, without pre-existing heart disease. CCM was first termed in 1989 [1]. Since then, this entity has been investigated by many studies, which led to the first definition in 2005 at the World Congress of Gastroenterology in Montreal [2] called the WCG criteria. Following advances in imaging technology, the diagnostic criteria were redefined by a multidisciplinary expert group, resulting in the Cirrhotic Cardiomyopathy Consortium criteria (CCC criteria) [2]. The WCG criteria emphasize the blunted cardiac response to exercise, volume challenge, or pharmacological stimuli, whereas the CCC criteria are based on contractile function at rest (Table 1). Nevertheless, the newer CCC criteria appear superior [3,4] and should be used going forward.

Table 1. Diagnostic criteria systems for cirrhotic cardiomyopathy.

| Criteria | Systolic Dysfunction | Diastolic Dysfunction |
|---|---|---|
| WCG criteria (2005) | LVEF < 55%<br>Or<br>Blunted increase in contractility on stress testing | E/A ratio < 1.0<br>Or<br>DT > 200 ms<br>Or<br>IVRT > 80 ms |
| CCC criteria (2019) | LVEF $\leq$ 50%<br>Or<br>GLS < 18% | $\geq$3 of the followings<br>E/e′ ratio $\geq$ 15<br>e′ septal < 7 cm/s<br>TR velocity > 2.8 m/s<br>LAVI > 34 mL/m$^2$ |

WCG: World Congress of Gastroenterology; CC: cirrhotic cardiomyopathy consortium; LVEF: left ventricular ejection fraction; E/A: E-wave to A-wave ratio; DT: mitral deceleration time; IVRT: isovolumetric relaxation time; GLS: global longitudinal strain, absolute value; TR: tricuspid regurgitation; LAVI: left atrial volume index.

Although CCM has been investigated extensively, the management is still not standardized because there is no clearly-accepted specific treatment. Due to the significant baseline peripheral vasodilatation in cirrhotic patients, vasodilators such as angiotensin-converting enzyme (ACE) inhibitors are unfeasible to treat CCM, as further vasodilation may lower the mean arterial pressure below the cutoff value (approximately 65 mmHg) that induces kidney injury. Patients with cirrhosis are not sensitive to cardiac glycosides; thus, these drugs cannot be used.

However, there are general supportive measures, current potentially useful therapies, and the future possibility of specific treatments for CCM, which will be summarized in the present review.

## 2. Clinical Relevance

CCM is clinically significant because when the cirrhotic heart is challenged, the subclinical dysfunction becomes overt. These challenges include exercise, transjugular intrahepatic portosystemic shunt (TIPS) insertion, drugs, and liver transplantation [5]. Regarding the last, during the transplantation procedure, intravenous fluids augment cardiac preload, and postoperatively, systemic vascular resistance is raised, which increases cardiac afterload. All these challenges significantly aggravate any pre-existing CCM. It was demonstrated that cardiovascular complications are the third-leading cause of death in patients after liver transplantation, accounting for 7–21% of deaths [6]. Even without any challenges, CCM plays an essential role in mortality. Premkumar and coworkers demonstrated that the mortality rates within 2 years of cirrhotic patients were parallel to the grades of left ventricular diastolic dysfunction (LVDD grade 1, 10.5% mortality; grade 2, 22.5%; and grade 3, 40%) [7]. Furthermore, LVDD also correlates with the incidence of acute kidney injury (OR 6.273, $p < 0.05$) and hepatic encephalopathy (OR 5.6, $p < 0.05$).

## 3. Pathogenic Mechanisms

Cirrhosis is defined as hepatic architectural damage characterized by nodular regeneration and diffuse fibrosis; these features lead to liver dysfunction and portal hypertension. Mechanisms underlying CCM and portal hypertension have been recently reviewed in detail [2]. Liver dysfunction impacts cardiac molecules, for example, decreased density of β-adrenergic receptors [8], an increased cholesterol-to-phospholipid ratio of the cardiomyocyte sarcolemmal plasma membrane [9], and abnormal contractile filaments, such as a myosin heavy chain (MHC) shift from the stronger α-MHC to the weaker and slower β-MHC isoform [10]. Portal hypertension causes intestinal vascular congestion, which results in bacterial translocation and endotoxemia. Under stimulation by lipopolysaccharides, pro-inflammatory cytokines, such as TNFα and interleukin (IL)-1β, are increased. These cytokines further augment nitric oxide and carbon monoxide, which inhibit cardiac contractility. Other cardiac contractile inhibitors include oxidative stress, apoptosis, and bile acids.

## 4. Management

There are currently no guidelines on the treatment of CCM. The general management of overt noncirrhotic heart failure usually requires oxygen and afterload and preload reduction [11]. Preload reduction includes water and sodium restriction and diuretics. Unfortunately, long-term diuretic application may cause electrolyte imbalances and renal injury [12]. Afterload reduction mainly consists of vasodilation. However, vasodilators are usually not suitable for treating heart dysfunction in cirrhosis because these patients often have significant vasodilatation and hypotension. Thus, there is a real risk that vasodilators may worsen a cirrhotic patient's clinical condition [5]. Therefore, ACE inhibitors or angiotensin receptor blockers are not feasible in patients with advanced cirrhosis. The potential therapies for CCM may include nonselective beta-blockers (NSBBs. Table 2), antioxidants, and anti-apoptotic and anti-inflammatory agents.

**Table 2.** Effects of beta-blockers.

| First Author (Ref.) | NSBB | Subjects | Effects |
|---|---|---|---|
| Poynard [13] | Propranolol, nadolol | Patients | Decreases bleeding, improves survival |
| Sersté [14] | Propranolol | Patients with refractory ascites | Decreases 1-year survival rate |
| Silvestre [15] | metoprolol | Patients with CCM | No change in stroke volume or diastolic function |
| Leithead [16] | Propranolol, carvedilol | Patients with refractory ascites | Improves survival |
| Mookerjee [17] | Propranolol | Patients with ACLF | Improves inflammation and survival |
| Premkumar [18] | carvedilol + ivabradine | Patients with CCM | Improves LVDD and survival |
| Zambruni [19] | Nadolol | Patients with cirrhosis | Decreases QTc in patients with prolonged QTc over 1–3 months |
| Henrikson [20] | Propranolol | Patients with cirrhosis | Decreases QTc over 90 min |

NSBB: non-specific beta-blocker. ACLF: acute-on-chronic liver failure. CCM: cirrhotic cardiomyopathy. LVDD: left ventricular diastolic dysfunction. QTc: corrected QT interval.

### *4.1. Non-Selective Beta-Blockers*

NSBBs have a long history in the therapy of cirrhotic patients. In 1980, Lebrec and colleagues [21] conducted a randomized clinical trial with propranolol, which demonstrated that it significantly decreased the portal pressure in patients with cirrhosis and portal hypertension, at doses reducing the heart rate by 25%. They speculated that propranolol might be valuable in preventing recurrent bleeding caused by esophageal varices based on the portal-hypotensive effect. Many subsequent studies have confirmed this initial speculation. NSBBs have thus been used in cirrhotic patients with portal hypertension for more than 40 years [21,22]. To the present date, NSBBs are still a standard of care to prevent variceal bleeding and rebleeding [23].

#### 4.1.1. Issues of NSBBs in Portal Hypertension

Since Lebrec and colleagues first explored the application of NSBBs in cirrhotic patients with portal hypertension [21], many studies have been conducted. Poynard et al. [13], in 1991, analyzed four randomized controlled trials. They concluded that patients with NSBBs not only had fewer first episodes of bleeding but also had improved survival rates. Almost 30 years later, Serste and colleagues, i.e., the Lebrec group [14], conducted a prospective study in 151 cirrhotic patients with refractory ascites. The patients were divided into two groups: one group received propranolol (n = 77), and the other group did not. They reported that the survival time in the propranolol group was shorter than that in the control group. Furthermore, the 1-year probability of survival was significantly lower in the propranolol group compared with controls. They concluded that NSBBs should not be used in cirrhotic patients with refractory ascites. However, a major problem of this study was that it was not randomized; patients were selected by their physician to receive NSBB or not, and thus, there was an inescapable likelihood of selection bias. Comparing the patients' basic characteristics, in the propranolol group, the presence of esophageal varices and total bilirubin levels were significantly higher than in the controls, and the systolic blood pressure was significantly lower. These important differences of these baseline parameters strongly suggest that the patients in the propranolol group had more severe liver failure.

After further studies on the safety of NSBBs in advanced cirrhosis, a "window hypothesis" was proposed. The window hypothesis contends that NSBBs are neither useful nor necessary in the early stages of cirrhosis and potentially hazardous in the later stages, such as those patients with refractory ascites [24]. The sympathetic nervous system activity is nearly normal in the early stages of cirrhosis, and therefore, NSBBs will exert only modest effects at this stage; at the later end stage, although the sympathetic system is highly active, NSBBs at this stage not only inhibit the sympathetic system but also decrease the cardiac contractility and arterial pressure [24], which may result in tissue hypoperfusion and death. NSBBs may therefore only be clinically beneficial within a narrow 'window period' of the clinical course of cirrhosis.

However, recent studies do not seem to support the "window hypothesis"; Chen et al. [25] examined the National Health Insurance Research Database of Taiwan. Patients with cirrhosis taking propranolol vs. those not on this drug (controls) were matched for gender and age. The mean survival of cirrhotic patients with refractory ascites was $34.3 \pm 31.2$ months in the propranolol group and $20.8 \pm 26.6$ months in the control group ($p < 0.001$). They concluded that compared with controls, propranolol treatment reduces mortality. Leithead and coworkers [16] also demonstrated that even with refractory ascites, NSBB treatment confers benefits to cirrhotic patients with end-stage liver disease on the waiting list for liver transplantation.

Further evidence that demonstrated the beneficial effect of NSBBs is a study of patients with acute-on-chronic liver failure (ACLF). Mookerjee et al. [17] examined the effect of NSBBs on systemic inflammation in patients with ACLF and found that NSBBs significantly reduced white cell count and the concentration of plasma C-reactive protein (CRP). The severity of inflammatory reaction was an independent predictor for the development of ACLF after enrollment and for ACLF-associated mortality. NSBB treatment downregulated the grades of ACLF and decreased mortality rates. In contrast, patients without NSBB treatment tended to show a worsening of ACLF during their hospital stay. Moreover, patients who discontinued NSBB treatment had significantly higher 28-day and 3-month mortality rates [17].

An important caveat to emphasize is that all the above studies and indeed all previous comparative studies of NSBBs and mortality in advanced cirrhosis are nonrandomized and thus inevitably suffer from probable selection bias. This limitation decreases the strength of any conclusions that can be drawn from these studies. Nevertheless, at present, the tentative conclusion based on the most recent evidence suggests that in cirrhotic patients, NSBBs should be stopped only if and when the mean arterial pressure drops below 65 mmHg, as that is the approximate cut-off value at which renal hypoperfusion occurs [26].

#### 4.1.2. NSBBs for CCM Treatment

There are two pathways by which NSBBs theoretically could exert a therapeutic effect on CCM. One pathway is by blocking direct cardiac damage due to an overactivated sympathetic nervous system. The other is the beneficial effect of the decrease in portal hypertension.

The pathway of direct cardiac damage of the overactivated sympathetic nervous system is unrelated to portal hypertension. In patients with cirrhosis, the sympathetic nervous system is overactivated, manifesting as persistent adrenergic activation and high circulating levels of catecholamines [27]. The heart is one of the target organs that can be damaged by high levels of circulating catecholamines. An animal study demonstrated that an increase in portal and hepatic sinusoidal pressure leads to the activation of sympathetic nerves to the heart [28]. It is well known that sympathetic overactivation plays an important role in a variety of pathophysiological processes in cardiovascular diseases [29]. Cao et al. [30] furthermore specified that it is the β-adrenergic receptor (β-AR) overactivation that is a major pathological factor mediating cardiac inflammatory injury and causing cardiac dysfunction. Cardiac inflammatory injury is a key mechanism underlying the development of cardiac diseases [31].

Regarding the portal hypertension-related pathway, in cirrhotic patients, the portal venous hypertension causes mesenteric congestion. The congested gut impairs bowel motility and consequently leads to increased intestinal permeability and bacterial overgrowth [32,33]. The bacterial overgrowth stimulates the production of endotoxin, and the increased intestinal permeability augments the absorption of endotoxin. Moreover the dysfunctional cirrhotic liver has reduced detoxication capability, and the presence of portosystemic collateral circulation enables endotoxin to directly enter the systemic circulation. All the above changes in patients with cirrhosis cause endotoxemia and systemic inflammation, a phenomenon termed the 'inflammatory phenotype'. In summary, an inflammatory phenotype seems to underlie disease severity in many cirrhosis-related complications, including CCM.

Portal hypertension-associated inflammation is an essential pathogenic event in CCM. NSBBs are demonstrated to decrease portal pressure; thus, this class of drugs theoretically could have a therapeutic effect on CCM. Another mechanism by which NSBBs could decrease endotoxemia is by increasing bowel motility and reducing intestinal permeability, thus decreasing bacterial translocation [34]. NSBBs have anti-inflammatory effects, which may be beneficial in CCM because this condition displays an inflammatory phenotype: inflammatory cytokines, such as TNFα and IL-1β, are increased in the cirrhotic heart. Furthermore, NSBBs improve both systolic and diastolic function in patients with non-cirrhotic chronic heart failure [35]. However, there is no solid evidence to date to demonstrate that NSBBs have clear therapeutic effects on CCM.

Because of the observations above, many centers have investigated the possible therapeutic effect of NSBBs on CCM. Although current theories suggest that NSBBs may exert therapeutic effects, the pertinent studies have not confirmed this.

Although theoretically it is rational to use NSBBs to treat CCM, there are some difficulties in the clinical application. First, although cardiac function at resting status is normal, i.e., left ventricular ejection fraction (LVEF) is preserved due to the vasodilatation, the contractile responsiveness is decreased, such as decreased global longitudinal train (GLS, <18%). Furthermore, diastolic function is also abnormal, manifested as a reduced relaxation velocity of ventricular muscle (diastolic mitral annular velocity for example). Unfortunately, NSBBs possess not only anti-inflammatory effects but also inhibit the contractility-stimulating $\beta_1$-AR, therefore potentially further inhibiting cardiac systolic and diastolic function.

In a recent randomized controlled trial, Premkumar et al. [18] enrolled 189 cirrhotic patients divided into 3 groups: carvedilol (an alpha- and beta-blocker) alone, carvedilol + ivabradine (a cardiac pacemaker current [$I_f$] blocker), and standard medical therapy (SMT) for 52 weeks. The targeted heart rate reduction (THR) was defined as heart rate reduction to 55–65 beats per minute. They observed that patients treated with carvedilol + ivabradine showed an improvement of LVDD and improved survival compared with the SMT group. Even the patients treated with carvedilol alone showed modest improvements in cardiac and clinical parameters. In patients who obtained THR with carvedilol treatment, the E/e′ was insignificantly decreased by 0.6%. In comparison, there was a 14.2% increase in E/e′ in the SMT group (0.6% vs. 14.2%, $p = 0.003$). These data confirmed a therapeutic effect of carvedilol on diastolic dysfunction. One issue to mention is that this study did not specifically report the therapeutic effect of carvedilol on cirrhotic patients with refractory ascites and Child–Pugh class C. The most promising results with the combination carvedilol + ivabradine therapy are encouraging and warrant further study in larger trials.

Silvestre and colleagues [15] performed a randomized, double-blinded, placebo-controlled trial to evaluate the effect of 6 months of metoprolol on CCM, randomizing 41 patients to the metoprolol group and 37 to a placebo group. Thirty-eight patients in the metoprolol group and thirty-five in the placebo group finished the study. The study did not show any significant differences in the improvement of stroke volume or diastolic dysfunction. Indeed, no echocardiography parameter or morphology was significantly different between the metoprolol and placebo groups. Furthermore, metoprolol treatment did not change the levels of noradrenaline, plasma renin activity, and troponin compared with the placebo group. Clinical events, such as hospitalization and mortality rates, were not different significantly between the two groups. Therefore, the authors concluded that six months of metoprolol treatment does not improve cardiac function and morphology in patients with CCM. However, randomization may have produced selection bias by chance: 19.5% of patients in the metoprolol group were Child–Pugh class C, whereas this percentage in the placebo group was only 8.1%. Thus, the different severity of cirrhosis in the two groups may have contributed to a type II error.

Another prospective study consecutively enrolled 403 patients, 213 with compensated cirrhosis and 190 with decompensated cirrhosis [36]. This study reported that NSBBs were more effective on the heart and less effective on portal pressure in patients with decompensated cirrhosis than in those with compensated cirrhosis. At baseline, decompensated patients were more hyperdynamic than compensated patients, with higher heart rate and cardiac output (CO), lower arterial pressure, and higher portal pressure. NSBBs had greater reductions in heart rate (15 $\pm$ 12 vs. 10 $\pm$ 11, $p < 0.05$) and CO (17 $\pm$ 15% vs. 10 $\pm$ 21%; $p < 0.01$) in decompensated patients. However, NSBBs induced less portal pressure decrease in decompensated patients than in compensated patients (10 $\pm$ 18% vs. 15 $\pm$ 12%; $p < 0.05$). Furthermore, the CO decrease was an independent predictor of mortality in decompensated patients: compared with survivors, NSBBs produced a greater decrease in CO in decompensated patients who died (21 $\pm$ 14% vs. 15 $\pm$ 16%; $p < 0.05$). Death risk was higher in decompensated patients with CO < 5 L/min than in those with CO > 5 L/min. Based on the data above, these authors concluded that NSBBs may be detrimental in patients with end-stage cirrhosis and latent cardiomyopathy because NSBBs further reduce the cardiac compensatory reserve.

A potential benefit is that NSBBs shorten the prolonged QTc interval and decrease the risk of ventricular arrhythmias [37,38]. There is no controversy regarding this effect.

Why are the NSBB study results so discrepant? Does a therapeutic window also exist in CCM treatment with NSBBs? The explanations may be due to patient and NSBB selection. Some patients might have disease progression during the time course of treatment, such as those with ACLF. Moreover, several other variables, such as the patients' nutritional status, and differences in other standard medical therapies may also play a role.

In terms of NSBB selection, several different drugs have been used in studies. These include propranolol, nadolol, and carvedilol, all of which are true NSBBs, exerting effects on both $\beta_1$ and $\beta_2$ receptors. In addition, carvedilol is also an $\alpha_1$-adrenergic blocker. On the other hand, metoprolol, also studied in CCM, is a selective $\beta_1$-receptor blocker. All these differences may contribute to the observed discrepant therapeutic effects in patients with CCM.

*4.2. Potential Therapies in CCM (Table 3)*

There is currently no accepted specific treatment for CCM. As detailed above, NSBB therapy is controversial. Other potential strategies could be suggested by the pathogenesis of CCM, such as antioxidants and anti-inflammatory and anti-apoptotic substances (Figure 1). The study by Taprantzia et al. [39] reported that compared with healthy controls, oxidative indicators, such as lipid peroxidation and malondialdehyde (MDA) levels, were significantly increased in cirrhotic patients, thus showing that oxidative stress is significantly augmented in cirrhosis [40]. Our previous studies demonstrated that cardiac inflammation, oxidative stress, and apoptosis play a significant pathogenic role in

CCM [41–43]. Agents active against oxidative stress, inflammation, and apoptosis may therefore have potential in clinical application to treat CCM.

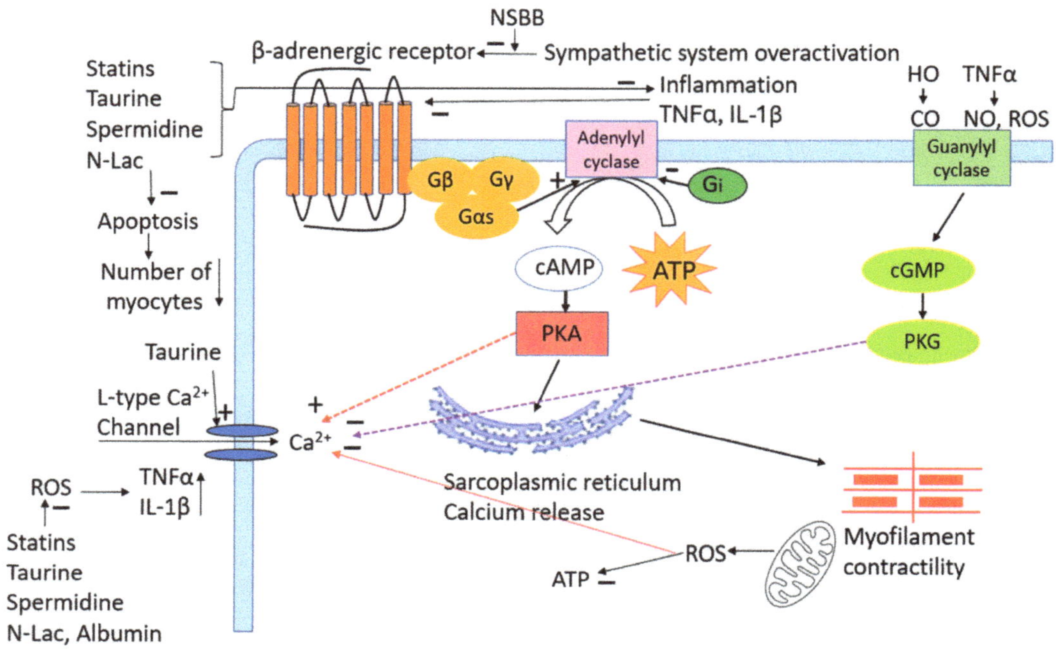


**Figure 1.** Schematic diagram of pathogenic mechanisms of CCM and therapeutic targets. + denotes positive or stimulatory effect. —denotes negative or inhibitory effect. NSBB: non-specific beta-blocker. TNFα: tumor necrosis factor alpha. IL-1β: interleukin 1beta. HO: hemo oxygenase. CO: carbon monoxide. NO: nitric oxide. ROS: reactive oxygen species. PKA: protein kinase A. PKG: protein kinase G. N-Lac: N-acetyllactosamine.

**Table 3.** Potential therapies in cirrhotic cardiomyopathy.

| First Author (Ref.) | Substance | Mechanism of Action | Species/Model | Effects |
|---|---|---|---|---|
| Bortoluzzi [44] | Albumin | Decreases inflammatory and oxidative stress | $CCl_4$-cirrhotic rats | Enhances systolic function |
| Fernandez [45] | Albumin | Reduces systemic inflammation | Patients with decompensated cirrhosis | Improves cardiac function |
| Mousavi [46] | Taurine | Reduces oxidative stress, protein carbonylation, improves mitochondrial function, and increases ATP levels | Bile duct-ligated cirrhotic rats | protects liver and heart from injury |
| Sheibani [47] | Spermidine | Decreases inflammatory and oxidative stress | Bile duct-ligated cirrhotic rats | Enhances systolic function, decreases QTc |
| Yoon [48] | Galectin-3 inhibitor (*N*-acetyllactosamine) | Decreases inflammation by inhibiting TNFα | Bile duct-ligated cirrhotic rats | Increases blood pressure; enhanced systolic and diastolic function |
| Niaz [49] | Statin (atorvastatin) | Decreases inflammation and oxidative stress | Bile duct-ligated cirrhotic rats | Increases chronotropic response to isoproterenol; decreases QTc interval. |
| Node [50] | Statin (simvastatin) | Attenuates systemic inflammation | Patients with dilated cardiomyopathy | Improves LVEF, NYHA classification |

$CCl_4$: carbon tetrachloride; QTc: corrected QT interval; TNFα: tumor necrosis factor alpha; LVEF: left ventricular ejection fraction; NYHA: New York Heart Association functional classification.

#### 4.2.1. Statins

Statins not only inhibit cholesterol synthesis and downregulate the serum cholesterol level but also possess antioxidant and anti-inflammatory effects. Bielecka-Dabrowa et al. [51] investigated the role of atorvastatin on the parameters of inflammation and left ventricular function in patients with dilated cardiomyopathy (DCM). They showed that atorvastatin significantly reduced the inflammatory cytokines in plasma, such as TNFα and IL-6. It also decreased the cardiac dysfunction marker N-terminal pro-brain natriuretic peptide (BNP) concentration. Atorvastatin significantly improved cardiac function as manifested by decreased left ventricular diastolic and systolic diameters. Furthermore, it significantly increased LVEF. Finally, this drug also significantly increased the probability of 5-year survival.

Niaz et al. [49] tested the effects of atorvastatin on cirrhotic hearts induced by bile duct ligation (BDL) in rats. They reported that the chronotropic responses of atria from BDL rats to isoproterenol were decreased compared with those from sham-operated controls. The response was increased in BDL rats treated with atorvastatin. Furthermore, the QTc interval and serum BNP, TNFα, and MDA levels were increased in BDL rats, and atorvastatin significantly decreased these parameters. In summary, atorvastatin improved the chronotropic hyporesponsiveness and downregulated the oxidative stress and inflammation in cirrhotic rats. From the evidence above, both in humans and animal models, statins seem to exert a therapeutic effect on cardiac function, mediated via the inhibition of inflammation, apoptosis [52], and oxidative stress. Therefore, statins are potentially useful therapeutic agents that need further study. Moreover, given its already proven excellent safety profile, it can safely be used in almost all patients with cirrhosis except perhaps those with severely decompensated liver function.

#### 4.2.2. Taurine

Taurine possesses multiple functions, including the modulation of protein phosphorylation, calcium ion regulation, membrane stabilization, bile acid conjugation, lipid metabolism, glucose regulation, antioxidation, anti-inflammation, and anti-apoptosis [53,54]. It is an abundant amino-sulfonic acid in many tissues, such as skeletal muscle, liver, platelets, and leukocytes, especially in electrically excitable tissues, such as the heart [54]. Low taurine serum levels have been closely associated with many oxidative stress-mediated pathologies, including hepatic disorders and cardiomyopathy [55]. It plays a significant role in reducing lipid peroxidation products [54], therefore protecting cells from tissue damage [56]. It also exerts a protective effect on oxidative stress-induced vascular dysfunction [57], which may also apply to the heart [55].

Pion et al. [58], in a feline model, showed that taurine depletion leads to cardiomyopathy. Another study [59], using a taurine transporter-knockout model in mice, showed that these animals naturally develop cardiac dysfunction. Beyranvand and coworkers [60] verified that taurine supplementation increases the exercise capacity in patients with heart failure, and this effect is partially due to the antioxidant role of taurine.

All these data suggest that taurine is essential for cardiovascular function. However, the role of taurine in CCM is not well studied. Since the biosynthesis of taurine is primarily in the liver [54], cirrhosis decreases the functional liver mass and consequently the synthesis of taurine [5].

It is known that the cardiac content of taurine is significantly decreased in the cirrhotic heart [61] and parallel to the decrease in taurine is the decreased antioxidant capacity in these hearts. Thus, the supplementation of taurine may be potentially beneficial. Taurine has been shown to reduce lipid peroxidation and protect cells from damage [56]. Liu and coworkers [62] created a model of transverse aortic constriction-induced heart failure in mice and demonstrated that taurine exerts a protective effect on cardiac function. The mechanisms are due to a decrease in myocyte oxidative stress, apoptosis, hypertrophy, and cardiac fibrosis.

The results obtained from non-cirrhotic heart failure may also be applicable to CCM. In the BDL rat model of cirrhosis, Mousavi et al. [46] showed that oxidative stress, including lipid peroxidation, reactive oxygen species, protein carbonylation, and the GSH/GSSG ratio, were significantly increased in the cirrhotic heart. Taurine administration significantly reduced tissue oxidative stress and increased the total antioxidant capacity and mitochondrial ATP content [46].

In summary, taurine decreased oxidative stress and improved mitochondrial function in the cirrhotic rat heart. Moreover, it also reduced creatine kinase MB (CK-MB), a surrogate marker of heart injury. Taurine is therefore a valuable candidate treatment that warrants further human studies in CCM.

#### 4.2.3. Spermidine

Spermidine, like taurine, also possesses antioxidant, anti-inflammatory, and anti-apoptotic effects [47,63]. Chen et al. [64] used a transverse aortic constriction model in mice to investigate the role of spermidine in heart failure (HF). They divided the animals into four groups: sham controls, HF, HF + spermidine, and HF + spermidine antagonist (trans-4-methylcyclohexylamine (4-MCHA)). They reported that spermidine significantly decreased the left ventricular mass. The most significant changes in echocardiographic parameters were in the HF mice treated with 4-MCHA. This group demonstrated further increases in left ventricular systolic and diastolic diameters, left ventricular end-diastolic, and diastolic volumes and further decreases in LVEF. Moreover, 4-MCHA significantly increased the cardiac content of galectin-3, an inhibitor of cardiac function [48]. Finally, the mice treated with 4-MCHA showed the greatest extent of cardiomyocyte apoptosis. These data demonstrated that spermidine inhibition worsened cardiac function and spermidine improved cardiac function in heart failure.

Omar et al. [65] evaluated the impact of spermidine in a rat model of acute myocardial infarction (AMI) induced by isoproterenol. Compared to the untreated group, spermidine significantly reversed electrocardiographic RR interval, QRS, QT intervals, and ST segments towards normal ranges. Furthermore, serum CK-MB and lactate dehydrogenase, the parameters of cardiac injury, were significantly reduced by spermidine. Furthermore, compared with the untreated AMI group, spermidine significantly rescued the reduced antioxidant capacity [65]. Martinalli et al. [66] administered spermidine to patients with peripheral artery disease and reported that it increased maximal walking distance and reduced oxidative stress in these patients. What is the effect of spermidine on CCM? Sheibani et al. [47] investigated the effects of spermidine in the BDL-induced cirrhotic rat. They showed that it significantly decreased the QTc interval, which is consistent with a study of Omar and colleagues [65].

Furthermore, compared with the control group, spermidine significantly reduced the cardiac oxidative stress and inflammation: decreased levels of malondialdehyde, increased superoxide dismutase and GSH, and decreased TNF$\alpha$ and IL-1$\beta$. Moreover, the contractility of isolated ventricular papillary muscles from the BDL + spermidine group was significantly increased compared with BDL controls. These studies give us hope that spermidine may one day be applicable to cirrhotic patients with cardiovascular dysfunction.

#### 4.2.4. Galectin-3 Inhibitor

Galectins are members of the lectin family. Galectin-3 is one of the 15 mammalian galectins identified to date [67]. Galectin-3 is widely distributed in the nucleus, cytoplasm, cell surface, extracellular space, and the blood circulation [68]. It is closely associated with CCM because (1) galectin-3 levels are significantly increased in cirrhotic patients [69] and animal models of liver fibrosis [70]. Moreover, galectin-3 is increased in the cirrhotic heart [48]. (2) It serves pleiotropic functions, including inflammation [71], oxidative stress, and apoptosis [72], which are pathogenic mechanisms of CCM. Galectin inhibitors therefore are theoretically attractive to investigate for CCM treatment.

We [48] investigated the role of galectin-3 in CCM pathogenesis, using N-acetyllactosamine (N-Lac) as a galectin-3 inhibitor. We divided rats into four groups, sham operated controls, sham + N-Lac, BDL, and BDL + N-Lac. In these animals, the left ventricular content of galectin-3, pro-inflammatory cytokine TNFα, BNP, the collagen I and III ratio, blood pressure, and cardiomyocyte contractility were measured. We demonstrated that galectin-3, TNFα, BNP, and the collagen I and III ratio were significantly increased in the hearts isolated from BDL rats compared with those from sham controls. Blood pressure and systolic and diastolic contractile velocities were significantly decreased in cardiomyocytes isolated from BDL rats. The galectin-3 inhibitor significantly decreased levels of galectin-3, TNFα, BNP, and the collagen I/III ratio in cirrhotic hearts and significantly increased the blood pressure and improved the cardiomyocyte contractile velocities of the BDL rats. N-Lac had no effect on sham controls. The galectin-3 inhibitor decreased the cardiac content of TNFα and improved the depressed contractility in the cirrhotic heart. With the data above, we concluded that the increase in galectin-3 in the cirrhotic heart plays an important role in the inhibition of cardiac contractility. This effect is mediated via TNFα.

#### 4.2.5. Albumin

Albumin is synthesized exclusively by the liver, so its serum levels are reduced in acute and/or chronic liver disease [73]. It may be a candidate for the treatment of CCM for the following reasons: (1) Hypoalbuminemia is common in patients with advanced cirrhosis. Thus, improving hypoalbuminemia should reduce ascites formation by increasing plasma colloid osmotic pressure. (2) Albumin decreases the protein expressions of $G\alpha i_2$, TNFα, and iNOS [44], which are known inhibitors of cardiac contractility. Albumin decreases TNFα via 2 mechanisms, binding serum TNFα and blunting the overexpression of TNFα in cardiac tissue. (3) It decreases oxidative stress [74], which is an important initiator of inflammation. Albumin binds many substances, such as NO, reactive oxygen species (ROS), and proinflammatory cytokines, which may be involved in the pathogenesis of both peripheral arterial vasodilatation and cardiac dysfunction in patients with cirrhosis. (4) Albumin increases adenylate cyclase 3, the enzyme that catalyzes ATP to cAMP [44], and is thus a key mediator of the ventricular-stimulatory pathway.

Bortulozzi and coworkers [44] used carbon tetrachloride ($CCl_4$) to induce cirrhosis and ascites in rats, subsequently infusing intravenous albumin to determine its effects on the cirrhotic heart. They demonstrated that the cardiac expression of TNFα, iNOS, and NAD(P)H-oxidase activity were significantly increased in the cirrhotic heart, and cardiac contractility was significantly decreased in cirrhotic rats compared to controls. Albumin infusion reversed the protein expressions of TNFα, iNOS, and NAD(P)H-oxidase to control levels, and the depressed cardiac contractility also reversed back to normal.

A clinical study also demonstrated the role of albumin in cardiac contractility in patients with cirrhosis. The Pilot-PRECIOSA study [45] demonstrated that patients who received a high albumin dose (1.5 g/kg weekly) showed improvement in systolic function with increases in cardiac index and left ventricular stroke work index.

Because the antioxidant and volume-expanding properties of albumin, regardless of any possible cardioprotective effects, are beneficial in almost all patients with cirrhosis and it lacks any significant downside, we believe this therapeutic agent is highly promising and could be considered at any stage of cirrhosis, not just those with advanced disease.

#### 4.2.6. Direct Antioxidants

Hydrogen is a direct antioxidant. The small size of the hydrogen molecule allows it to easily penetrate the cell membrane to the cytosol. It is naturally metabolized without residue, and therefore, there are no side effects [75]. Similar to taurine and spermidine, hydrogen has antioxidant [76], anti-inflammatory, and anti-apoptotic effects [77,78]. Jing et al. [79] tested the effect of hydrogen-rich saline on isoproterenol-induced myocardial infarction (MI) in rats. They reported that hydrogen-rich saline decreased MDA, increased superoxide dismutase, and decreased serum TNFα and IL-6 in the MI heart. Furthermore, hydrogen-rich saline decreased cardiac CK-MB levels in the MI rats compared to control rats. Hydrogen-rich saline pretreatment also reduced the infarct size, alleviated pathological changes in the left ventricle, and improved cardiac function.

Lee et al. [80] tested the effect of hydrogen on BDL-induced cirrhosis in rats, finding that hydrogen-rich saline significantly decreased thiobarbituric acid-reacting substances (TBARS) and MDA, markers of oxidative stress, and increased superoxide dismutase and GSH, which are markers of antioxidant reserves in BDL rats. Consistent with the study of Jing et al. [79], hydrogen-rich saline reduced pro-inflammatory markers, including TNF$\alpha$, IL-1$\beta$, and IL-6. The study of Lee and colleagues did not test the role of hydrogen-rich saline on direct cardiac function. Instead, they showed an improvement of hyperdynamic circulation. Qian et al. [81] used hydrogen-rich water (4 mL/kg orally three times a day) to treat patients with chronic graft-versus-host disease and demonstrated that it prolonged the survival time and increased the survival rate during 4 years of treatment. They speculated that these therapeutic effects were mediated via the antioxidant and anti-inflammatory effects of hydrogen. Accordingly, we suggest that hydrogen may improve cardiac function in CCM because hydrogen-rich saline attenuates oxidative stress and inflammation in subjects with cirrhosis, and these phenomena are pathogenic mechanisms of CCM.

### 4.3. Liver Transplantation

Liver transplantation remains the definitive 'cure' for cardiovascular anomalies of cirrhosis. A recent study showed that within one year after liver transplantation, 34% of CCM patients recovered according to the 2005 Montreal criteria and 57% according to the 2019 CCC criteria [82]. However, the recovery process is challenging, and the overall cardiovascular system experiences both risks and benefits. As stated above, the benefits accrue over a longer term, whereas many of the risks occur during the perioperative and short-term postoperative states. During the procedure of liver transplantation, the hemorrhage and clamping of the major blood vessels may cause hypovolemia, whereas aggressive fluid resuscitation may cause volume overload. Perioperative hemodynamic fluctuations significantly affect cardiac function. Other factors, such as acidosis, hypothermia, and electrolyte disturbances, may impair cardiac contractility [83]. Citrated blood transfusion may cause hypocalcemia [84], which further depresses cardiac contractility.

After liver transplantation, the peripheral vascular resistance immediately increases, as does the blood pressure, which raises both cardiac preload and afterload. These challenges may result in overt cardiac failure in patients with CCM [85]. Another challenge of liver transplantation is the shortage of donor organs, which limits its application. The high cost, the complexity of the procedure, the need for the long-term use of immunosuppressants, and complications such as infections and rejection also limit its widespread clinical application. In many economically underdeveloped global regions, liver transplantation is simply not available.

## 5. Future Possibilities

Since the traditional therapeutic strategies for non-cirrhotic cardiac dysfunction, such as vasodilators, are not applicable in CCM, other potential treatments have been investigated over the past decade. In particular, therapies aimed at correcting the pathogenesis-related targets, such as antioxidants and anti-inflammatory and anti-apoptotic agents, may be beneficial to patients with CCM. To date, these strategies are mostly limited to animal research, so these agents need to be validated in well-designed clinical trials.

Another therapeutic potential agent is NSBBs. Theoretically, NSBBs should have therapeutic effects on CCM. However, the results from different studies are inconsistent. Currently, NSBBs are only a standard of care for the prevention of primary and secondary bleeding caused by gastroesophageal varices. Borrowing from the therapeutic concept of systemic hypertension, which needs lifelong treatment, we may also need to treat portal hypertension lifelong rather than just administering NSBBs when variceal bleeding forces us to do so. Because portal hypertension is an important pathogenic factor underlying CCM, treating portal hypertension may lead to the improvement of CCM.

**Funding:** This review article received no external funding.

**Data Availability Statement:** Not applicable.

**Conflicts of Interest:** The authors declare no conflict of interest.

## References


1. Lee, S.S. Cardiac abnormalities in liver cirrhosis. *West J. Med.* **1989**, *151*, 530–535.
2. Liu, H.; Nguyen, H.H.; Yoon, K.T.; Lee, S.S. Pathogenic Mechanisms Underlying Cirrhotic Cardiomyopathy. *Front. Netw. Physiol.* **2022**, *2*, 849253. [CrossRef]
3. Liu, H.; Naser, J.A.; Lin, G.; Lee, S.S. Cardiomyopathy in cirrhosis: From pathophysiology to clinical care. *JHEP Rep.* **2024**, *6*, 100911. [CrossRef]
4. Liu, H.; Lee, S.S. Diagnostic Criteria of Cirrhotic Cardiomyopathy: Out with the Old, in With the New? *Hepatology* **2021**, *74*, 3523–3525. [CrossRef]
5. Yoon, K.T.; Liu, H.; Lee, S.S. Cirrhotic Cardiomyopathy. *Curr. Gastroenterol. Rep.* **2020**, *22*, 45. [CrossRef]
6. Liu, H.; Jayakumar, S.; Traboulsi, M.; Lee, S.S. Cirrhotic cardiomyopathy: Implications for liver transplantation. *Liver. Transpl* **2017**, *23*, 826–835. [CrossRef]
7. Premkumar, M.; Devurgowda, D.; Vyas, T.; Shasthry, S.M.; Khumuckham, J.S.; Goyal, R.; Thomas, S.S.; Kumar, G. Left Ventricular Diastolic Dysfunction is Associated with Renal Dysfunction, Poor Survival and Low Health Related Quality of Life in Cirrhosis. *J. Clin. Exp. Hepatol.* **2019**, *9*, 324–333. [CrossRef]
8. Lee, S.S.; Marty, J.; Mantz, J.; Samain, E.; Braillon, A.; Lebrec, D. Desensitization of myocardial beta-adrenergic receptors in cirrhotic rats. *Hepatology* **1990**, *12*, 481–485. [CrossRef]
9. Ma, Z.; Meddings, J.B.; Lee, S.S. Membrane physical properties determine cardiac beta-adrenergic receptor function in cirrhotic rats. *Am. J. Physiol.* **1994**, *267*, G87–G93. [CrossRef]
10. Honar, H.; Liu, H.; Zhang, M.L.; Glenn, T.K.; Ter Keurs, H.; Lee, S.S. Impaired myosin isoform shift and calcium transients contribute to cellular pathogenesis of rat cirrhotic cardiomyopathy. *Liver Int.* **2020**, *40*, 2808–2819. [CrossRef]
11. Chayanupatkul, M.; Liangpunsakul, S. Cirrhotic cardiomyopathy: Review of pathophysiology and treatment. *Hepatol. Int.* **2014**, *8*, 308–315. [CrossRef]
12. Nobbe, A.M.; McCurdy, H.M. Management of the Adult Patient with Cirrhosis Complicated by Ascites. *Crit. Care Nurs. Clin. N. Am.* **2022**, *34*, 311–320. [CrossRef]
13. Poynard, T.; Cales, P.; Pasta, L.; Ideo, G.; Pascal, J.P.; Pagliaro, L.; Lebrec, D. Beta-adrenergic-antagonist drugs in the prevention of gastrointestinal bleeding in patients with cirrhosis and esophageal varices. An analysis of data and prognostic factors in 589 patients from four randomized clinical trials. Franco-Italian Multicenter Study Group. *N. Engl. J. Med.* **1991**, *324*, 1532–1538.
14. Serste, T.; Melot, C.; Francoz, C.; Durand, F.; Rautou, P.E.; Valla, D.; Moreau, R.; Lebrec, D. Deleterious effects of beta-blockers on survival in patients with cirrhosis and refractory ascites. *Hepatology* **2010**, *52*, 1017–1022. [CrossRef]
15. Silvestre, O.M.; Farias, A.Q.; Ramos, D.S.; Furtado, M.S.; Rodrigues, A.C.; Ximenes, R.O.; de Campos Mazo, D.F.; Zitelli, P.M.Y.; Diniz, M.A.; Andrade, J.L.; et al. Beta-Blocker therapy for cirrhotic cardiomyopathy: A randomized-controlled trial. *Eur. J. Gastroenterol. Hepatol.* **2018**, *30*, 930–937. [CrossRef]
16. Leithead, J.A.; Rajoriya, N.; Tehami, N.; Hodson, J.; Gunson, B.K.; Tripathi, D.; Ferguson, J.W. Non-selective beta-blockers are associated with improved survival in patients with ascites listed for liver transplantation. *Gut* **2015**, *64*, 1111–1119. [CrossRef]
17. Mookerjee, R.P.; Pavesi, M.; Thomsen, K.L.; Mehta, G.; Macnaughtan, J.; Bendtsen, F.; Coenraad, M.; Sperl, J.; Gines, P.; Moreau, R.; et al. Treatment with non-selective beta blockers is associated with reduced severity of systemic inflammation and improved survival of patients with acute-on-chronic liver failure. *J. Hepatol.* **2016**, *64*, 574–582. [CrossRef]
18. Premkumar, M.; Rangegowda, D.; Vyas, T.; Khumuckham, J.S.; Shasthry, S.M.; Thomas, S.S.; Goyal, R.; Kumar, G.; Sarin, S.K. Carvedilol Combined with Ivabradine Improves Left Ventricular Diastolic Dysfunction, Clinical Progression, and Survival in Cirrhosis. *J. Clin. Gastroenterol.* **2020**, *54*, 561–568. [CrossRef]
19. Zambruni, A.; Trevisani, F.; Di Micoli, A.; Savelli, F.; Berzigotti, A.; Bracci, E.; Caraceni, P.; Domenicali, M.; Felline, P.; Zoli, M.; et al. Effect of chronic beta-blockade on QT interval in patients with liver cirrhosis. *J. Hepatol.* **2008**, *48*, 415–421. [CrossRef]
20. Henriksen, J.H.; Bendtsen, F.; Hansen, E.F.; Moller, S. Acute non-selective beta-adrenergic blockade reduces prolonged frequency-adjusted Q-T interval (QTc) in patients with cirrhosis. *J. Hepatol.* **2004**, *40*, 239–246. [CrossRef] [PubMed]
21. Lebrec, D.; Nouel, O.; Corbic, M.; Benhamou, J.P. Propranolol—A medical treatment for portal hypertension? *Lancet* **1980**, *2*, 180–182. [CrossRef]
22. Hillon, P.; Lebrec, D.; Munoz, C.; Jungers, M.; Goldfarb, G.; Benhamou, J.P. Comparison of the effects of a cardioselective and a nonselective beta-blocker on portal hypertension in patients with cirrhosis. *Hepatology* **1982**, *2*, 528–531. [CrossRef] [PubMed]
23. de Franchis, R.; Baveno, V.I.F. Expanding consensus in portal hypertension: Report of the Baveno VI Consensus Workshop: Stratifying risk and individualizing care for portal hypertension. *J. Hepatol.* **2015**, *63*, 743–752. [CrossRef] [PubMed]

24. Krag, A.; Wiest, R.; Albillos, A.; Gluud, L.L. The window hypothesis: Haemodynamic and non-haemodynamic effects of beta-blockers improve survival of patients with cirrhosis during a window in the disease. *Gut* **2012**, *61*, 967–969. [PubMed]
25. Gupta, N.; Bhat, S.N.; Reddysetti, S.; Afees Ahamed, M.A.; Jose, D.; Sarvepalli, A.S.; Joylin, S.; Godkhindi, V.; Rabaan, A. Clinical profile, diagnosis, treatment, and outcome of patients with tubercular versus nontubercular causes of spine involvement: A retrospective cohort study from India. *Int. J. Mycobacteriol.* **2022**, *11*, 75–82. [PubMed]
26. Tellez, L.; Ibanez-Samaniego, L.; Perez Del Villar, C.; Yotti, R.; Martinez, J.; Carrion, L.; Rodriguez de Santiago, E.; Rivera, M.; Gonzalez-Mansilla, A.; Pastor, O.; et al. Non-selective beta-blockers impair global circulatory homeostasis and renal function in cirrhotic patients with refractory ascites. *J. Hepatol.* **2020**, *73*, 1404–1414. [CrossRef] [PubMed]
27. Henriksen, J.H.; Moller, S.; Ring-Larsen, H.; Christensen, N.J. The sympathetic nervous system in liver disease. *J. Hepatol.* **1998**, *29*, 328–341. [CrossRef]
28. Kostreva, D.R.; Castaner, A.; Kampine, J.P. Reflex effects of hepatic baroreceptors on renal and cardiac sympathetic nerve activity. *Am. J. Physiol.* **1980**, *238*, R390–R394. [CrossRef] [PubMed]
29. Li, L.; Hu, Z.; Xiong, Y.; Yao, Y. Device-Based Sympathetic Nerve Regulation for Cardiovascular Diseases. *Front. Cardiovasc. Med.* **2021**, *8*, 803984. [CrossRef]
30. Cao, N.; Wang, J.J.; Wu, J.M.; Xu, W.L.; Wang, R.; Chen, X.D.; Feng, Y.N.; Cong, W.W.; Zhang, Y.Y.; Xiao, H.; et al. Glibenclamide alleviates beta adrenergic receptor activation-induced cardiac inflammation. *Acta. Pharmacol. Sin.* **2022**, *43*, 1243–1250. [CrossRef]
31. Ruparelia, N.; Chai, J.T.; Fisher, E.A.; Choudhury, R.P. Inflammatory processes in cardiovascular disease: A route to targeted therapies. *Nat. Rev. Cardiol.* **2017**, *14*, 133–144. [CrossRef]
32. Seo, Y.S.; Shah, V.H. The role of gut-liver axis in the pathogenesis of liver cirrhosis and portal hypertension. *Clin. Mol. Hepatol.* **2012**, *18*, 337–346. [CrossRef]
33. Madsen, B.S.; Havelund, T.; Krag, A. Targeting the gut-liver axis in cirrhosis: Antibiotics and non-selective beta-blockers. *Adv. Ther.* **2013**, *30*, 659–670. [CrossRef]
34. Reiberger, T.; Mandorfer, M. Beta adrenergic blockade and decompensated cirrhosis. *J. Hepatol.* **2017**, *66*, 849–859. [CrossRef]
35. Hole, T.; Froland, G.; Gullestad, L.; Offstad, J.; Skjaerpe, T. Metoprolol CR/XL improves systolic and diastolic left ventricular function in patients with chronic heart failure. *Echocardiography* **2004**, *21*, 215–223. [CrossRef]
36. Alvarado-Tapias, E.; Ardevol, A.; Garcia-Guix, M.; Montanes, R.; Pavel, O.; Cuyas, B.; Graupera, I.; Brujats, A.; Vilades, D.; Colomo, A.; et al. Short-term hemodynamic effects of beta-blockers influence survival of patients with decompensated cirrhosis. *J. Hepatol.* **2020**, *73*, 829–841. [CrossRef]
37. Yoon, K.T.; Liu, H.; Lee, S.S. beta-blockers in advanced cirrhosis: More friend than enemy. *Clin. Mol. Hepatol.* **2021**, *27*, 425–436. [CrossRef]
38. Lee, W.; Vandenberk, B.; Raj, S.R.; Lee, S.S. Prolonged QT Interval in Cirrhosis: Twisting Time? *Gut. Liver* **2022**, *16*, 849–860. [CrossRef]
39. Taprantzi, D.; Zisimopoulos, D.; Thomopoulos, K.C.; Spiliopoulou, I.; Georgiou, C.D.; Tsiaoussis, G.; Triantos, C.; Gogos, C.A.; Labropoulou-Karatza, C.; Assimakopoulos, S.F. Propranolol reduces systemic oxidative stress and endotoxemia in cirrhotic patients with esophageal varices. *Ann. Gastroenterol.* **2018**, *31*, 224–230. [CrossRef] [PubMed]
40. Liu, H.; Nguyen, H.H.; Hwang, S.Y.; Lee, S.S. Oxidative Mechanisms and Cardiovascular Abnormalities of Cirrhosis and Portal Hypertension. *Int. J. Mol. Sci.* **2023**, *24*, 16805. [CrossRef] [PubMed]
41. Yang, Y.Y.; Liu, H.; Nam, S.W.; Kunos, G.; Lee, S.S. Mechanisms of TNFalpha-induced cardiac dysfunction in cholestatic bile duct-ligated mice: Interaction between TNFalpha and endocannabinoids. *J. Hepatol.* **2010**, *53*, 298–306. [CrossRef]
42. Nam, S.W.; Liu, H.; Wong, J.Z.; Feng, A.Y.; Chu, G.; Merchant, N.; Lee, S.S. Cardiomyocyte apoptosis contributes to pathogenesis of cirrhotic cardiomyopathy in bile duct-ligated mice. *Clin. Sci.* **2014**, *127*, 519–526. [CrossRef]
43. Liu, L.; Liu, H.; Nam, S.W.; Lee, S.S. Protective effects of erythropoietin on cirrhotic cardiomyopathy in rats. *Dig. Liver Dis.* **2012**, *44*, 1012–1017. [CrossRef] [PubMed]
44. Bortoluzzi, A.; Ceolotto, G.; Gola, E.; Sticca, A.; Bova, S.; Morando, F.; Piano, S.; Fasolato, S.; Rosi, S.; Gatta, A.; et al. Positive cardiac inotropic effect of albumin infusion in rodents with cirrhosis and ascites: Molecular mechanisms. *Hepatology* **2013**, *57*, 266–276. [CrossRef]
45. Fernandez, J.; Claria, J.; Amoros, A.; Aguilar, F.; Castro, M.; Casulleras, M.; Acevedo, J.; Duran-Güell, M.; Nuñez, L.; Costa, M.; et al. Effects of Albumin Treatment on Systemic and Portal Hemodynamics and Systemic Inflammation in Patients with Decompensated Cirrhosis. *Gastroenterology* **2019**, *157*, 149–162. [CrossRef]
46. Mousavi, K.; Niknahad, H.; Ghalamfarsa, A.; Mohammadi, H.; Azarpira, N.; Ommati, M.M.; Heidari, R. Taurine mitigates cirrhosis-associated heart injury through mitochondrial-dependent and antioxidative mechanisms. *Clin. Exp. Hepatol.* **2020**, *6*, 207–219. [CrossRef] [PubMed]
47. Sheibani, M.; Nezamoleslami, S.; Mousavi, S.E.; Faghir-Ghanesefat, H.; Yousefi-Manesh, H.; Rezayat, S.M.; Dehpour, A. Protective Effects of Spermidine Against Cirrhotic Cardiomyopathy in Bile Duct-Ligated Rats. *J. Cardiovasc. Pharmacol.* **2020**, *76*, 286–295. [CrossRef]
48. Yoon, K.T.; Liu, H.; Zhang, J.; Han, S.; Lee, S.S. Galectin-3 inhibits cardiac contractility via a tumor necrosis factor alpha-dependent mechanism in cirrhotic rats. *Clin. Mol. Hepatol.* **2022**, *28*, 232–241. [CrossRef] [PubMed]
49. Niaz, Q.; Tavangar, S.M.; Mehreen, S.; Ghazi-Khansari, M.; Jazaeri, F. Evaluation of statins as a new therapy to alleviate chronotropic dysfunction in cirrhotic rats. *Life Sci.* **2022**, *308*, 120966. [CrossRef]

50. Node, K.; Fujita, M.; Kitakaze, M.; Hori, M.; Liao, J.K. Short-term statin therapy improves cardiac function and symptoms in patients with idiopathic dilated cardiomyopathy. *Circulation* **2003**, *108*, 839–843. [CrossRef]
51. Bielecka-Dabrowa, A.; Mikhailidis, D.P.; Rizzo, M.; von Haehling, S.; Rysz, J.; Banach, M. The influence of atorvastatin on parameters of inflammation left ventricular function, hospitalizations and mortality in patients with dilated cardiomyopathy—5-year follow-up. *Lipids Health Dis.* **2013**, *12*, 47. [CrossRef] [PubMed]
52. Liao, J.K. Statin therapy for cardiac hypertrophy and heart failure. *J. Investig. Med.* **2004**, *52*, 248–253. [CrossRef] [PubMed]
53. Tang, H.Y.; Huang, J.E.; Tsau, M.T.; Chang, C.J.; Tung, Y.C.; Lin, G.; Cheng, M.L. Metabolomics Assessment of Volume Overload-Induced Heart Failure and Oxidative Stress in the Kidney. *Metabolites* **2023**, *13*, 1165. [CrossRef] [PubMed]
54. Baliou, S.; Adamaki, M.; Ioannou, P.; Pappa, A.; Panayiotidis, M.I.; Spandidos, D.A.; Christodoulou, I.; Kyriakopoulos, A.M.; Zoumpourlis, V. Protective role of taurine against oxidative stress (Review). *Mol. Med. Rep.* **2021**, *24*, 605. [CrossRef]
55. Schaffer, S.W.; Jong, C.J.; Ramila, K.C.; Azuma, J. Physiological roles of taurine in heart and muscle. *J. Biomed. Sci.* **2010**, *17* (Suppl. S1), S2. [CrossRef]
56. Goodman, C.A.; Horvath, D.; Stathis, C.; Mori, T.; Croft, K.; Murphy, R.M.; Hayes, A. Taurine supplementation increases skeletal muscle force production and protects muscle function during and after high-frequency in vitro stimulation. *J. Appl. Physiol.* **2009**, *107*, 144–154. [CrossRef] [PubMed]
57. Ozsarlak-Sozer, G.; Sevin, G.; Ozgur, H.H.; Yetik-Anacak, G.; Kerry, Z. Diverse effects of taurine on vascular response and inflammation in GSH depletion model in rabbits. *Eur. Rev. Med. Pharmacol. Sci.* **2016**, *20*, 1360–1372.
58. Pion, P.D.; Kittleson, M.D.; Rogers, Q.R.; Morris, J.G. Myocardial failure in cats associated with low plasma taurine: A reversible cardiomyopathy. *Science* **1987**, *237*, 764–768. [CrossRef] [PubMed]
59. Ito, T.; Oishi, S.; Takai, M.; Kimura, Y.; Uozumi, Y.; Fujio, Y.; Schaffer, S.W.; Azuma, J. Cardiac and skeletal muscle abnormality in taurine transporter-knockout mice. *J. Biomed. Sci.* **2010**, *17* (Suppl. S1), S20. [CrossRef]
60. Beyranvand, M.R.; Khalafi, M.K.; Roshan, V.D.; Choobineh, S.; Parsa, S.A.; Piranfar, M.A. Effect of taurine supplementation on exercise capacity of patients with heart failure. *J. Cardiol.* **2011**, *57*, 333–337. [CrossRef]
61. Najibi, A.; Rezaei, H.; Manthari, R.K.; Niknahad, H.; Jamshidzadeh, A.; Farshad, O.; Yan, F.; Ma, Y.; Xu, D.; Tang, Z.; et al. Cellular and mitochondrial taurine depletion in bile duct ligated rats: A justification for taurine supplementation in cholestasis/cirrhosis. *Clin. Exp. Hepatol.* **2022**, *8*, 195–210. [CrossRef] [PubMed]
62. Liu, J.; Ai, Y.; Niu, X.; Shang, F.; Li, Z.; Liu, H.; Li, W.; Ma, W.; Chen, R.; Wei, T.; et al. Taurine protects against cardiac dysfunction induced by pressure overload through SIRT1-p53 activation. *Chem. Biol. Interact.* **2020**, *317*, 108972. [CrossRef] [PubMed]
63. Jiang, D.; Wang, X.; Zhou, X.; Wang, Z.; Li, S.; Sun, Q.; Jiang, Y.; Ji, C.; Ling, W.; An, X.; et al. Spermidine alleviating oxidative stress and apoptosis by inducing autophagy of granulosa cells in Sichuan white geese. *Poult. Sci.* **2023**, *102*, 102879. [CrossRef] [PubMed]
64. Chen, Y.; Guo, Z.; Li, S.; Liu, Z.; Chen, P. Spermidine Affects Cardiac Function in Heart Failure Mice by Influencing the Gut Microbiota and Cardiac Galectin-3. *Front. Cardiovasc. Med.* **2021**, *8*, 765591. [CrossRef] [PubMed]
65. Omar, E.M.; Omar, R.S.; Shoela, M.S.; El Sayed, N.S. A study of the cardioprotective effect of spermidine: A novel inducer of autophagy. *Chin. J. Physiol.* **2021**, *64*, 281–288. [CrossRef] [PubMed]
66. Martinelli, O.; Peruzzi, M.; Bartimoccia, S.; D'Amico, A.; Marchitti, S.; Rubattu, S.; Chiariello, G.A.; D'ambrosio, L.; Schiavon, S.; Miraldi, F.; et al. Natural Activators of Autophagy Increase Maximal Walking Distance and Reduce Oxidative Stress in Patients with Peripheral Artery Disease: A Pilot Study. *Antioxidants* **2022**, *11*, 1836. [CrossRef]
67. Dumic, J.; Dabelic, S.; Flogel, M. Galectin-3: An open-ended story. *Biochim. Biophys. Acta* **2006**, *1760*, 616–635. [CrossRef]
68. Krzeslak, A.; Lipinska, A. Galectin-3 as a multifunctional protein. *Cell. Mol. Biol. Lett.* **2004**, *9*, 305–328. [PubMed]
69. Wanninger, J.; Weigert, J.; Wiest, R.; Bauer, S.; Karrasch, T.; Farkas, S.; Scherer, M.N.; Walter, R.; Weiss, T.S.; Hellerbrand, C.; et al. Systemic and hepatic vein galectin-3 are increased in patients with alcoholic liver cirrhosis and negatively correlate with liver function. *Cytokine* **2011**, *55*, 435–440. [CrossRef]
70. Abu-Elsaad, N.M.; Elkashef, W.F. Modified citrus pectin stops progression of liver fibrosis by inhibiting galectin-3 and inducing apoptosis of stellate cells. *Can. J. Physiol. Pharmacol.* **2016**, *94*, 554–562. [CrossRef]
71. Dong, R.; Zhang, M.; Hu, Q.; Zheng, S.; Soh, A.; Zheng, Y.; Yuan, H. Galectin-3 as a novel biomarker for disease diagnosis and a target for therapy (Review). *Int. J. Mol. Med.* **2018**, *41*, 599–614. [CrossRef] [PubMed]
72. Al-Salam, S.; Kandhan, K.; Sudhadevi, M.; Yasin, J.; Tariq, S. Early Doxorubicin Myocardial Injury: Inflammatory, Oxidative Stress, and Apoptotic Role of Galectin-3. *Int. J. Mol. Sci.* **2022**, *23*, 12479. [CrossRef]
73. Trebicka, J.; Garcia-Tsao, G. Controversies regarding albumin therapy in cirrhosis. *Hepatology* **2023**, *7*, 10–97. [CrossRef] [PubMed]
74. Mapanga, R.F.; Joseph, D.E.; Saieva, M.; Boyer, F.; Rondeau, P.; Bourdon, E.; Essop, M.F. Glycation abolishes the cardioprotective effects of albumin during ex vivo ischemia-reperfusion. *Physiol. Rep.* **2017**, *5*, e13107. [CrossRef] [PubMed]
75. Saengsin, K.; Sittiwangkul, R.; Chattipakorn, S.C.; Chattipakorn, N. Hydrogen therapy as a potential therapeutic intervention in heart disease: From the past evidence to future application. *Cell Mol. Life Sci.* **2023**, *80*, 174. [CrossRef] [PubMed]
76. Graves, J.; Mason, M.; Laws, D. A case of orthodeoxia platypnoea in a patient with adult polycystic kidney and liver disease with a patent foramen ovale. *Acute Med.* **2007**, *6*, 126–127. [CrossRef] [PubMed]
77. Ohta, S. Molecular hydrogen as a preventive and therapeutic medical gas: Initiation, development and potential of hydrogen medicine. *Pharmacol. Ther.* **2014**, *144*, 1–11. [CrossRef] [PubMed]

78. Yao, L.; Chen, H.; Wu, Q.; Xie, K. Hydrogen-rich saline alleviates inflammation and apoptosis in myocardial I/R injury via PINK-mediated autophagy. *Int. J. Mol. Med.* **2019**, *44*, 1048–1062. [CrossRef] [PubMed]
79. Jing, L.; Wang, Y.; Zhao, X.M.; Zhao, B.; Han, J.J.; Qin, S.C.; Sun, X.J. Cardioprotective Effect of Hydrogen-rich Saline on Isoproterenol-induced Myocardial Infarction in Rats. *Heart Lung Circ.* **2015**, *24*, 602–610. [CrossRef]
80. Lee, P.C.; Yang, Y.Y.; Huang, C.S.; Hsieh, S.L.; Lee, K.C.; Hsieh, Y.C.; Lee, T.Y.; Lin, H.C. Concomitant inhibition of oxidative stress and angiogenesis by chronic hydrogen-rich saline and N-acetylcysteine treatments improves systemic, splanchnic and hepatic hemodynamics of cirrhotic rats. *Hepatol. Res.* **2015**, *45*, 578–588. [CrossRef]
81. Qian, L.; Liu, M.; Shen, J.; Cen, J.; Zhao, D. Hydrogen in Patients with Corticosteroid-Refractory/Dependent Chronic Graft-Versus-Host-Disease: A Single-Arm, Multicenter, Open-Label, Phase 2 Trial. *Front. Immunol.* **2020**, *11*, 598359. [CrossRef] [PubMed]
82. Ali, S.A.; Arman, H.E.; Shamseddeen, H.; Elsner, N.; Elsemesmani, H.; Johnson, S.; Zenisek, J.; Khemka, A.; Jarori, U.; Patidar, K.R.; et al. Cirrhotic cardiomyopathy: Predictors of major adverse cardiac events and assessment of reversibility after liver transplant. *J. Cardiol.* **2023**, *82*, 113–121. [CrossRef] [PubMed]
83. Myers, R.P.; Lee, S.S. Cirrhotic cardiomyopathy and liver transplantation. *Liver Transpl.* **2000**, *6*, S44–S52. [CrossRef] [PubMed]
84. Pandey, C.K.; Singh, A.; Kajal, K.; Dhankhar, M.; Tandon, M.; Pandey, V.K.; Karna, S.T. Intraoperative blood loss in orthotopic liver transplantation: The predictive factors. *World J. Gastrointest. Surg.* **2015**, *7*, 86–93. [CrossRef]
85. Rahman, S.; Mallett, S.V. Cirrhotic cardiomyopathy: Implications for the perioperative management of liver transplant patients. *World J. Hepatol.* **2015**, *7*, 507–520. [CrossRef]

MDPI AG
Grosspeteranlage 5
4052 Basel
Switzerland
Tel.: +41 61 683 77 34

*International Journal of Molecular Sciences* Editorial Office
E-mail: ijms@mdpi.com
www.mdpi.com/journal/ijms

www.ingramcontent.com/pod-product-compliance
Lightning Source LLC
Chambersburg PA
CBHW061602210926
51850CB00009B/2182
*9783725852956*